W9-CRB-274

Scaling Up

The Institution of Chemical Engineers and the Rise of a New Profession

Colin Divall   and   **Sean F. Johnston**
University of York   *University of Glasgow*

with a chapter by

James F. Donnelly
University of Leeds

KLUWER ACADEMIC PUBLISHERS
DORDRECHT/BOSTON/LONDON

Chemistry Library

A C.I.P. catalogue record for this book is available from the Library of Congress.

ISBN 0-7923-6692-1

Published by Kluwer Academic Publishers,
PO Box 17, 3300 AA Dordrecht, The Netherlands.

Sold and distributed in North, Central and South America
by Kluwer Academic Publishers,
101 Philip Drive, Norwell, MA 02061, USA

In all other countries, sold and distributed
by Kluwer Academic Publishers,
PO Box 322, 3300 AH Dordrecht, The Netherlands

Printed on acid-free paper

All rights reserved
© 2000 Kluwer Academic Publishers
No part of the material protected by this copyright may be reproduced
or utilized in any form or by any means, electronic, mechanical,
including photocopying, recording or by any information storage and
retrieval system, without written permission from the copyright owners.

Printed and bound in Great Britain by Antony Rowe Limited. ac

Scaling Up

The Institution of Chemical Engineers
and the Rise of a New Profession

Chemists and Chemistry

VOLUME 20

A series of books devoted to the examination of the history and development of chemistry from its early emergence as a separate discipline to the present day. The series will describe the personalities, processes, theoretical and technical advances which have shaped our current understanding of chemical science

The titles published in this series are listed at the end of this volume.

T P
1
D48
2000
CHEM

TABLE OF CONTENTS

LIST OF FIGURES

LIST OF TABLES

ACRONYMS AND ABBREVIATIONS USED IN THE TEXT

Organisations

ABCM	Association of British Chemical Plant Manufacturers
ACostE	Association of Cost Engineers
AIChE	American Institute of Chemical Engineers
AIEE	American Institute of Electrical Engineers
AIMME	American Institute of Mining and Metallurgical Engineers
ANC	Australian National Committee
ASCE	American Society of Civil Engineers
ASME	American Society of Mechanical Engineers
A&W	Albright & Wilson Ltd.
BAC	British Association of Chemists
BCECA	British Chemical Engineering Contractors Association
BCPMA	British Chemical Plant Manufacturers Association
BEA	British Engineering Authority
BEP	Board for the Engineering Profession
BER	Board for Engineering Regulation
BES	Bureau of Engineering Surveyors
BNEC	British Nuclear Energy Congress
BNES	British Nuclear Energy Society
CAPITB	Chemical and Allied Products Industry Training Board
CAT	College of Advanced Technology
CBI	Confederation of British Industry
CBMPE	Council of British Manufacturers of Petroleum Equipment
CEG	Chemical Engineering Group of the Society of Chemical Industry
CEI	Council of Engineering Institutions
CIA	Chemical Industries Association
CND	Campaign for Nuclear Disarmament
CS	Chemical Society
CSTI	Council of Science and Technology Institutes
DECHEMA	Deutsche Gesellschaft für Chemisches Apparatewesen, Chemische Technik und Biotechnologie e.V

DEUA	Diesel Engine Users' Association
DSIR	Department of Scientific and Industrial Research
EFCE	European Federation of Chemical Engineering
EIJC	Engineering Institutions Joint Council
EngC	Engineering Council
EUSEC	Conference of Engineering Societies of Western Europe and the USA
FBI	Federation of British Industries
FEANI	Fédération Européene d'Associations Nationales d'Ingenieurs
HSE	Health and Safety Executive
HEOD	High Explosives Operating Department
IChemE	Institution of Chemical Engineers
ICE	Institution of Civil Engineers
ICI	Imperial Chemical Industries
IEAust	Institute of Engineers, Australia
IEnergy	Institute of Energy
IEE	Institution of Electrical Engineers
IERE	Institution of Electronic and Radio Engineers
IGasE	Institution of Gas Engineers
IICE	Indian Institute of Chemical Engineers
IMat	Institute of Materials
IMatE	Institution of Materials Engineers
IMechE	Institution of Mechanical Engineers
IMet	Institute of Metals
IMinE	Institution of Mining Engineers
IMM	Institution of Mining and Metallurgy
IPlantE	Institution of Plant Engineers
IProdE	Institution of Production Engineers
IStructE	Institution of Structural Engineers
KCL	King's College, London
MIT	Massachusetts Institute of Technology
MLNS	Ministry of Labour and National Service
NACEIC	National Advisory Council on Education for Industry and Commerce
NECTAR	National Council for Technological Awards
NEDC	National Economic Development Council
NPL	National Physical Laboratory
OEEC	Organisation for European Economic Co-operation
OPEC	Organisation of Petroleum Exporting Countries
PITB	Petroleum Industry Training Board
PPA	Process Plant Association
PRO	Public Record Office
RIC	Royal Institute of Chemistry
RSC	Royal Society of Chemistry
SCI	Society of Chemical Industry
SCOTEC	Scottish Technical Education Council

TEC	Technician Education Council
TPC	Technical Personnel Committee
TUC	Trades Union Congress
UCL	University College, London
UGC	University Grants Committee
UKAEA	United Kingdom Atomic Energy Authority
UMIST	University of Manchester Institute of Science and Technology
WFEO	World Federation of Engineering Organisations

Periodicals

AIChE Trans.	*American Institute of Chemical Engineers Transactions*
Am. J. Sociology	*American Journal of Sociology*
Ann. Sci.	*Annals of Science*
Brit. J. Hist. Sci.	*British Journal for the History of Science*
Chem. Age	*Chemical Age*
Chem. News	*Chemical News*
Chem. & Indus.	*Chemistry and Industry*
Chem. Eng. & Works Chemist	*Chemical Engineering & the Works Chemist*
Chem. Trade J.	*Chemical Trade Journal*
Chem. Trade J. & Chem. Eng.	*Chemical Trade Journal & Chemical Engineer*
Econ. Hist. Rev.	*Economic History Review*
J. BNEC	*Journal of the British Nuclear Energy Conference*
J. BNES	*Journal of the British Nuclear Energy Society*
J. Inst. Petroleum	*Journal of the Institute of Petroleum*
J. Soc. of Arts	*Journal of the Society of Arts*
J. Soc. Hist.	*Journal of Social History*
J. SCI	*Journal of the Society of Chemical Industry*
Proc. CEG	*Proceedings of the Chemical Engineering Group*
Quart. Bull. IChemE	*Quarterly Bulletin of the Institution of Chemical Engineers*
Sci. Stud.	*Science Studies*
Soc. Stud. Sci.	*Society for the Study of Science*
Tech. & Culture	*Technology & Culture*
TCE	*The Chemical Engineer*
Trans. Faraday Soc.	*Transactions of the Faraday Society*
Trans. IChemE	*Transactions of the Institution of Chemical Engineers*
Trans. Newcastle Chem. Soc.	*Transactions of the Newcastle Chemical Society*

Archives

AF	Application Form for membership, 1923–1999, IChemE.
BCA	Birmingham City Archives.
Brennan archive	First General Secretary John Basil Brennan papers, IChemE.
Carr papers	F. H. Carr papers, Imperial College Archives.
CM	Council Minute book (16 volumes, 1923–99), IChemE.
CRO	Cheshire Record Office.
Davis collection	George E. Davis book collection, IChemE.
Davis papers	George E. Davis papers, Science Museum, London.
Evans archives	Fourth General Secretary Trevor Evans papers, IChemE.
Gayfere	'Gayfere St' archives, IChemE.
Hartley papers	Churchill College Archive Centre, University of Cambridge.
ICA	Imperial College Archives.
ICE CM	Council Minutes, Institution of Civil Engineers.
IEE CM	Council Minutes, Institution of Electrical Engineers.
IMechE COP	Council Papers, Institution of Mechanical Engineers.
IMechE Hinton	Hinton collection, Institution of Mechanical Engineers.
MEC	Minutes of the Education Committee, IChemE.
MPDC	Minutes of the Professional Development Committee (PDC), IChemE.
MCEC	Minutes of the Chemical Education Committee, King's College, London.
MCSC	Minutes of the Chemistry Sectional Committee, UMIST.
MLNS	Ministry of Labour and National Service.
MSUL	Minutes of the Senate of the University of London.
Nathan papers	Sir Frederic Lewis Nathan papers, IChemE.
PRO	Public Record Office.
AB, CAB, BT, LAB, EDUC, WO	Papers at the Public Record Office from the Atomic Energy Authority, Cabinet, Board of Trade and Ministries of Labour, Education and Works, respectively.
SCI CM	SCI Council Minute Book.
SRO	Scottish Record Office.
Swindin papers	Norman Swindin papers, Loughborough University Archives.
UCLA	University College London Archives.
ULA	University of London Archives.
UMISTA	University of Manchester Institute of Science and Technology Archives

Fifty years hence the historian may ask, "But who were these founders?" True, three have in due course become Presidents and their life histories are therefore recorded in our proceedings, but what of the others? Some of them will no doubt be occupying this chair before long but the Provisional Council does not by any means include all who bore the heat and burden of these pioneer days. Before it is too late, therefore, I should like to have the records of all our pioneers placed among our archives and I promise that this will be done.

William Cullen, IChemE President, 1938

[It wasn't.]

[For the 50th anniversary celebrations in 1972] the content should be of a forward looking nature rather than a retrospective.

IChemE Council Minutes, 26 Oct 1966

Council gave support to the proposal that a history of the Institution/profession in the UK should be progressed as an IChemE millennium project ... It was believed that the ability to prepare such a history in the future, based on any kind of contemporary archive, would be prejudiced by not beginning work in the next period.

IChemE Council Minutes, 14 Mar 1996

FOREWORD

Precursors of the modern chemical industry began to emerge in Northern Europe in the middle of the eighteenth century. The Industrial Revolution boosted activities such as soap-making, glassmaking and textiles production, which required increasing quantities of chemical products. The Lead Chamber process for the manufacture of sulphuric acid, required for the production of dye, was developed in the 1740s by John Roebuck then based in Birmingham. Production of this key commodity rose steadily. By the 1820s, British annual production had reached 10 000 tons of 100% acid. By 1900, Britain was producing one quarter of the world's output with an annual production approaching one million tons.

Demand for alkalis for glassmaking and soap-making, for textile dyes and for bleach was also growing rapidly in the second half of the eighteenth century, and it became clear that existing sources of these materials would not be sufficient. In response to a prize established by the *Académie des Sciences*, Nicholas Leblanc had devised by 1791 a method for converting common salt into soda ash, which was to become the central operation of the world alkali industry for about one hundred years.

By the 1860s, a substantial industry had developed in Britain based on these technologies. In 1862, the British Leblanc industry consumed 250 000 tons of salt, generated revenues of £2 500 000 and employed some 10 000 workers. A survey in 1864 identified 83 Leblanc factories in operation in Britain, the majority located in Lancashire and the North East of England. Between thirty and forty percent of the annual production of bleaching powder was transported to North America via the ports of Liverpool and Glasgow, accompanied by up to 60 percent of the total soda output, There were also significant exports to countries of Northern Europe from factories in the North-East of England.

In parallel with these industrial developments, the conceptual basis of the science of chemistry was being established. Lavoisier, in his *Traité Elémentaire de Chimie* published in 1789, enunciated the principle of conservation of mass. The atomic theory was developed in the first half of the nineteenth century by John Dalton (1766–1844), Jöns Jakob Berzelius (1779–1848), and others. By 1850, chemical scientists had at their disposal workable assumptions regarding the structure of matter, and laws to describe observed chemical phenomena. A growing and profitable industry combined with discoveries from the young science of chemistry led to the exploration of a wide range of potential industrial products based on chemical manufacture (synthetic dyes and pigments, fertilisers, explosives and pharmaceuticals among others).

By the middle of the nineteenth century, the industry in Britain faced a number of challenges. Significant levels of atmospheric pollution (largely hydrochloric acid gas) and solid waste (calcium sulphide) were being generated. Indeed, the Leblanc system has been characterised as turning out one product and wasting two in the process. The environmental problems associated with the industry were the subject of public concern, litigation, and later of legislative action. A petition in 1839 complained that *the gas from these manufactories is of such a deleterious nature as to blight everything within its influence, and is alike baneful to health and property.* The first Alkali Act of 1863 required manufacturers to absorb 95% of the by-product gaseous hydrogen chloride. As a result, the gas emitted from the Leblanc process was treated before discharge to atmosphere with water in Gossage towers (first designed by William Gossage in 1836) to absorb the hydrogen chloride. Although improved atom economy for the process was achieved, since part of the hydrochloric acid was then used in the manufacture of bleaching powder, nevertheless a significant amount of the acid was discharged into nearby streams and rivers. However, a more robust long-term solution was emerging. Ernest Solvay and his brother developed in the 1860s a commercial ammonia-soda process based on chemistry discovered as early as 1811 by Auguste Fresnel. This continuous process was imported to Britain by Ludwig Mond in 1874, and marked a key transition point of the industry to more modern methods of manufacture.

The use of coal gas for street lighting was well established by the mid-nineteenth century, and most cities had gas works. It became difficult to dispose of the large quantities of tarry residue left when the coal gas was manufactured. It was discovered that coal tar was a rich source of aromatics, but demand for these materials did not initially absorb the supply. This situation was to change with the development towards the end of the nineteenth century of a synthetic dyestuffs industry, a forerunner of the modern organic chemicals industry.

The early development of the chemical industry in Britain (as of the Industrial Revolution more generally) had not been the result of a systematic application of scientific knowledge nor a product of the formal education system of the country. Rather its growth was the result of emerging technologies applied in an *ad hoc* manner. A Government committee investigating the state of the chemical industry in the 1860s found that board members were uninterested in science and unable to think in scientific terms. Concerns that Britain was losing its industrial supremacy, through less than adequate education in comparison with competitors, were gaining currency. Lyon Playfair's famous letter to the Times in 1867 prompted the Government to set up a Select Committee to *inquire into the provisions for giving instruction in theoretical and applied sciences to the industrial classes.*

It is against this background that a number of pioneers began to argue in the second half of the nineteenth century in Britain for the recognition of a new kind of professional properly equipped to meet the needs of the chemical industry. No doubt, to its proponents, the case for the profession of chemical engineering and for a society to foster its development was overwhelming. But an Institution of Chemical Engineers was not founded in the UK until 1922, and it was really only after the Second World War that a flourishing profession of chemical engineering became fully established in this country. It is with the struggle to establish the profession and the Institution from roots in the chemical industry of the nineteenth century to the current position as one of the 'Big Four' engineering professions with members serving a very wide variety of industries that this book is concerned. It presents a rich case study, of the development over the course of the twentieth century of an important technical profession. As such, the book will be of interest not only to chemical engineers but also to historians, sociologists and members of the science studies community. Anyone wishing to know how chemical engineering came to be will find a mine of useful information and important insights in this valuable and fascinating book.

John Perkins
President of the Institution of Chemical Engineers, 2000–2001
Courtaulds Professor of Chemical Engineering, Imperial College London, 2000–

PREFACE

The representation of chemical engineering as a skilled specialism, an academic subject and an acknowledged profession is scarcely a century old. Its multiple identities followed different national trajectories. In Britain – shaped by institutions, firms and governments – the emergence of recognised chemical engineers was the unique result of professional aspirations and contingency.

This book is not framed as an inward-looking or congratulatory institutional history. Nor does it attempt to catalogue the minutiae of every accomplishment, individual or committee ever associated with the Institution of Chemical Engineers. We have, instead, tried to paint with some precision the evolution of an important specialism against a rich backdrop of technical, economic and political events, and from a historical and sociological perspective. Neither a festschrift nor an exposé, our study charts the difficult and complex emergence, stabilisation and maintenance of an occupational identity. Not only a careful and systematic answer to the question 'What happened?' but also 'Why did events turn out as they did, and not otherwise?' and 'What can this tell us?'. This approach highlights the importance of scale, so central to the practice of chemical engineering: how a series of seemingly mundane decisions and singular episodes combined with long-term professional goals and enduring historical changes to create a peculiarly national variety of technical expert. But while avoiding both sentimentality and scepticism, it documents the remarkable rise of the profession due in no small part to the energetic actions of the Institution of Chemical Engineers.

The 'chemical engineer' was a type constructed both co-operatively and by conflict between interest groups. The product of a particular time and place, he (and more recently, she) was moulded by a changing social and technical environment. The collection of attributes that characterise this species of specialist has evolved over a century, and continues to evolve today. For this reason, professional identity is an important theme of this book. First used with some regularity as a description in 1880s Britain, the label 'chemical engineer' became variously defined: as an 'engineer of industrial efficiency' by the government after the first world war, as an affordable hybrid 'jack of all trades' by small chemical firms, as a non-existent or undesirable entity by large

firms preferring teams of chemists and mechanical engineers or, after the second
world war, as a theory-based scientific professional (indeed, the most 'scientific'
of engineers) by academics. Chemical engineers present a particularly interesting
problem for the historian because – unlike electrical engineering, for example,
where one can speak of communities of telegraphists or radio engineers – the
notion of a chemical engineering 'community' is punctured by the pervasive
secrecy that operated in the chemical industry, which organised workers along
the lines of industries (e.g. alkali production) and, even more insistently, by the
cloaking traditions of firms.

    Chemical engineers, and the specialism that they created, had an ephemeral
identity for decades. From the transitory ghost of an idea shaped by a conjunction of interests, chemical engineering evolved into a concept buoyed up by
growing consensus to compete with prevailing industrial cultures. It took on
an identity in Britain quite different from that in America and Western Europe.
Because of this fractured and mutating representation, we cannot trace backwards our present or historical notions of chemical engineering before the late
nineteenth century, any more than we can validly seek the origins of Art Deco
in ancient Egypt. Something akin to the tasks taken on by chemical engineers
arguably existed before the late nineteenth century, but chemical engineers
most assuredly did not. Consequently, the book is organised chronologically
to span twelve decades, although some chapters focus on specific aspects of
this history.

    The purpose of this book is thus to provide a critical history and analysis
of an idea and its consequences: the evolution to the end of the twentieth
century of an identity associated with the occupation, discipline and especially
the profession of chemical engineering in Britain, and the crucial role of the
Institution of Chemical Engineers in that development.

    The evolution of the profession is a topic unjustly neglected by historians
and sociologists. Until fairly recently, most historical studies of chemical engineering have been written by former practitioners. The majority of these, in
common with the few studies by professional historians, has focused on the
emergence of the profession in the United States of America, where the subject
had its most unproblematic success. Such works understandably have tended
to assume that the development of the intellectual foundations of the discipline
in Britain was strongly influenced by American experiences. By contrast, the
present study highlights the national differences in the cognitive content and
scope of the subject.

    The history of this engineering specialism reveals a wealth of insights concern-
ing the rise of the technical professions in Britain, their relevance in an indu-
strialised economy and polity, the role of the universities in the emergence of
academic disciplines, and changing definitions of the relationship between
science and technology. We therefore hope that it will be of interest to practising
chemical engineers, historians, sociologists, and the science studies community.

    The major funding for this study was provided by the Institution of Chemical
Engineers, which very generously allowed unhindered access to archives, and
provided considerable assistance in gaining access to historical materials and

interviewees. Funding for earlier research was provided to one of us (Divall) by the ESRC through a grant held by Dr. Jon Harwood, Prof. John V. Pickstone and Prof. Richard Whitley of the University of Manchester and by Manchester Metropolitan University; the National Railway Museum generously allowed time away from other duties during the latter stages of the project. Some material in Chapters 3, 4, 6, 7 and 8 has appeared previously in papers published in *Contemporary British History, Journal of Contemporary History, Technology & Culture, Social Studies of Science* and *Minerva*.

Archives and their guardians were crucial in providing historical material. We would like to thank Birmingham City Archives; the University of Birmingham; the University of Cambridge; the Cheshire Record Office, Chester; the University of Glasgow; Heriot-Watt University; Imperial College, London; the Institution of Chemical Engineers; the Institution of Civil Engineers; the Institution of Electrical Engineers; the Institution of Mechanical Engineers; King's College, London; Loughborough University; the Public Record Office, Kew; the Scottish Record Office; the Society of Chemical Industry; University College, London; the University of Manchester Institute of Science and Technology (UMIST); the Senate House, University of London; the University of Sheffield; and, the University of Strathclyde, for providing access to their collections. Bodies that provided helpful responses to direct enquiries included the British Chemical Engineering Contractors' Association, the Canadian Society for Chemical Engineering, the Indian Institute of Chemical Engineers and the IChemE in Australia.

Experienced practitioners were another valuable source of information. Those interviewed or who provided personal accounts included Mr Hugh D. Anderson, Mr. W. E. Bryden, Dr. Peter G. Caudle, Mr. A. Cluer, Mr. G. A. Dummett, Mr. Kenneth L. Emler, Dr. Trevor Evans, Prof. Sir Hugh Ford, Prof. Donald C. Freshwater, Mr. George H. Hill, Mr. C. H. Hopkins, Mr. Reg Mason, Mr. C. S. H. Munro, Mr. C. Alan Noble, Mr. Derek J. Oliver, Mr. George M. Ossipoff, Mr. Y. K. Raghunatha Rao, Prof. Jack F. Richardson, Prof. Peter N. Rowe, Mr. George Sachs, Mr. B. R. Scott, Dr. David H. Sharp, Prof. S. R. Siemon, Mr. John Solbett, Dr. Philip H. Sykes, Prof. S. Russell Tailby, Mr. Dennis R. Twist, Prof. Sir Frederick Warner and Mr. Neville A. Whiffen. Mrs. N. Valentin provided access to the files of her late husband, Dr. F. H. H. Valentin.

Academics who assisted with local information, besides some of the interviewees mentioned above, include Prof. Barry Azzopardi, University of Nottingham, Prof. P. J. Bailes, University of Bradford; Dr. Alistair Nicoll, University of Paisley; Dr. Robert Edyvean, University of Sheffield; Dr. A. Foord and Dr. H. J. Briers, Loughborough University; Dr. Ron G. Hill, Heriot-Watt University; Mr. J. Satchmel, University of Newcastle; Prof. Leslie W. Shemilt, McMaster University; and Dr. Eric L. Smith, Aston University.

The writing of a history is a collaborative affair, but we have perhaps benefited more than usual from parallel research by others. We are grateful to the historians who have contributed information and/or assistance to this project, particularly Dr. James F. Donnelly for Chapter 2 on the 'prehistory' of chemical engineering; Dr. Gerrylynn Roberts and Dr. Robin Mackie of the Open University, who shared data on membership they had collected laboriously from application forms at the IChemE; Dr. Robert Bud of the Science Museum, London; Dr. Steve Sturdy at the University of Edinburgh; Dr. Paolo Palladino at the University of Lancaster, and Dr. Clive Cohen.

The progress of the project was greatly assisted by an Advisory Group whose members included:

Dr. Robert Bud, Science Museum, London
Dr. Gerrylynn K. Roberts, Open University
Professor Emeritus Donald Freshwater
Professor Emeritus Sir Frederick Warner
Professor John Garside, UMIST
Dr. Trevor Evans, General Secretary, IChemE
Dr. Peter Varey, Director of Publications, IChemE.

The responsibility for the content and conclusions in the book nevertheless lies with us, and the analysis and views we present are not necessarily those of any of the persons mentioned above.

Finally, we should like to thank our partners and families, Karen Hunt, and Libby, Daniel and Samuel Johnston, for putting up with the fallout from this project over several years.

Colin Divall and Sean Johnston
Hayfield and York, September 2000

CHAPTER 1

INTRODUCTION:
AN ELUSIVE PROFESSION

How did 'chemical engineers' acquire a professional identity, and what was their role in inventing chemical engineering itself? These terms became increasingly common from the late nineteenth century to describe certain work practices in the chemical manufacturing industries – principally the design, adaptation and operation of chemical plant and processes. A body of knowledge with that name was being taught regularly in a handful of American and British colleges by the first decade of the twentieth century.[1] From a meagre presence in Britain before the first world war, chemical engineering became, by the end of the century, one of the 'big four' engineering professions, and a major contributor to the British economy. Yet this 'success story' is not a mere parallel of its better known American counterpart. Its sources are dissimilar and complex. In Britain, different industries harboured the malcontents who promoted the specialism; the competition of established technical professions was more obstructive; the role of the state was considerably more explicit; industrial cultures were a more heterogeneous mixture of home-grown, European and American traditions; and educational provision evolved more centrally, if episodically. In this quagmire of competing factors, the would-be profession struggled for an identity. The role of the Institution of Chemical Engineers (IChemE) proved central to this evolution, articulating a public identity while remaining alert to the exploitation of new opportunities.

Until the second world war, the nascent profession grew in the shadow of that of chemistry and, to a lesser extent, those of civil, mechanical and electrical engineering. Chemical engineers wished to take over from these professions the tasks of scaling up manufacturing processes from the laboratory to the industrial level, and activities concerning chemical plant. In 1922 the foundation of the Institution of Chemical Engineers gave an organisational focus for these

---

[1] We will adopt a semantic difference between the terms *'chemical engineer'* and *chemical engineer*. The former (in quotes) is someone identified from outside (e.g. by contemporary non-practising observers or later historians) as performing certain occupational tasks; the latter is a self-conscious individual who promoted the project of professionalisation.

claims. It institutionalised these ideas, not least by contrasting them with opposing visions of chemical and process specialists. The small association was, however, no match for the might of the Institute of Chemistry, which commanded the loyalties of the majority of professional chemists working in industry. But working between the world wars in association with a tiny number of teachers in the universities and elsewhere, the IChemE defined a distinctive form of academic training. This made it clear that the chemical engineer was not to be regarded merely as a hybrid of a chemist and an engineer.

A novel conceptual framework – based on what came to be called 'unit operations' – understood the manufacturing of chemicals as a series of discrete physical operations. The principal tasks of the chemical engineer were to ensure the containment of chemicals during the manufacturing process, to secure their movement from one stage of the manufacturing process to another, and to provide the physical conditions that would permit chemical reactions to work efficiently and economically on the large scale. All of this required a knowledge of chemistry (particularly physical chemistry) greatly in excess of that required of other kinds of engineer. But the 'unit operations' distanced chemical engineers intellectually from chemists, and suggested that the new profession might have more in common with the older engineering disciplines. After the second world war, this tentative intellectual connection with the established branches of engineering was strengthened at the organisational level. In the 1950s, the IChemE was gradually accepted as a kindred body by the principal associations of professional engineers; while the Institution did not abandon its links with chemists, it did not develop them so assiduously. By the 1960s the IChemE was a member of the Council of Engineering Institutions, unlike its one time rival, the Institute of Chemistry (by now the Royal Institute of Chemistry, RIC). By contrast, when in the late 1960s the RIC started to canvass support for a similar federation of chemical associations, the IChemE had little to do with the scheme. By the end of the century, the IChemE was one of the most important bodies relating to the Engineering Council – the chemistry associations, by contrast, had nothing to do with the organisation.

The history of this subject is clearly of some interest to its growing number of practitioners – some 25 000 in the UK at the end of the twentieth century.[2] But there are good reasons for wider attention. The subject had an intimate involvement with many British and international events during the twentieth century. It therefore illuminates that history, albeit from an unusual perspective: the story of chemical engineering reveals the 'underbelly' of British science and technology. A conventional history of intellectual discovery and technical advancement would fail to give prominence to the institutions, professional interactions, government policies, workplace categorisations and industrial pressures that were so important to changes in chemical engineering. And there

---

[2]   The IChemE in 1999 had about 21 000 members of all classes in the UK. The fraction of non-member practitioners is not known accurately but is estimated to be between 10% and 30% of all practitioners.

are also deeper motivations for scholarly interest, which the remainder of this chapter will introduce.

Precisely because of its tortuous evolution, British chemical engineering is of considerable historical and sociological interest. The tribulations and regional detours of the subject demonstrate that it was in no sense 'destined to be'. The profession was not a natural or inevitable consequence of technological progress. Its history is therefore much more than a linear sequence of dates, discoveries and developments. What, then, can its troubled growth reveal? British chemical engineering is ideal for examining the balance between professional aspirations and historical contingency in what historical sociologist Andrew Abbott has called 'the ecology of the professions'.[3] Its identity was defined perpetually by its neighbours: between chemistry and engineering, between science and engineering, was its 'proper' identity that of a hybrid, a convenient compromise, or an unique specialism? In its gradual insinuation as a sort of 'Goldilocks profession' – neither too big nor too small, neither so weak as to fail nor powerful enough to command authority, and not entirely convincing as either a 'theoretical discipline' or 'indispensable occupation' – this staking of the middle ground was long-lasting and characteristic.

SURVEY OF ANALYTICAL STUDIES

Abbott's insight that professions must be understood as co-evolving in a changing environment is near the theoretical centre of this book. He dismisses earlier claims by historians and sociologists that the attainment of professional status – 'professionalisation' – follows a regular sequence of, for example, ethical codes of practice, academic training programmes, entry examinations, vocational qualifications and licensing or, alternatively, that it can be interpreted as reflecting a straightforward strategy of the consolidation of social and economic power. Indeed, he argues that the emergence and development of professions cannot be understood at all adequately as isolated movements; instead they must be analysed in their particular historical contexts as parts of evolving systems of interdependent yet competing occupational specialisms. Within this social ecology, Abbott urges an initial focusing on groups that undertake common work rather than on the separate ways they might organise institutionally: only then should we shift the focus of our analysis to discover how the link between an occupational group and 'its' work is created and anchored by formal and informal social structures, practices and discourses in such a way that the group comes to gain the degree of social and economic authority characteristic of a 'profession'.

Abbott's key argument is that the historical development of professions hinges on 'jurisdictional disputes' between occupational groups; jurisdictional claims over 'professional' tasks in the workplace motivate and shape subsequent organisational developments. Survival in the competitive system of the profes-

---

[3]   Andrew Abbott, *The System of the Professions: An Essay on the Division of Expert Labour* (Chicago: Chicago University Press, 1988).

sions is promoted by the particular tactics adopted by practitioners to strengthen their collective claims to authority. The history of chemical engineering as a profession supports the view that the achievement and maintenance of jurisdiction over technical tasks may require the endorsement of several social groups, including, for example, employers and government.

Yet sociologists of the professions such as Abbott and Keith MacDonald have thus far treated the engineering professions cursorily.[4] Historians, for their part, have long been concerned to understand the politics of organised interest groups that has characterised the workings of the British state. But even the most important work has virtually ignored the part played by the professional institutions of the technical occupations. Keith Middlemas's magisterial three volume study, *Power, Competition and the State* does not appear to contain a single reference to the engineering institutions, although it mentions on numerous occasions the Engineering Employers' Federation and the manual engineers' trades unions.[5] Middlemas's work does, however, draw our attention to the shifting alliances and tensions that exist between different parts of the state and government. Perhaps it is even more surprising that Harold Perkin's *The Rise of Professional Society: England Since 1880* is almost as neglectful of technical professions.[6]

There is as yet no comprehensive study of the interaction between the various parts of the state, the associations of professional engineers and related scientific workers, and engineering employers in Britain. The politics of those technical occupations that lay claim to professional status remains a surprisingly neglected area of the historiography of modern Britain. Nor have there been studies by analysts of historical and sociological processes dealing with the emergence of 'sub-professions' – particularly important in Britain – such as nuclear engineering, which for a time after the second world war was seen as a logical territory for expansion by chemical engineers. It is significant that engineering professions since the second world war, led by the chemical engineers, have been increasingly dominated by a scientific perspective. The evolution of explicitly scientific professions has attracted the attention of some historians of science and technology.[7] Yet the failure of researchers in 'mainstream' history to engage this issue of the gradual but nearly continuous shifting of the balance between technical 'art' and 'science' may explain the absence of substantive work on these newer engineering specialisms.

THE CENTRALITY OF IDENTITY

We attempt to redress these deficiencies through a detailed study of chemical engineering from a particularly fruitful perspective: that of individual, profes-

4   Keith M. MacDonald, *The Sociology of the Professions* (London: Sage Publications, 1995).
5   Robert Keith Middlemas, *Power, Competition and the State* (Basingstoke: Macmillan, 1990).
6   Harold Perkin, *The Rise of Professional Society: England Since 1880* (London: Routledge, 1989).
7   See, for example, C. A. Russell, Noel G. Coley and G. K. Roberts, *Chemists by Profession: The Origins and Rise of the Royal Institute of Chemistry* (Milton Keynes: Open University Press, 1977).

sional and institutional identity. Such an approach is timely in two respects. First, identity has increasingly served as the starting point for a wide variety of investigations in cultural history and sociology. And second, a self-conscious awareness and promotion of identity has been a phenomenon of modern times, as argued by Anthony Giddens.[8] The extension of the professional identity of chemical engineers from the workplace and university successively to regional, national and international institutions is mirrored by larger-scale changes in society.[9]

As suggested by the capsule history above, and developed as the underlying theme in the following chapters, chemical engineers have assumed multiple identities through their history. These characterisations have alternately been claimed by the practitioners themselves and imposed upon them by others. While seeing themselves as a social or professional 'group', others nevertheless relegated them to a mere 'category' of worker, if indeed they were singled out at all. Indeed, the more common practice of chemical firms in the early years was to promote a 'corporate' or 'industrial' identity – attaching employees to a particular firm or chemical process for their entire working lives. Hence the identity of the 'chemical engineer' could not be established unilaterally. As Richard Jenkins has discussed, identity is the result of negotiation or agreement between parties.[10] Nascent 'chemical engineers' had to work out not only in what respects they were similar to each other, but how they all differed as a group from others.

Different identities have also been serial and concurrent. The definition of the 'chemical engineer' evolved episodically in the eyes of industry and the state, yet was simultaneously different for various engineering and scientific communities. This heterogeneity and malleability of these identities was influential in the ultimate success of the profession.

The profession's identity had several dimensions which delimited its frontiers. The chemical engineering profession adopted a succession of positions along the science/engineering axis, for example. Another distinctive attribute in the profile of working chemical engineers was their particular educational background, which had an enduring relationship with social class. During the past quarter century, too, gender has become a significant variable refashioning their professional identity. And the content of 'chemical engineering' practice has been strongly circumscribed by local industrial conditions, hence the importance of considering regional variations. Regionalism has also delineated the profession by introducing tensions between the organisational centre of the IChemE and its peripheries in Britain and the Commonwealth, and between the IChemE and American and European institutions. Abbott's metaphor of professional jurisdiction as territorial competition draws explicitly on this geographical dimension for good reason.

---

[8] Anthony Giddens, *Modernity and Self-Identity: Self and Society in the Late Modern Age* (Cambridge: Polity, 1991).

[9] See Jonathan Friedman, *Cultural Identity and Global Process* (London: Sage, 1994).

[10] Richard Jenkins, *Social Identity* (London: Routledge, 1996).

Similarly, certain aspects of identity have been advanced by particular tactics. The cognitive identity of the discipline of chemical engineering was strengthened by the innovative concept of unit operations. The courting of patronage from government departments and industrial associations advanced the validation of the profession; the organisers explicitly recognised a political dimension. So, too, were the affinities of professional chemical engineers strengthened by links (at various times) with other professional engineering and scientific societies. By contrast, an *occupational* identity was asserted with difficulty, given the established employment categories of 'engineer' and 'chemist' favoured by industry and state institutions alike.

In concert with such tactics went the invention of a professional image, which included the elaboration of legends of pioneering antecedents and critical events to buttress a sometimes fluid identity.[11] Such self-conscious image building even employed potent symbolic elements, utilising the award of medals based on founding fathers, the iconography of institutional seals and the rhetoric of Presidential addresses and institutional mottos. Engagement with the past, however, varied through the century, as reasons altered for praising or neglecting past events and representations. Considering such constituents, the history of this specialism bears notable parallels with that of some national and ethnic groups.[12] Just as Gerard Delanty has written of Europe, 'the European idea emerged and was sustained more by conflict and division than by consensus and peace' and arguing that it was 'a contested concept ... about exclusion and the construction of difference based on norms of exclusion', so Andrew Abbott contends that professions evolve by competition and territorial definitions.[13] More generally, the analogy of professions as struggling nations is strengthened by the imprecision of their definitions. Hugh Seton-Watson's observation that 'no "scientific definition" of the nation can be devised; yet the phenomenon has existed and exists', is equally apt for professions.[14] And just as for nationalism and nations, professionalisation is not necessarily a process of formalising pre-existing and natural groups of specialists, but rather the invention and maintenance of such groups. Questions of authority and representation are at the heart of the creation of professions.

As suggested by this brief discussion, our point of departure is a view of the identity of chemical engineers as 'non-essentialist', that is, as not having a fixed, authoritative meaning. Their identity has always been subjective, contested and shaped by their relationships with 'others'. As such, it reveals much about not only those who became 'chemical engineers', but of those who did not.

---

[11]   To speak of 'invention' is not to imply any cynical promotion, or to dispute the importance of the subject and its reality to practitioners and beneficiaries, but to stress that it is a product of history and culture as much as a 'natural' technological category.

[12]   See, for example, Benedict Anderson, *Imagined Communities: Reflections on the Origin and Spread of Nationalism* (London: Verso, 1991); Gerard Delanty, *Inventing Europe: Idea, Identity, Reality* (Basingstoke: Macmillan, 1995); and, Murray G. Pittock, *The Invention of Scotland* (London: Routledge, 1991).

[13]   Delanty, op. cit, pp. vii and 1.

[14]   Hugh Seton-Watson, *Nations and States* (London: Methuen, 1977), p. 5.

THE IMPORTANCE OF THE CHEMICAL ENGINEERING PROFESSION

Our work aims to tie together previously isolated empirical data and disparate analytical approaches. A contextual history of a British engineering specialism can be considerably more than the sum of its parts, disclosing as it does the interactions and linkages between players that are as important as the individual professions themselves. A similar objective pertains for the bases of our analysis. The sociology of the professions has for too long presumed a simple model of scientific and technical expertise, taking it as universal, progressive and uncontroversial.[15] Sociologists of scientific knowledge, on the other hand, while more sophisticated in their treatment of such evidence, have tended to neglect the organised social structures – the professions – often responsible for and underlying its generation. To fully explain the nature of these entities in the British context, we therefore consider professional bodies, their members, their work and their productions as equally important components in an historical milieu. The third fertile research tradition that must be incorporated is the flourishing history of technology, which recently has brought new perspectives for understanding the technological aspects of society. Several writers acknowledge the success with which science and technology have been harnessed to the task of modernising the British economy. A fine-grained study of the historical development of one of the major professions could not be more propitious.

There are other questions that a study of chemical engineering history can illuminate. It is often said, for example, that the performance of the British economy is damaged by the influence of political structures and occupational organisations dating from the earliest days of industrialisation. In particular, a good deal of criticism has been levelled at the organisation of professional engineers. Some commentators point out that the engineering associations – established from the early nineteenth century on the model of the self-governing bodies of the legal and medical professions – are unusually distanced from the concerns of business and the state. Critics compare this state of affairs unfavourably with those among Britain's industrial rivals in Europe, North America, the Pacific Rim economies, and elsewhere; there, it is argued, engineers are much better integrated with wealth-producing institutions and structures. In this context, chemical engineers are of particular interest since they tend to work in one of the few industrial sectors – chemical and allied manufacturing – where Britain's economic record clearly bears comparison with that of its competitors.

An appreciation of the IChemE's relations with other professional groups is thus important for this book. 'Manpower' policy, for example, was a vital domain in which the IChemE had to persuade the various governmental, educational and industrial authorities of the distinctive character and value of the chemical engineer if the profession were to thrive. The Institution achieved this goal, particularly after the second world war, partly by mobilising support

---

[15]   E.g. Peter Whalley, *The Social Production of Technical Work: The Case of British Engineers* (Basingstoke: Macmillan, 1986).

among groups of industrialists, politicians and high officials who were not
persuaded of the adequacy of the provision made by chemists. The analysis of
how this was done suggests that there has been a greater measure of agreement
between the IChemE and certain industrial employers than one would expect
from the arguments of the critics of the engineering associations.

This study is also of significance for the literature on the role of corporatism
in British politics. In an important series of articles, Kevin McCormick, for
instance, has analysed the development since 1939 of new forms of state power
intended to recognise, legitimate and incorporate organised interest groups.[16]
He argues that corporatist structures should be conceived as lying along a
continuum: at one end are those forms of organisation involving a high degree
of state intervention, centralisation and coercion of the incorporated bodies
('state corporatism'); at the other, those in which relatively autonomous, representative bodies come together in voluntary association ('societal corporatism').
McCormick suggests that the degree of state intervention in a particular domain
of policy has historically depended upon two factors. First, the changeable
perception of industrialists' interests by politicians and different parts of
Whitehall; and, secondly, on the degree of co-ordination between departments
of state. He concludes that attempts to create durable corporatist institutions
at the national level have foundered on the lack of corporate organisations at
lower levels, including that of industrial employers, and on the tendency of the
groups that are incorporated to pursue their own interests in their own way.

The history of the professional organisation of chemical engineers is grist to
the mill for all of these points. The degree to which the IChemE became
incorporated into formalised state structures of 'manpower' planning varied
considerably through the century. This was at least partly a result of the
changing perception of employers' 'needs' by parts of the state. But it is
important to realise that under certain circumstances, the IChemE played a
large part in shaping the state's perceptions of these 'needs'. The institution
was most successful when it functioned within the chemical and process industries as a kind of corporatist body of the 'societal' kind – it secured significant
policy concessions when it was able to demonstrate to high officials a substantial
measure of agreement among a representative body of industrial employers.

Our work suggests that it is necessary to attend to the particularities of
historical episodes if we are to understand the circumstances under which a
voluntary association like the IChemE can secure a consensus among industrial
interests.

THE MAKING OF CHEMICAL ENGINEERS

We devote considerable attention to the history of chemical engineering education. Our focus on the professional aspects of this process has an important
bearing on the literature concerning the role played by universities and aca-

[16] K. McCormick, 'Engineers, British culture and engineering manpower reports: The historical
legacy revisited', *Manpower Studies* (1981), 131–5.

demic knowledge in the formation of technical experts. As discussed above, recent historians and sociologists have largely turned away from trying to agree on the characteristics that define a professional ideal type and instead have concentrated on the ways in which certain occupational groups struggle to achieve social and economic authority as 'professions'. There seems to be agreement among many commentators that a crucial stage in the making of any profession is the founding of a means of producing specialist, formalised knowledge. Simply put, control over the production of such knowledge is held to be *a* cause – if not *the* cause – of 'professionalisation'. One particularly influential version of this thesis holds that universities have become increasingly central to professional identity as practitioners have based their claims to social status on technical expertise underpinned by codified knowledge.

We do not seek wholly to dissent from this kind of analysis, which might be called the 'academic account of professionalisation'. But we agree with those analysts who suggest that it can be fruitful to consider more carefully the role played by universities in the production of formal knowledge of practice. Historically, the codification of technical expertise is interesting because so often it has been the chief point of conflict in Abbott's 'jurisdictional' disputes between occupational groups. Yet theorists rarely acknowledge, other than in passing, that the loci and practices of the production and transmission of such knowledge are historically contingent and culturally specific. In particular, the growth of vocational knowledge and learning within the universities can only be properly understood if one considers the attempts by certain groups of academics to gain authority within the academy. In other words, academic accounts of professionalisation often turn out to rest on accounts of academic professionalisation that are themselves poorly grasped.

Through the case of chemical engineering we seek to illuminate what Abbott aptly describes as 'the embarrassing British case' for academic accounts of professionalisation. Following Abbott, we agree that for much of the nineteenth century the association between the universities and, in particular, the evolving engineering professions, was not particularly strong. But we differ from him in his implication that this was more or less a constant state of affairs. In fact the universities became increasingly important from the middle of the nineteenth century, even for those branches of engineering that had already achieved a high degree of social status and economic authority without the benefit of a close association with the academy. This shaped the nature of the later relationship between the universities and the professions of civil, mechanical and electrical engineering. But with chemical engineering, matters were very different. The occupation emerged as an industrial specialism somewhat later, and the universities and codified knowledge played a very much more marked role in the struggles of the early practitioners in their jurisdictional disputes with cognate experts. The dynamics, and the eventual resolution, of the tensions between university academics and practitioners in the realm of chemical engineering also contrasted quite markedly with those in the other branches of engineering, and we attribute such variation to differences in the wider social,

economic and institutional contexts of both academic and occupational
practice.

The very success of these initiatives also raised problems concerning the
appropriate mix of academic education and practical training. The production
of codified knowledge and its transmission to would-be chemical engineers in
the academy was as much a kind of professional work as the forms of industrial
practice that they underpinned. By analysing chemical engineering academics'
efforts to assert their authority as professionals within the university, we lend
weight to the more general claim that jurisdictional negotiations between
practitioners and academics are an important, and probably inevitable, aspect
of the making of any occupation once it becomes associated with 'a body of
relatively abstract knowledge, susceptible of practical application'.[17]

SCOPE

Our study thus attempts a contextual history of the chemical engineering
profession by drawing on economic, technological, cultural and sociological
aspects. The essence of the story is the recognition of the 'chemical engineer'
as a distinct type of specialist; attempts to claim intellectual and occupational
tasks from chemists; and, the consolidation of these jurisdictional ties in the
peculiar environments of twentieth century Britain.

As suggested above, this book concentrates on the social history of chemical
engineering as a profession in Britain, and particularly the part played by the
IChemE in its growth. The time period consequently focuses on the period
from about 1880, when the first attempts to found such an organisation were
made and when the expanding chemical industry began increasingly to employ
such specialists, to the end of the twentieth century. We are not so insular as
to suggest that indigenous developments were solely important, however.
Comparative aspects of the subject, such as the intellectual and professional
connections with chemical engineering in the USA and developments in
Commonwealth countries, are treated where relevant but do not form our
central thrust; we concentrate on the deciding factors for the British profession.

The evolution of chemical engineering is studied as an organised *occupational
activity*, as an *academic discipline* and (most intensively) as a *profession*. The
activities examined include technical practice, working environment and social
interactions. We explore the practical scope and demands of a career in chemical
engineering – as an employee, designer, plant supervisor, consultant, academic
and Institution council member. In addition, the interplay between chemical
engineers and their peers, and with society at large, is highlighted. We nevertheless recognise that writing a balanced social history of the occupation is
hampered by scattered and incomplete primary sources. The 'view from the
coal face' was little documented in official records. Practitioners' reminiscences
can suggest merely the variability and uniqueness of each job, firm and activity

---

[17]   MacDonald, op. cit.

over the century. A representation of what it meant to practise chemical engineering in past decades cannot adequately be grasped from anecdotes.

The *discipline*, however, can more faithfully be mapped. We elucidate the conceptual attributes defined by chemical engineers, by educators and by their contemporaries, targetting the intellectual ideas that played a role in distinguishing chemical engineering from other academic subjects. These ideas included 'unit actions', 'unit processes', costing, and mass and energy transport. Vaunted in the period after the first world war, such conceptual entities fell largely outside the domain of practising chemists and mechanical engineers. This intellectual framework therefore distanced chemical engineers from chemists (and particularly from the closely related occupations of industrial chemist and chemical technologist), and suggested that the new profession might have more in common with the older engineering disciplines.

The investigation of *professional* aspects includes the social definition of chemical engineers as specialists. We have studied their visibility, status and perceived importance relative to other professionals. The standards of qualification defined by the IChemE were crucial to these questions, as were the continuing interactions with government and industry for recognition.

The context in which these aspects of chemical engineering evolved is highly relevant. We account for the role of the IChemE as a focus for a professional identity, as an activist for a disciplinary definition, and as a liaison between government, industrialists, practitioners and educators. And the study does more than explain the past: by exploring the causes of the trajectory of British chemical engineers, it also reveals constraints on their future course.

CHAPTER 2

'... THAT DOUBTFUL AND INDESCRIBABLE PERSON, THE CHEMICAL ENGINEER ...'

INTRODUCTION

In 1839, Andrew Ure, during a discussion of the lead chamber process in his *Dictionary of Arts Manufactures and Mines*, made the following remark about the steam nozzle in the chamber:

> It deserves to be noted that the incessant tremors produced in this pipe by the escape of steam, cause the orifice to contract ... Provision should therefore be made against this event, by the chemical engineer.[1]

So far as is known this is the first use of the term chemical engineer in print. It did not appear in subsequent editions of the book, and stands as an intriguing but somewhat anachronistic early reference. Yet it set the tone for what was to follow. Ure was a chemist, said to be the first man to earn his living as a consultant analytical chemist in the United Kingdom.[2] His comment related to the heavy chemical industry. And the context, together with other remarks by Ure, suggest that he was using the term 'chemical engineer' as a shorthand title for a mechanical engineer working in the chemical industry.

After Ure's early statement, references to chemical engineering were, to say the least, intermittent. The Oxford English Dictionary gives no 19th century exemplar use of 'chemical engineering' or 'chemical engineer'. However, by the mid-1870s a few examples can be found. While it is difficult to recover this early history of the title, it seems likely that there was a link with mechanical engineering. In 1966 Hardie and Pratt claimed that in the 1870s 'process inventions ... generally had more engineering in them than they had chemistry', though the basis of their claim is obscure.[3] Works were of course employing

---

[1] Andrew Ure, *A Dictionary of Arts Manufactures and Mines* (London: Longman, Orme, Brown, Green & Longman, 1839), p. 1220.

[2] For this comment (p. 314) and more on Ure (1778–1857) see W. V. Farrar, 'Andrew Ure, F.R.S., and the Philosophy of Manufactures', *Notes and Records of the Royal Society 27* (1972–3), pp. 299–324.

[3] D. W. F. Hardie and J. D. Pratt, *A History of the Modern British Chemical Industry* (Oxford: Pergamon, 1966), p. 40.

steam engines. Purpose-built revolving furnaces had appeared in the Leblanc industry from 1853, together with mechanisation of the earlier 'decomposing' of the sodium chloride with sulphuric acid. The finishing process, too, began to undergo mechanisation. The advantage of such mechanisation appears often to have been less any improved technical quality of the product, and perhaps not even greater throughput, but rather 'the dispensing with skilled labour'.[4] It seems likely that the Leblanc industry, which was relatively advanced mechanically and operated on a large scale (one-half a million tons of salt were decomposed annually by the 1870s), was the first to make extended use of steam power, and perhaps of specialist engineers.[5] The semi-continuous Solvay process extended this mechanisation. Smaller and less capital intensive chemical industries no doubt had less motivation for mechanisation and consequently less obvious call for the 'engineer'. But all of this is somewhat speculative since there has been little systematic study of the machinery associated with the chemical industry. The industry was notoriously heterogeneous and attention has been focused mainly on the chemistry of processes. The interplay between machinery and chemical process runs through much of what follows, but that is not to say that chemical engineering was some kind of fusion between mechanical engineering and chemistry. Indeed it was precisely in refusing such a perspective that the field had its origins as a distinct discipline. Despite all of this, at base we are simply unprovided with evidence as to why a few men in the 1870s began to use the term chemical engineer to denote something different from a species of mechanical engineer working in a chemical factory.

As we have seen, Andrew Abbott sees the ultimate ground of the trajectory and relationships of professions in the attempt to gain what he calls jurisdiction over a particular class of tasks. Chemical engineering appears firmly grounded in the attempt to create and define such a class of tasks: the control of the chemical and physical processing of bulk materials on an industrial scale. While they might not have framed it in quite this way, those who first used the term do appear to have had something like this jurisdictional claim in mind. If this point is accepted the generic aspirations of chemical engineering begin to appear implicit from the beginning. But, as Abbott recognises, the social and institutional circumstances of the attempt to establish a jurisdiction can itself be constitutive of the task domain. He distinguishes 'objective' and 'subjective' influences on tasks and jurisdictional claims. Chemical engineering is set within a technical sphere, which has its share of objective influences. But the emergence of chemical engineering was subject to a range of subjective influences, although 'contingent' or 'social' might be better adjectives. It is by no means clear that any major degree of jurisdiction (or even clarity about the viability and boundaries of the task domain over which jurisdiction was sought) had been established, even when the field began to develop the academic and institutional

---

4   J. Lomas, *A Manual of the Alkali Trade. including the Manufacture of Sulphuric Acid. Sulphate of Soda and Bleaching Powder* (2nd. ed. London: Lockwood, 1886), p. 182.
5   L. F. Haber, *Chemical Industry during the Nineteenth Century* (Oxford: Oxford University Press, 1958), p. 59.

trappings of professionalism in the third decade of the twentieth century. The domain was still less clear during the period with which this chapter is concerned, and the jurisdiction barely even embryonic.

SOME STATISTICAL BACKGROUND

Though they emerged only in the 1870s, the terms chemical engineer and chemical engineering were relatively common by the end of the nineteenth century. They were already the focus of disagreement, and experiencing some hostility from competitors, particularly chemists. A few statistics will give some sense of the extent to which the title was used. Of the 300 or so members of the Society of Chemical Industry at the time of its first General Meeting, in June 1882, fifteen (about 5%) described themselves as chemical engineers.[6] Even this small proportion was to decrease during subsequent decades. In 1900 only 42 of the total British membership of about 2300 described themselves as chemical engineers, that is, about 1.8%. By 1917 the number was 65, showing a moderate rise to about 2.2%. However when the Chemical Engineering Group of the Society was established in 1918, it attracted some 400 members. Chemical engineering appears then to have been acknowledged as an activity occurring within chemical manufacturing, and even to have attracted some limited affiliation. But it was not understood as a well-defined occupation, let alone one possessing the trappings or power associated with a profession. Chapter 3 will examine the effect of the first world war in catalysing the beginning of the transition.

The new field had connections with both chemistry and mechanical engineering. Yet, while many of the early discussions appeared to see mechanical engineering as the parent activity, with chemical manufacturing merely the context, the self-descriptions of the first members and associates of the IChemE tell a different story. In 1917 the Society of Chemical Industry published one of its last membership lists. How those who were to become members and associates of the IChemE gave their occupations at that time is summarised in Table 2-1.

Table 2-1 Self-descriptions of future members of the IChemE in 1917

Self-description	Number
Chemist	39
Manager	14
Chemical engineer	13
Engineer	9
Manufacturer	5
Superintendent	2
Academic	2
Other or none	5

---

[6] *J. SCI 1* (1882), 250.

When it is recalled that managers frequently had a background in chemistry, it is apparent that the genealogy of chemical engineering, while it encompassed both chemistry and mechanical engineering, favoured the former. This asymmetry was reflected in the stances adopted towards the two fields by the key protagonists of chemical engineering. While relations with chemistry, or more especially academic and laboratory chemistry, received careful, though not necessarily hostile, attention, particularly at an institutional level, those with mechanical engineering were less significant.

TWO EARLY CHEMICAL ENGINEERS

We return now to the origins of the term. The two most prominent users of the title 'chemical engineer' in the 1870s and 1880s were John Morrison and, a little later, George E. Davis. Both men were involved with the heavy chemical industry, Morrison on what were later known as Merseyside and Tyneside, Davis mainly in the Manchester area. Both had begun work in industrial laboratories and both had a background in chemistry. Little more of direct relevance to this study is known about Morrison, beyond his apparently very enthusiastic use of the title chemical engineer. It was included in his membership entry for the Chemical Society, where occupations were not commonly given. How his usage of the term was intended, or interpreted by his contemporaries, is difficult to judge. He wrote a few papers on matters connected with the chemical industry, but did not give chemical engineering any explicit attention. Nevertheless, in those few papers some of the characteristic concerns of the chemical engineer are already visible, most notably an emphasis on the centrality of scaling up, after working first with 'an experimental plant on a manufacturing scale'.[7] In no sense however, as we shall see, could Morrison be seen as a mechanical engineer working in a chemical context. Further, the idea that manufacturing chemistry was some unproblematic 'application' of the laboratory work of the academic chemist – so common in the rhetoric of early Professors of Chemistry such as Edward Frankland (Owens College), Alexander Williamson (University College) and even Henry Roscoe (Owens College) – was dismissed. Morrison also clearly expressed the situation of technical employees in the chemical industry, in a way that reflected important structural aspects of employment in the chemical industry.

> Most of us are of laboratory descent – men who, with a strong love to the calling of our choice, have early yielded to the discovery that we could never look to analytical chemistry other than as a sort of *pis-aller* – a kind of out-of-elbows trade – forming simply a stepping-stone to an indefinite something better. That at first undefined something, however, speedily resolved itself into a managership, and a managership, therefore, became henceforth the summit of our most ardent hopes.[8]

This movement, from laboratory chemist to manager (the latter term loosely

---

[7]   J. Morrison, 'On the manufacture of salt-cake', *J. Soc. of Arts 24* (1875–6), 639–45. See also 'On the manufacture of caustic soda', *Trans. Newcastle Chem. Soc. 3* (1874), 27–55.

[8]   Morrison, *ibid.*

understood) appears to have dominated the trajectory of technical workers in the chemical industry during the period with which this section is concerned. By contrast, few examples of movement from mechanical engineering to works management are to be found.[9] Chemical engineering would struggle to find a location within this structure.

George E. Davis (1850–1907) was an altogether more visible figure than Morrison. He occupied an important place in the process by which chemical engineering came to be recognised as a distinct occupational specialism, though he died nearly twenty years before the formation of the IChemE. Born in Eton and educated at Slough Mechanics' Institute and the Royal School of Mines, he apparently took no formal qualification at the latter. He was employed from 1871 as an analyst at the bleach works of Richard Bealey & Co. in Radcliffe. Over the next few years he moved between several works in Cannock Chase, St. Helen's and Runcorn, taking increasing responsibility for the processes proper. In May 1878 he was appointed to the Alkali Inspectorate, a post from which he resigned in July 1884.[10] He established a successful chemical engineering consultancy practice, eventually joining his brother Alfred R. Davis (1862–1936), to form the firm of Davis Bros. In 1887 he founded the main commercial periodical of the chemical industry, the *Chemical Trade Journal*.[11]

Davis has three claims to a place in the history of British chemical engineering: his attempt to constitute what became the Society of Chemical Industry as a Society of Chemical Engineers, the lecture series which he gave in Manchester in 1888 and his textbook on chemical engineering, published in 1901–4. Yet when he died, in 1907, chemical engineering still occupied an enigmatic place in the intellectual and professional landscape of the chemical industry. We will have more to say about him below.

CHEMICAL ENGINEERING AND THE SOCIETY OF CHEMICAL INDUSTRY

The Society of Chemical Industry (SCI) originated from a sequence of meetings which occurred in the Merseyside area between 1879 and 1881. It was preceded by a few small-scale organisations for industrial chemists and manufacturers in that region and on Tyneside (the Newcastle Chemical Society, 1868, the

---

[9]   J. F. Donnelly, 'Consultants, managers, testing slaves: changing roles for chemists in the British alkali industry, 1850–1920', *Tech. & Culture* 35 (1994), 100–28. Idem., 'Structural locations for chemists in the British alkali industry', in E. Homburg, A. S. Travis and H. G. Schröter (eds.), *The Chemical Industry in Europe, 1850–1914. Industrial Growth, Pollution, and Professionalization* (Dordrecht: Kluwer, 1998), 203–20.

[10]   These details are taken from the Davis papers, Science Museum Library, London, and supplemented by D. C. Freshwater, 'George E. Davis. Norman Swindin, and the empirical tradition in chemical engineering', in W. F. Furter (ed.), *History of Chemical Engineering* (Washington, D.C.: American Chemical Society, 1980), pp. 97–111 (though corrected at some points) and N. Swindin, 'The George E. Davis Memorial Lecture', *Trans. IChemE* 31 (1953), 187–200, together with some of Davis's contemporary publications such as G. E. Davis, 'On the estimation of nitrous acid in nitro-sulphuric acid', *Chem. News* 24 (1871), 257–59 and 'A new process for the production of carbonate and caustic soda, without the formation of any noxious waste, and the recovery of the sulphur', *J. Soc.Arts 25* (1877), 633–42.

[11]   The first edition was published on 27 May 1887. Davis Papers, Science Museum, DAV 6/4.3.

Tyne Social Chemical Society, 1873, and the Faraday Society, based in Widnes, St. Helen's and Liverpool, 1874).[12] At meetings of these bodies, matters of general technical or commercial interest were discussed, some broadly within the remit of chemical engineering, though the term was rarely used. There is little record that any of the societies aspired to any fuller sense of professional identity, and what there is points towards chemistry as the central affiliation. George Davis, who had been Honorary General Secretary of the Faraday Society, later wrote that his then employer David Gamble, hearing of the Society,

> ... told me that he had heard I was connected with a secret society and desired to know whether this was the case or not. I told him that whether it was called a secret society or not, it was a combination of chemists having no interest but their own profession, and no adversaries so far as they were concerned.[13]

In 1872 the Tyne Social Chemical Society fiercely rejected a move to join with its contemporary, which served a more elevated clientele, stating in its response that it was 'a society of managers and chemists only, and not of manufacturers'.[14] As Morrison so clearly described, the standard trajectory for such men (including both himself and George Davis) was from employment in an analytical laboratory in the works to managerial responsibility. Davis would join the Institute of Chemistry in 1878, while Morrison was, as has already been mentioned, a member of the Chemical Society.

The first attempt to establish a larger scale organisation orientated towards industrial chemistry occurred around 1880, and led to the formation of the SCI as a national society. In the period between the earliest recorded meeting in this process, in November 1879, and the inaugural meeting of the SCI proper, in April 1881, the character of the proposed society altered. Whatever stirrings there may have been towards some form of organisation with broadly professional concerns amongst the predecessors of the SCI, the process led in practice to a society dominated by owners and academics and orientated towards technical and commercial concerns.[15] More significantly, from our point of view, there were several attempts during this period to focus the society on chemical engineering, at least at a titular level. In April 1880, at a meeting in Owens College, two chemical manufacturers – Hugh Lee Pattinson, of Tyneside, and Eustace Carey, of Merseyside – suggested that the proposed organisation take the title 'Society of Chemical Engineers'. William Hunt, of Wednesbury, argued for the title 'Institution of Chemical Engineers'.[16] Carey's

---

[12]   J. Donnelly, 'Defining the industrial chemist in the United Kingdom, 1850–1921', *J. Soc. Hist.* 29 (1996), 779–96.

[13]   *J. SCI 34* (1915), 749. On Sir David Gamble (1823–1907) see *J. SCI 26* (1907), 245.

[14]   *Chem. News* (1872), 57. C. Russell, *Science and Social Change, 1700–1900* (London: Macmillan, 1983), pp. 210–11.

[15]   Society of Chemical Industry, Minute Book of the Preliminary Meetings (held at the SCI offices in London); *Chemical News 43* (1881), 164. Donnelly, 'Defining the industrial chemist', op. cit.

[16]   The most prominent of these men was Eustace Carey (1835–1915), manager of the Leblanc firm Gaskell Deacon and later Secretary of the United Alkali Co. *J. SCI 34* (1915), 261.

suggestion was carried unanimously. Reporting the meeting, *Chemical News* noted that this title was

> ... very largely supported on the ground that most of the difficulties in the application of chemical science arose on the question of apparatus and physical condition involved.[17]

It is necessary to be careful about the significance attributed to the proposed title. It may have been no more than an expression of the view that manufacturing chemistry was very different from laboratory chemistry, which was a commonplace within the meetings of the embryonic society. D. B. Hewitt, of Brunner, Mond, told the inaugural meeting:

> ... there was not a sufficient supply of men of engineering skill also versed in the arts of manufacturing chemistry. They could obtain plenty of men capable of carrying through processes in the laboratory, but not competent to apply these on the large scale.[18]

The proposed name was ratified in December 1880, and was used in communications to the press, with the further comment that:

> It may indeed afterwards prove desirable to found a distinct branch of the Engineering Profession.[19]

By this time George Davis was the Secretary of the Society.

However, when it was eventually inaugurated, in April 1881, the Society adopted the title which it still retains. In 1895 Davis wrote as follows about this turn of events:

> ... the idea which ran through the minds of all concerned was the formation of a Society to represent chemical engineering ... (but) when the meeting was held one or two of the professorial type foresaw that if it was a Society of Chemical Engineers it was more than probable they would be left out in the cold.[20]

He himself claimed to have suggested the more anodyne title, in order to reconcile the various groups. Davis's comment may perhaps be interpreted to mean that he did indeed see the term chemical engineering as distinguishing the practice of industrial chemistry from academic chemistry, rather than envisaging the possibility of its existence as an academic discipline. Despite its absence from the Society's title, chemical engineering still appeared prominently in its Bye-laws, representing one of its two primary aims:

> To promote the acquisition and practice of that species of knowledge which constitutes the profession of a Chemical Engineer.[21]

But this nominal emphasis appeared to make little concrete impact. Davis later remarked that a local manufacturer had turned to him after the first vote on

---

[17]   *Chem. News* 43 (1881), 164.
[18]   *The Chemist and Druggist* 23 (1881), 182–4.
[19]   Minute Book of the Preliminary Meetings; *Chem. News* 43 (1881), 81 and 154.
[20]   *J. SCI* 34 (1915), 750. See also *ibid.*, 14 (1895), 935.
[21]   *Ibid.*, 1 (1882), 288.

the name and said that he 'did not know that there was an animal of that genus in existence; I have heard of civil engineers, electrical engineers etc., but never of a chemical engineer'.[22]

Chemical engineering was, then, a term with some contemporary currency, but its significance is far from clear. It is not difficult for a small group to influence such events, and what its members intended by the term remains obscure. Nevertheless, despite its obscurity, over the ensuing decades the visibility of the term grew. This growth might be explained, at an ideological level, by the strong association between engineering and 'practical' knowledge and at a more concrete level by the increasing use of machinery within the heavy chemical industry. But, whatever influences were operative, they were insufficient to supply 'chemical engineering' with a coherent identity, cognitively, academically or institutionally. Creating even the beginnings of such an identity was to occupy the best part of four decades, and, even with the establishment of the IChemE, was far from complete. Within this process, none of the several sites at which such an identity might be sustained (notably in academe, in a public statutory system of control or in the workplace itself) appears to have been successfully colonised by chemical engineering as an independent field. Jurisdiction was an even more distant prospect.

A DISCURSIVE EXISTENCE: 1881–1909

After the failure to ground the SCI in chemical engineering, the embryonic discipline occupied an indeterminate existence. There were no attempts to set up any kind of professional organisation, and, as we have seen, chemical engineering was rarely used as a self-description by members of the SCI. A few courses in educational institutions adopted it as a title for all or part of their work.[23] The most prominent of these was probably the Central Institution of the City and Guilds of London, under the somewhat unwilling leadership of H. E. Armstrong, but the interpretation to be placed on its 'Diploma of Chemical Engineer' (sic) which existed from 1885 to 1891 is at best uncertain.[24] Another course existed at the Glasgow and West of Scotland Technical College (1887, formerly Anderson's University, later the University of Strathclyde).[25] But for many years these courses had difficulty in defining any characteristic curricular emphasis. Usually they consisted of a mixture of descriptive accounts of chemical processes, mechanical engineering and analytical chemistry, as

---

[22]   *Chem. Trade J.* 7 (1888), 306.

[23]   For a list of these early courses, somewhat loosely described as chemical engineering, see F.R. Whitt, 'Early teachers and teaching of chemical engineering', *TCE* (1969), CE357–360; (1971), 370–4. The list is inevitably somewhat incomplete, and it seems likely that several technical colleges, certainly that at Bradford, also offered courses with the title.

[24]   City and Guilds of London Institute for the Advancement of Technical Education, *Programme of the Central Institution* (1885), 24–6. Emphasis in the original.

[25]   On the Central Institution see J. F. Donnelly, 'Chemical engineering in England, 1880–1922', *Ann. Sci.* 45 (1988), 555–90. On the Glasgow and West of Scotland Technical College see *idem.*, 'Chemical education and the chemical industry during the late nineteenth and early twentieth centuries', PhD thesis, University of Leeds, 1988, chapter 4.

appears also to have been the case in the early chemical engineering courses at institutions in the USA such as MIT.[26] Looking more broadly, chemical technology as a generalised study in academe made little headway in British academic institutions, and several courses failed, for reasons which have been discussed elsewhere.[27]

Nevertheless, when the subject of education for the chemical industry was raised, as it frequently was, the term chemical engineering increasingly made an appearance. A call for a Department of Chemical Engineering at Glasgow University in 1889 was typical. During these years chemical engineering appeared as a promising title, in need of something of which to be the title: its usage was only loosely related to the field which was eventually created. In a debate on the subject in 1889 at the Manchester section of the SCI, one participant commented:

> I do not think that more than 50% of the chemists are able to make a scaled drawing from which chemical engineers could make plant ...[28]

Here the sense of the chemical engineer as a mechanical engineer working in a chemical context is prominent.

Around this time George Davis gave his well-known course of lectures on chemical engineering at the Manchester Technical School. They were, so he claimed, 'the first lectures on the subject that have been delivered in the native tongue to the English-speaking race'.[29] Davis's account of chemical engineering displayed his characteristic concerns: the mechanisation of processes ('the application of machinery and plant to the utilisation of chemical action on the large scale'); the gap between laboratory and plant-scale operations; and the 'technical experiment', carried out on a scale intermediate between that of the laboratory and the plant. Davis's argument can be construed as a response to the greatly exaggerated claims about the technical utility of academic chemistry which had been in circulation since the mid-century. But it also shifted the reference of chemical engineering away from constructing machinery in a chemical context towards engineering chemical processes on an industrial scale. Davis began to explore the manner in which the new field might walk an independent line between affiliation to chemistry and to mechanical engineering.

---

[26] 'The course is so arranged that the student will receive a suitable general training in mechanical engineering, and at the same time will devote a portion of his time to the study of the applications of chemistry to the arts ...' *Chem. Trade J.* 2 (1888), 408–9. J. W. Westwater, 'The beginnings of chemical engineering in the USA', in Furter, *op. cit.*, 140–52.

[27] J. F. Donnelly, 'Getting technical: the vicissitudes of academic industrial chemistry in nineteenth century Britain', *History of Education* 26 (1997), 125–43.

[28] *J. SCI* 8 (1889), 341.

[29] *Chem. Trade J.* 2 (1888), 290, 306–8 etc. The first lecture was delivered on 11 January 1888. Davis Papers, Science Museum, DAV 6/4.3. Clive Cohen has sought, somewhat ungenerously, to demote Davis within the origins of chemical engineering. Without wishing to underestimate the gap between Davis's descriptive approach and the subsequent development of unit operations, it seems perverse to deny that his formulation served a significant historical purpose. C. Cohen, 'The early history of chemical engineering: a reassessment', *Brit. J. Hist. Sci.* 29 (1996), 171–95.

The circumstances surrounding Davis's activities at this time highlight under-lying issues of secrecy within chemical technology. In 1884 an advertisement for his chemical engineering consultancy practice in the *Journal of the Society of Chemical Industry* apparently made reference to his work in the Alkali Inspectorate (it has not been possible to trace a copy of the original advertise-ment). The advertisement provoked anger among members of the Society's Council, notably Alexander Chance, and may even have precipitated Davis's resignation from the Inspectorate.[30] In 1888 Davis (as Editor of the *Chemical Trade Journal*) paraphrased the criticism of his lecture series on chemical engineering in the following terms:

> It is all very fine for Davis after having the *entrée* of all the chemical works in the country to now go and lecture about them.[31]

Whether his target was again Chance is uncertain, but Davis's response was scathing:

> This little speech shows the absolute ignorance of the speaker upon the subject of chemical engineering. The science of chemical engineering does not consist in hawking about trade secrets.

The idea that chemical engineering resolved tensions caused by the tendency towards 'secret working' within the chemical industry will be returned to below. Davis's course did not become established, though another with the same title was reintroduced under E. L. Rhead in 1896.[32]

Davis's contemporaries during the late 19th century displayed divergent attitudes towards the idea of chemical engineering. For some it was apparently a key term, denoting competence in the specific task of designing and operating chemical works.[33] Though he did not use the term chemical engineering, Norman Tate, Chairman of the Liverpool Section of the SCI, offered a compara-ble view to that of Davis in 1890. His emphasis was on generality:

> ... expositions of general operations, such as would deal with the operations of fuel-burning, ebullition and evaporation, solution, crystallisation, fusion etc. These could be made useful in connexion with the work of almost any manufactory ... instead of some of the courses on special processes, courses on general operations would often prove more useful.[34]

[30] Society of Chemical Industry, Council Minutes, 21 November 1884, 1 December 1884. On Chance (1844–1917) glassmaker and alkali manufacturer. D. W. F. Hardie, *A History of the Chemical Industry in Widnes* (London: ICI, 1950), pp. 126–7.
[31] *Chem. Trade J. 2* (1888), 290.
[32] *J. SCI, 18* (1899) 437–9. On E. L. Rhead (1864–1940) see *JPIC* (1940), 341.
[33] R. Galloway, *Education, Scientific and Technical* (London: Trübner & Co.1881), pp.380–1. I. Levinstein, 'Observations and suggestions on the present position of the British chemical industries, with special reference to coal-tar derivatives', *J. SCI 5* (1886), 351–7. Robert Galloway (1822–1906) was Professor of Practical Chemistry at the Government School of Science, Dublin. Ivan Levinstein (1845–1916) was perhaps the foremost U.K. manufacturer of synthetic dyestuffs at that time. *J. SCI 35* (1916), 458.
[34] A. N. Tate, 'On some aids to the further development of chemical industry' *J SCI 9* (1890), 1010–12. Tate (1837–1892) was an analyst and consultant chemist (fitting the pattern of many of the early promoters of chemical engineering) who had also operated a private school of chemistry in Liverpool. *JCS 63* (1893), 764–5.

Arthur Smithells, Professor of Chemistry at the Yorkshire College of Science, suggested somewhat ambivalently in 1894 that chemical engineering was close to being construed as the general title for competence in the operation of chemical plant. The demand for immediate usefulness after training was 'at the bottom of all these schemes for producing chemical engineers'. But such competence could only be learnt in the works: 'there is no royal road to chemical engineering'.[35] Academic chemists, with exceptions, such as the physical chemist Frederick Donnan, remained sceptical of technological courses within universities for many years.[36] Despite this, Donnan later appeared to seek to appropriate chemical engineering to his own emergent specialism of physical chemistry.[37]

Contemporary discussion suggests that the currency of chemical engineering during the closing two decades of the century as a term was substantial, but its position in any other respect was scarcely a secure one. Despite this, Britain evidently had some reputation abroad in the area. The German industrial chemist Carl Duisberg, a key figure in Bayer and later IG Farben, could write in 1896:

> Nothing in my opinion, is worse than to make of a chemist an *ingenieur-chimiste*, as is done in France, or chemical engineer, as is very often done in England. Division of labour is here absolutely necessary'.[38]

In 1899 an extended discussion on the subject of chemical engineering took place in the SCI, under pressure from the then President George Beilby. The discussions revealed that there had been little development of view over the preceding two decades, though Beilby himself had some interesting remarks to make. He suggested that the difficulty in defining a generalised education in chemical technology was a significant problem, and that chemical engineering appeared as a possible resolution, with types of apparatus representing a possible organising structure. Despite the difficulty in defining the precise function of the chemical engineer, an emphasis on apparatus appeared central.

> Apparatus is generally the property of the whole trade, or it is patented ... Processes on the other hand are much more difficult to protect by patents and are often worked secretly.[39]

[35] A. Smithells, Chairman's Address to the Yorkshire Section, *J. SCI 13* (1894), 18–21. On Smithells (1860–1939) Professor of Chemistry at the Yorkshire College of Science see DNB.

[36] This is especially reflected in the various discussions which occurred in the Institute of Chemistry, notably Institute of Chemistry, *Conference of Professors of Chemistry* (London, 1913); and *idem.*, *Conference on the Place of Applied Chemistry in the Training of Chemists* (London: Institute of Chemistry, 1925).

[37] F. G. Donnan, 'The university training of technical chemists', *J. SCI 28* (1908), 275–80; 'The training of technical chemists', *Chem. Trade J.* (1915), 519–20. Comments in *J. SCI 35* (1916), 1190. On Donnan (1870–1956) the first Professor of Physical Chemistry in Great Britain (at Liverpool University and later University College, London) see DNB.

[38] C. Duisberg, 'The education of chemists', *J. SCI 15* (1896), 427–32. On Duisberg (1861–1935) see *Chemist and Industry 13* (1935), 392–4.

[39] G. T. Beilby, 'The relations of the Society to chemical engineering and to industrial research', *J. SCI 18* (1899), 333–40. The discussion was followed up at *ibid.*, pp. 437–43. On Beilby (1850–1924) a director of the Cassel Cyanide Co. see DNB.

There is evidence from his comments that Beilby was influenced by the views of George Davis (he referred to Davis's yet-unpublished book), though Davis himself was not particularly supportive of Beilby's views in his own remarks.

CONSULTANCY, SECRECY AND GENERALITY

This slow and institutionally unsupported development reached a kind of climax when George Davis began to publish his *Handbook of Chemical Engineering* just after the turn of the century. The book drew heavily for its structure on the 1888 lectures, but was much larger. Davis offered a framework for the knowledge base of the chemical industry, distinguishing applied chemistry and chemical engineering as concerned respectively with chemical processes and the operation of industrial scale plant, and again stressing the need for the 'technical experiment'.[40] His book was followed by another in 1906 by Jacob Grossmann.[41] Grossmann's book also placed a heavy emphasis on scaling up, and the need for intermediate scale experiments in order to achieve this.

These two texts were both produced by consultants, and it is possible to argue, if somewhat speculatively, that these early accounts of chemical engineering recognised a distinctive potential knowledge base and jurisdictional domain, and that this recognition owed something to the peculiar role of the consultant. That role, as Davis had implied in response to the criticism of his original lecture course, involved moving between works in a somewhat privileged position. By contrast, employees within works were often required to sign a commitment to secrecy, and forbidden to take other employment within specified time periods.[42] The sensitivity within the industry to commercial secrecy has already been touched upon, and other examples could be given. In 1885, William Perkin, speaking in connection with the training of technical chemists, suggested that 'processes which are *publicly* known and taught, are more or less antiquated'. Nearly 40 years later the relative emphasis on 'secret working' within the industry continued to be noted.[43] The City and Guilds, which had administered technical examinations in fields such as dyestuffs and alkali manufacture since 1875, had difficulties when a particular examination was criticised by one manufacturer for supposedly including a question which sought to obtain details of a secret process on the manufacture of -naphthol and -naphthalinine [sic]. This was said to be a 'fishing' question:

---

[40]   G. E. Davis, *A Handbook of Chemical Engineering* (Manchester: Davis, 1901–2).
[41]   J. Grossmann, *The Elements of Chemical Engineering* (London: Griffin, 1906). On Grossmann (1854–1920) see *PIC* (1920), 247.
[42]   Brunner, Mond is an interesting example of the most advanced heavy chemical firm of its day which was yet still very concerned about secret working. See Cheshire County Record Office. Managing Directors' Minute Book, BM3/2. Norman Swindin commented that Georg Lunge was banned from the Northwich works after publishing a full account of the process in his well known textbook. N. Swindin, *Engineering Without Wheels. A Personal History* (London: Weidenfeld & Nicholson, 1962), p. 37.
[43]   *J. SCI 4* (1885), 437. H. E. Potts, *Patents and Chemical Research* (Liverpool: Liverpool University Press, 1921).

If employers get the idea that there is a danger of their particular processes or methods being revealed ... I can hardly imagine anything more disastrous'.[44]

Such comments referred mainly to the specifically chemical aspects of processes. By contrast, as Beilby noted, machinery had a much more public and standardised character. Indeed to some degree machinery was a commodity in itself. During the first world war an editorial in the *Chemical Trade Journal* claimed that those who manufactured and sold such machinery were appropriating the title 'chemical engineer'.[45] Abbott has noted that the embodiment of a profession's knowledge base (and problem solving strategies) within commodities is a threat which many professions have faced.[46]

It would be an exaggeration to suggest that Davis's and Grossmann's textbooks articulated a well-defined model of chemical engineering, or that their appearance represented any radical breakthrough in the creation of the discipline. What they did was gesture towards a distinctive conceptualisation of the field grounded in the physical environment of the chemical (and physical) processes involved.[47] To the extent that this conceptualisation allowed a generalisation across plants and processes it is apparent how it reflected the very specific working situation of the consultant. But too little is known of the working practices of consultants for this to be more than an interesting hypothesis.

At any rate, throughout this time the title chemical engineer (and by implication the occupation understood as a specialism) received little recognition in works, and was only claimed with any enthusiasm by consultants such as Davis and Grossmann. The point is given support by the comment of Raphael Meldola in 1909 that the

few expert practitioners (in chemical engineering) who could be named are probably better off as private consultants than they would be if they transferred their services to an educational institution.[48]

THE NEW CENTURY: PROFESSIONAL INSTITUTIONS AND INDUSTRIAL CHEMISTRY

The issues of generality and secrecy (which might be construed, in academic terms, as requiring the creation of a publicly accessible conceptualisation of chemical technology) can also be found in academe and to a degree in professional organisations. Several attempts to establish courses in chemical technol-

---

[44]   City and Guilds of London. Minutes of the Board of Examiners. 30 June 1896. (Held at the Guildhall Library, London.)

[45]   *Chem. Trade J. 57* (1915), 517–8.

[46]   Abbott, *op. cit.*, 146.

[47]   Hornix has argued similarly that standardised apparatus developed in the organic sector pointed towards the concept of 'unit operations'. W. J. Hornix, 'From process to plant: innovation in the early artificial dye industry', *Brit. J. Hist. Sci. 25* (1992), 65–90 (89).

[48]   R. Meldola, 'Education and research in applied chemistry', *J. SCI 28* (1909), 555–77 (572). On Meldola (1849–1915) Professor of Chemistry at Finsbury Technical College see J. Marchant (ed.), *Raphael Meldola* (London: Williams & Norgate, 1916).

ogy in the United Kingdom experienced such difficulties.[49] The problem was also experienced by the Institute of Chemistry, when, in 1903, it sought to extend its authority into the field of industrial chemistry, by establishing a qualifying examination. The Institute had long drawn its core membership from consultant analytical chemists and academe: industrial chemists were more marginal. By the turn of the century the growth of employment in industry was well underway, and the attempt to establish such hegemony predictable. But identifying and certifying the heterogeneous knowledge demands of the chemical industry proved much more difficult than doing so for the well-established protocols of analytical chemistry.[50]

In 1893 the Institute had offered 'Elementary mechanics, steam, and general chemical engineering' as one of the optional 'courses of instruction' which an applicant might have attended, 'it being provided that Theoretical and Practical Chemistry are to be regarded as principal subjects'. However the detailed account makes clear that it was primarily understood to mean mechanical engineering.[51] The 1903 Committee began its task by seeking to

... prepare a full definition of Chemical Engineering including the formulation of a detailed syllabus of the subjects which a course in Chemical Engineering should embrace.[52]

Why chemical engineering should be have been identified as the primary focus of a committee the explicit brief of which concerned chemical technology is unclear. It may at the least be taken as a further indicator of the currency which the term had built up. In the committee's subsequent meetings, and the many drafts of its reports, which were produced over a period of two years, concerns over generality and secrecy were dominant. Both were emphasised in the first report, and they were linked through the requirement that no question should involve

the disclosure by Candidates of processes and plant peculiar to particular works.[53]

The Committee made several attempts to resolve these problems within the regulations for the proposed examinations. One formulation, in October 1904, stated that the focus of the examination should be on

Such industries (as) are established in or near any large town and are, as a rule, carried on without secrecy ...[54]

The Institute eventually placed the emphasis in its examinations on 'General

[49] Donnelly, 'Getting technical', op. cit.
[50] C. A. Russell, N. G. Coley and G. K. Roberts, Chemists by Profession: The Origins and Rise of the Royal Institute of Chemistry (Milton Keynes: Open University Press/Royal Institute of Chemistry, 1977), chapter 10.
[51] Institute of Chemistry, Regulations for Admission to Membership, (1893), 9.
[52] Institute of Chemistry. Committee Minute Book. Special Committee re. Technological Examinations. 16 October 1903.
[53] Institute of Chemistry Council Minutes. 11 December 1903.
[54] Ibid., 20 October 1904.

Chemical Technology'.[55] For whatever reason the initial stress on chemical engineering had vanished. It may perhaps have represented too great a shift from the core emphasis of the Institute: and the potential power of the new formulation may already have been sensed. In any event chemical engineering was not even mentioned in the regulations governing the examinations as established. As matters turned out the examinations attracted very few candidates, and the Institute turned its attention simply to a *de facto* recruitment of industrial employees. No doubt it would have argued that a generalised and publicly accessible approach to chemical technology was already available – chemistry.

The second decade of the new century saw the first attempts to establish organisations for industrial chemists since the foundation of the Society of Chemical Industry, to some extent as a direct response to the failure of the Institute of Chemistry to appeal to industrial chemists. The first such organisation was the Association of Chemical Technologists, established in March 1911.[56] Neither the Association nor its successor from June 1914, the Institution of Chemical Technologists, showed much direct interest in chemical engineering. Both directed their attention towards attacking the dominance of academic chemistry, and chemists. Yet, oddly, the Association's 'official organ' was called *Chemical Engineering and the Works Chemist*[57] and its second Council included J. W. Hinchley who was, as will be seen, a very significant figure in the history of chemical engineering and the IChemE. The Institution was the precursor of other organisations (the British Association of Chemists and the National Association of Industrial Chemists, both established in 1917). All of these bodies were concerned primarily with confronting the authority-claims of academic chemistry over chemical technology and technical chemists within works. The BAC had some success in its confrontation with the Institute of Chemistry, but none was able, or perhaps even sought, significantly to replace the hegemony of chemistry in general.

In 1915 a proposal for an Institution of Chemical Engineers was put forward by one of the self-styled chemical engineers in the SCI, Arthur Lymn, who remarked that the abuse of the title was 'little short of a scandal'.[58] He envisaged a qualifying association, certifying competence in a specialist field. Lymn's proposal, and attendant discussion on the training of technical chemists, provoked considerable interest and support, though it met with hostility from some correspondents. The most notable of these was Charles Carpenter, at

---

[55]   *Proceedings of the Institute of Chemistry* (November, 1906), 35–41. This examination was supplemented by a specialist topic chosen by the candidate.

[56]   Donnelly, 'Defining the industrial chemist', op. cit..

[57]   *Chem. Eng. & Works Chemist 1* (1911–2), 1. The Chemical Trade Journal also added 'and Chemical Engineer' to its title in the early years of the century.

[58]   *Chem. Trade J. & Chem. Engineer 57* (1915), 546. Lymn (b.1875) had studied at the University of Zurich and been a consultant chemical engineer since 1910 (membership records of the IChemE). It is perhaps significant that he changed his title in the SCI lists to chemical and gas engineer in 1912, just after he moved from works management to consultancy.

that time President of the Society of Chemical Industry.[59] For Carpenter, chemical engineers were 'first and foremost practitioners of mechanical engineering'.

Carpenter followed up this response by presenting a paper to the Society of Chemical Industry, in which he argued for the operation of works with 'two distinct staffs'.[60] His remarks provoked mixed responses. J. W. Hinchley appeared for the first time as a promoter of the new field, which he was already teaching at Imperial College. He was predictably hostile to Carpenter's argument:

> ... he wished to put in an appeal for the chemical engineer. It was absurd to talk about the chemist appealing to the engineer unless they defined what sort of engineer was meant. The ordinary mechanical engineer was quite untrained in the particular points which the chemical manufacturer had to handle.[61]

Hinchley's remarks in turn did not go unchallenged. Chemical engineering was gradually coming to be perceived as title with wide-ranging implications. Bertram Blount attacked the provision 'ad hoc of the chemical engineer'. Arthur Smithells remained unconvinced, as he had been over twenty years before:

> ... they must not talk too much about that doubtful and indescribable person, the chemical engineer, being trained for that particular vocation in life.

He went to claim that, though demands for chemical engineers had been made for many years 'almost louder than any other in the industrial community', these (chemistry and engineering) were 'two distinct shades of talent and capacity'. Unsurprisingly Carpenter's argument received the support of the *Journal* in a subsequent editorial.

These arguments increasingly focused on the training of chemical engineers, but were accompanied by explicit denials that such a specialism existed. Davis's manufacturer of 1881 would now at least have heard of the 'genus', though whether he would have judged it real or artefactual is more doubtful. The argument was also becoming more complex. Speaking broadly but crudely, the competitors for jurisdiction appeared on the one hand the chemist and mechanical engineer and, on the other, the specialist in particular areas of chemical technology.

It was the former tension which dominated public debate, and reached a kind of culmination in 1917, with an extended discussion at the Faraday Society in which some of the major protagonists (e.g. Donnan, Bielby and Hinchley) took part.[62] The terms of the discussion focused on the extent to which chemists

[59]   *Ibid* 563–4, 585; *58* (1916), 74.
[60]   C. Carpenter, 'Chemistry and engineering', *J. SCI 35*(1916), 1185–91. On Carpenter (1858–1938) see *J. SCI* special issue (1931), 83. It will be seen in a later chapter that Carpenter appeared sufficiently to change his judgement about chemical engineering to become Vice President of the IChemE.
[61]   On John William Hinchley (1871–1931) see E. M. Hinchley, *John William Hinchley, Chemical Engineer: A Memoir* (London: Lamley & Co., 1935).
[62]   'The training and work of the chemical engineer. A general discussion', *Trans. Faraday Soc.* 13 (1917–8), 61–118.

should be taught engineering and engineers chemistry. John Hinchley, without involving himself in the more abstract statements of his fellow participants, began to sketch a domain of activity constitutive of a recognisable, if somewhat crudely defined, specialism. In the previous decade Hinchley had been laying some significant, if small scale, academic foundations for creating such a specialism.

THE BEGINNINGS OF A RESOLUTION

The difficulties of courses in chemical technology have already been alluded to. Among these difficulties was a tendency to fragment around particular industrial sectors. There was a need for a curricular means of avoiding this sensitivity to the specifics of particular processes. In the 1909 Presidential Address to the SCI already cited Raphael Meldola claimed that

> ... we are, I think, in a position to face that bugbear with a certain class of teachers – chemical technology in educational institutions. What does it really mean? ... It means *generalised chemical engineering* ... , a knowledge of the chemical, physical, and mechanical principles underlying the construction and working of the machinery and plant in general use under various modifications in all branches of chemical industry.[63]

This view was expressed by others.[64] In the same year as Meldola's address Hinchley had developed a course in chemical engineering at Battersea Polytechnic which has as strong a claim as any to be the first to be organised around the physical processes and apparatus of the chemical industry. It operated at a level of sophistication beyond that of George Davis, though not dramatically so.[65] The new course understood the field as an independent specialism. That at least was the perception of the *Chemical Trade Journal*, which commented:

> It is gratifying to see that our educational institutions are at last recognizing that chemical engineering cannot be taught successfully by digressions in cognate subjects. It is now a complete and separate subject, ... and its elements can no more be taught by studying those parts of chemistry, chemical technology, and engineering which entrench on each other.[66]

Hinchley began lecturing at Imperial College in 1911, and entered into a bruising campaign to constitute chemical engineering as the central discipline of the College's Department of Chemical Technology. Hinchley had a long battle with William Bone, Professor of Chemical Technology, described further in Chapter 4. Bone saw his department as an amalgamation of specialisms,

[63] R. Meldola, 'Education and research in applied chemistry', *J. SCI 28* (1909), 555–77 (571).
[64] E. C. C. Baly, 'The future position and prospects of the British chemical trade and the question of concerted action by manufacturers', *J. SCI 34* (1915), 53–5, comments of A. J. Allmand, 54.
[65] W. C. Peck, 'Early chemical engineering', *Chem. & Indus.* (1973), pp. 511–7. H. Arrowsmith, *Pioneering in Education for the Technologies: The Story of Battersea College of Technology. 1891–1962* (Guildford: University of Surrey, 1966).
[66] *Chem. Trade J. 45* (1909), pp. 307–8.

and was hostile to Hinchley's view that chemical engineering was foundational within it.[67] Hinchley also sought to place the field in carefully studied relationships with the supposed underlying sciences. He intended students to determine

> factors of construction or factors relating to design, which cannot be deduced from physical or chemical knowledge ... (apparatus should) not attempt to eliminate causes of disturbance which are present in factories, but make (...) it possible for the student to appreciate their value.[68]

The battle between Bone and Hinchley continued for many years. But it is perhaps possible to sense an anticipation of defeat in Bone's comment in 1926, after the foundation of the Chair of Chemical Engineering at University College, London, which Bone had done his best to sabotage, and for which Hinchley had applied. Since the establishment of that Chair and department Bone remarked:

> ... in some quarters, the term 'Chemical Engineering' is being used as synonymous with 'Chemical Technology'.[69]

The battle which Hinchley found himself engaged in could be seen as a microcosm of the wider battle fought by chemical engineering over the next eighty years to establish itself as the generic discipline for industrial processes involving bulk chemical transformations. In the shorter term – the period 1911–26 which is the focus of the next chapter – chemical engineering was slowly establishing its claim to being able to resolve the difficulties of secrecy and generality which manufacturing chemistry had experienced in academe. Faced with this, chemical engineering's stress on generality and the operation of physical plant was both its strength and its weakness. It distanced the field from the dominance of chemistry, but invited a common objection: that it possessed no distinctive abstract conceptualisation, that it was a mere *ad hoc* treatment of chemical plant, requiring the skills of both chemist and mechanical engineer.

Buchholz has claimed that the emphasis on plant was a necessary theoretical regression from the nineteenth century foundational claims of chemistry. But this is a somewhat teleological view.[70] The approach gained momentum mainly from the failure of other ostensibly more promising approaches, such as a naturalistic categorisation by industrial sector, or the more abstract approach founded in physical science. During the period with which this chapter has been concerned, chemical engineering did not draw on a clearly identifiable task domain which could form the basis of a professional practice: it certainly did not seem so to most contemporaries. It was judged by some as a crude programmatic approach to the technological work within the chemical industry,

---

[67] M. de Reuck, 'History of the Department of Chemical Engineering and Chemical Technology' typescript in Imperial College archives, 1960. On Bone (1871–1938) see DNB.

[68] Letter Hinchley to Edward Thorpe, 14 December 1911, Imperial College London Archives KCT 10/3/1956.

[69] Board of Studies Minutes, Imperial College of Science and Technology, 1 June 1926.

[70] K. Buchholz, '*Verfahrenstechnik* (chemical engineering) – its development, present state and structure', *Sci. Stud. 9* (1979), 33–62.

and by others as little more than the opportunistic use of an attractive title. On its behalf a few protagonists claimed authority over a notionally generalised (or in Abbott's more flattering term, abstract) space of problems. But it would require extensive intellectual and institutional work to change that prototype into a coherent and recognisable practice which might form the basis of an occupation, discipline and profession. This work could be construed as conjuring chemical engineering into existence from a range of possible approaches to chemical technology, rather than uncovering an objectively existing task domain. Such a suggestion seems less far-fetched than might appear when the very different history of chemical engineering in Germany is recalled.

In the early 20th century chemical engineering was poised for growth. But it was sustained only by a loose network of individual practices, linked by a shared rhetoric. It was weakly conceptualised and institutionalised, and thus vulnerable to contingency. It is the influence of one of these, the first world war, that dominates the next chapter.

CHAPTER 3

CATALYSING AN IDENTITY, 1915–1925

Creating a viable profession demanded more than the zeal of a few promoters. Indeed, the activities of George E. Davis, much later cited as the progenitor of the British chemical engineering profession, were little mentioned in the second decade of the century.[1] This chapter focuses on a much more immediate source for subsequent folklore: the role of war in the coalescence of representative bodies that came to promote the occupation, discipline and profession in Britain.

As we have seen, by the 1910s chemical engineering in Britain had evolved

---

[1] The writings of G. E. Davis were not seen as the basis of a unifying narrative until after the second world war. There is a clear discontinuity in his role in foundational myths. Organisers of chemical engineering appear to have made no public mention of any perceived links between Davis and their movement, despite the continued existence of his periodical, the *Chemical Trade Journal* (retitled *The Chemical Trade Journal and Chemical Engineer* after his death in 1907). The IChemE Provisional Committee, in connection with an 'investigation they had made in connection with the previous attempts to organise chemical engineers in Great Britain', recorded seeing Davis's son Keville (1881–1934) erroneously identified as 'Kerfoot', but made no acknowledgement; no other mention of Davis is to be found in the Minutes or speeches of the period. The legend of G. E. Davis gained steam some two decades later. Norman Swindin's publications and lectures from the 1950s, and autobiography of 1962, publicised the work of his sometime employer; the IChemE instituted a decennial 'George E. Davis Memorial Lecture' in 1953 (a 'Hinchley Memorial Lecture' having been started four years earlier); Silver Jubilee (1933) retrospectives and histories of the AIChE do not mention Davis, and that for the Golden Jubilee (1958) provides only a sentence; Eric N. Davis (b.1888) found 'those in power' at the IChemE to be 'uncooperative, to put it mildly' to the notion of a memorial to his father in 1959, and finally offered his 'personal relics' to the IChemE in 1965, when the IChemE created the 'George E. Davis Award'; the new IChemE headquarters constructed in the mid 1970s gained the name the 'George E. Davis Building'; the Science Museum archive, at least partly the collection of E. N. Davis, was acquired in 1984. CM 21 Dec 1921; N. Swindin, 'Some memories of George E. Davis, Chemical Engineer', *Birmingham University Chemical Engineer 1* (1950), 98–101; *idem.*, 'The George E. Davis Memorial Lecture', *Trans. IChemE 31* (1953), 187–200; J. C. Olsen, 'Chemical engineering as a profession: origin and early growth of the American Institute of Chemical Engineers', *AIChE Trans. 28* (1932), 299–329; Sidney D. Kirkpatrick (ed.), *Twenty-Five Years of Chemical Progress* (New York: AIChE, 1933); F. J. Van Antwerpen & Sylvia Fourdrinier, *High Lights* (New York: AIChE, 1968); letter, E. N. Davis to N. Swindin, 23 Oct 1959, Swindin Papers box 74; Swindin, 'George E. Davis as I knew him', *Chem. Trade J.* 17 Jun 1965, 753–4; CM 21 Jul 1965; Davis papers.

from an individualistic practice into an occupation having an increasing collective awareness. Attempts to organise within the chemical industry proliferated. But in this potential-filled environment, the timing and composition of new professional bodies was undetermined. The vulnerable nascent profession vied with other professions for space on a constantly shifting terrain.

Several influences contributed to organisation. The immediate postwar period was one of overextended firms, increased international competition and economic uncertainty. Active organisation, training and accreditation became prominent issues for engineers, for the chemical industry, and for the government as a means of stabilising employment and economic well-being. The identification and subsequent expansion of the profession of chemical engineering in Britain – if only during this brief period – was critically dependent on the context of increasing collectivism. The results of this environment were new groupings of practitioners: the Chemical Engineering Group (CEG) of the Society of Chemical Industry in 1918–19, and the Institution of Chemical Engineers (IChemE) in 1922–23, along with two closely-linked manufacturers' bodies, the Association of British Chemical Manufacturers (ABCM) and British Chemical Plant Manufacturers Association (BCPMA).

The organisers of these new bodies certainly vaunted the wartime connections; indeed, the first article in the *Proceedings of the Chemical Engineering Group* in 1919 cites 'the stress of the European War' as the reason chemical engineering came to prominence.[2] Four years later, the *Transactions of the Institution of Chemical Engineers* attributed the renewed interest 'on the question of chemical plant, its efficiency, design, construction and working' to the war, and credited Lord Moulton, head of the department of explosives supply, with foreshadowing the Institution.[3]

Yet the reasons for the creation of these new organisations were more complex than their organisers portrayed them. As expounded by Abbott, professional 'jurisdiction' over technical tasks may require the combination of several mechanisms. A trend towards organisation of chemical practitioners had been growing steadily before the war. And the Ministry of Munitions played a key part in legitimating chemical engineering. It undertook a sweeping reorganisation of the British chemical industry – a restructuring that surpassed the aims of pre-war groups. But where earlier calls for organisation had created schisms and ineffectual special interest groups, the Ministry's actions fed into a growing

---

2   'History of the formation of the Group', *Proc. CEG 1* (1919), 3.

3   'The history of the formation of the Institution of Chemical Engineers', *Trans. IChemE 1* (1923), vii. Hinchley, Secretary of both bodies, is the probable author of these two texts. He had altered the message only slightly a decade later: 'The profession of the chemical engineer is of recent growth which has been accelerated by the experience of the Great War and by the subsequent competitive development of industry all over the world', he told delegates at an international engineering congress [J. W. Hinchley, 'The training of the chemical engineer', World Engineering Congress, Tokyo (1929)]. For Moulton's promotion of chemical engineering see, for example, his lecture at UCL 19 Mar 1920 ['The training and functions of the chemical engineer', reprinted *Trans. IChemE 17* (1939), 186–91].

postwar consensus.[4] Like a seed crystal, the war triggered crystallisation in the pre-existing super-saturated chemical environment.

The sources of this synoptic mood were diverse. The sheer numbers of industrial chemists and chemical engineers brought together for Ministry work emphasised their collective situation; some of them, identifying weaknesses in the system, promoted the achievements of the most prominent plant designer and manager as a model of practice; the Ministry itself evaluated and publicised its economic and administrative experiment in a detailed postwar analysis;[5] and, more generally, the political terrain occupied by the coalition government supported a degree of communality in industrial strategy across the political spectrum. The personal and institutional connections between campaigners for organisation nurtured a cohesive set of bodies in the postwar decade.

The war changed the context, transforming the inter-relationship of state, industry and engineers. New structures provided new opportunities for these specialists to press their professional claims. Nevertheless, they were still small in number; the identity of the chemical engineer existed more in the minds of a handful of individuals than among a unified practising collection of recognised workers.

WARTIME CHEMICAL INDUSTRY

At the beginning of the war the British government was unprepared for the required scale of explosives production. High explosives had never been pro-duced by the national ordnance factories, and there was little commercial capacity for TNT. For other materials such as cordite (the principal propellant used by the British military, and consisting principally of a mixture of nitroglyc-erine and nitrocellulose extruded in thick cords) 'there appeared to be no prospect of a large-scale demand after the War which would induce existing manufacturers to extend their works'.[6]

At the end of November 1914, the Defence of the Realm (Consolidation) Act permitted the government to commandeer commercial plants, but these were clearly inadequate to meet the demand. Following a request from the War

---

[4] See, for example, W. J. Reader, *Imperial Chemical Industries: A History* (Oxford: Oxford University Press, 1970), Vol. 1, Chapters 11 and 15.

[5] *The History of the Ministry of Munitions*, commissioned by Lloyd George early in the war, was distributed to large libraries in 12 volumes comprising over 5000 pages.

[6] D. Lloyd George, *War Memoirs* (London: Ivor Nicholson & Watson, 1933), p. 575. For the pantheon of available explosives, their constituents and manufacturing processes, see H. Fletcher Moulton, *The Life of Lord Moulton* (London: Nisbet, 1922), chap. 7, and R. P. Ayerst, M. McLaren & D. Liddell, 'The role of chemical engineering in providing propellants and explosives for the U.K. armed forces', in: W. F. Furter (ed.), *History of Chemical Engineering* (Washington D.C.: American Chemical Society, 1980), 367–92. There had been, however, moves three years earlier to expedite cordite production by Nobel's Explosives in preparation for the probability: see letter, F. J. Shand to F. L. Nathan 23 Aug 1911, 'F.L.N. 3' folder, Nathan papers, IChemE, and W. Hornby, *Factories and Plant* (London, 1958), pp. 83–4. Moreover, the state had favoured cordite firms from the 1890s: see R. C. Trebilcock, 'A "Special Relationship" – Government, Rearmament and the Cordite Firms', *Econ. Hist. Rev. 19* (1966), 364–79.

Office, the Board of Trade recommended Lord Moulton, then aged 69, as organiser of explosives supply.[7] Apparently based on the suggestion of Sothern Holland, Secretary of the Explosives Committee he set up in mid November, Moulton recruited Kenneth Bingham Quinan, the American manager of a de Beers explosives factory in South Africa, to oversee the development of new chemical plants.[8]

Following his arrival in January 1915, Quinan oversaw the construction of a new state TNT plant at Oldbury, and a TNT and gun-cotton plant at Queen's Ferry, near Chester, started that May.[9] The following month, in the wake of a disastrous offensive and munitions shortage dubbed 'the Great Shell Scandal', David Lloyd George was appointed to head the new Ministry of Munitions, and incorporated Lord Moulton's Committee as a Department of Explosives Supply within it.[10]

By July the new Ministry had established four more national explosives factories, the largest of which, a cordite production factory at Gretna in south-west Scotland, covered a 9000 acre expanse and employed over 16000 workers.[11] Seventeen others were founded within a year, and by the end of war there were thirty-two H. M. explosives factories.[12]

---

[7] Three-times Liberal MP, Lord Chief Justice and Lord of Appeal before WWI, John Fletcher Moulton (1845–1921) had acted in explosives patent-cases following an education in mathematics and law. See *J. SCI 40* (1921), 137R-138R; *Trans. IChemE 17* (1939), 184–5.

[8] Chief Advisor and Controller in the Explosives Supply Department and later responsible for poison gas production, Quinan (pronounced Kwī'-nan, although other branches of the family pronounced the name Kwĕnn'-an, Kwĕnn'-an or Kwēēn'-an) had developed his practical experience in explosives manufacture at an older relative's factory in California. For biographical details, see IChemE Application Form (henceforth AF) 109; obituary, *The Times* 28 Jan 1948, p. 4; C. S. Robinson, 'Kenneth Bingham Quinan, C.H., M.I.Chem.E., 1878–1948'; *TCE*, Nov 1966, CE290-CE322; and A. P. Cartwright, *The Dynamite Company* (London: MacDonald, 1964). See also L. F. Haber, *The Poisonous Cloud: Chemical Warfare in the First World War* (Oxford: Clarendon Press, 1986), pp. 148–9. Quinan remained General Manager of the Cape Explosives Company *in absentia* during the war. For his wartime records, see the extensive Public Record Office holdings, PRO SUPP 10. For genealogical information we thank Christopher Baker-Carr and William R. Quinan.

[9] For more on the wartime changes in the industry, see D. W. F. Hardie & J. Davidson Pratt, *A History of the Modern British Chemical Industry* (Oxford: Pergamon Press, 1966), pp. 98–115.

[10] See Moulton, *op. cit.* For accounts emphasising the administrative, rather than technical, history of the Ministry, see R. J. Q. Adams, *Arms and the Wizard: Lloyd George and the Ministry of Munitions, 1915–1916* (London: Cassell, 1978), and C. Wrigley, 'The Ministry of Munitions: an innovatory department', in: K. Burk (ed.), *War and the State: The Transformation of British Government* (London: George Allen & Unwin, 1982), 32–56. The DES was based at the Storey's Gate offices of the Institution of Mechanical Engineers.

[11] *The History of the Ministry 8* (London, 1922), Part II.

[12] Twenty-one were constructed by the state, the remainder being nationalised commercial factories. Initiated as late as Spring 1918, at least two were never completed. Gretna and Queen's Ferry (or Queensferry) had a capital cost exceeding £12.7 million, almost equal to that of the all other national explosives factories combined. See *History of the Ministry 8* (London, 1922), and R. MacLeod, 'The chemists go to war: the mobilisation of civilian chemists and the British war effort, 1914–1918', *Ann. Sci. 50* (1993), 455–81.

CHEMICAL ENGINEERING DURING THE FIRST WORLD WAR

The supply of chemical munitions was central to the waging of the war, and a direct reason for the formation of the coalition government in 1915. The 'chemists' war' eventually mobilised well over a thousand technical workers with chemical and engineering backgrounds.[13] While Quinan's staff in the Ministry was 'drawn mostly from South Africa',[14] the chemical factories were populated mainly with chemists, few of whom had any experience with explosives manufacture. Moulton recorded that many were withdrawn from military service or seconded from academic posts, 'drafted off to the new factories that were continually coming into being'.[15] The wartime curriculum vitae of the best such workers show advancement from 'trainee' or 'shift chemist' to 'chemist-in-charge' to 'plant designer' or Ministry Assistant, resident for only months in each post.[16]

K. B. QUINAN AS ARCHETYPAL CHEMICAL ENGINEER

Such explosives plant designers and operators were soon identified by the Ministry as essential to its goals. Quinan, as their chief, became the most conspicuous example in Britain, and his activities were employed by others to promote a new self-image for the subject. Yet Quinan himself never referred to chemical engineers in his factory organisations, dividing his factory staff instead into the traditional categories of engineer and chemist.[17] Ironically, too, his residency was brief and his influence short-lived; while serving as a convenient exemplar for those concerned with organising the profession, he participated only marginally in such activity himself.[18]

Quinan was represented as a model to emulate chiefly by William Macnab, an acknowledged authority on explosives and Ministry colleague.[19] Macnab portrayed Quinan and his techniques as a symbol of what was missing in the British chemical industry. He credited him with introducing clear methods of analysing the problems of chemical design and production, crucial for initiating 'staff and workers largely without expert knowledge of the work they had to do'. Quinan's techniques of statistical control for administration also were

---

[13] See Haber, *op. cit.* and R. MacLeod, 'Chemistry for King and Kaiser: revisiting chemical enterprise and the European war', in: Homburg et al., op. cit. Haber estimates the figure at about 1500; MacLeod adds another 800 to 1000 workers in the explosives factories and other war-related chemical occupations.

[14] *History of the Ministry*, *op. cit.*, p. 74, and Robinson, *op. cit.*

[15] H. F. Moulton, *op.cit.*, p. 216.

[16] See IChemE AF, e.g. W. B. Davidson (No. 23), H. F. Hill (43), W. S. Milne (60), A. Cottrell (135), H. W. Cremer (203).

[17] K. B. Quinan, 'Memo on organisation of staffs at Queensferry', PRO SUPP 10/286.

[18] A 'corresponding Vice President' of the IChemE, Quinan never attended Council meetings or communicated regularly with the Institution.

[19] Macnab (1858–1941) spent his early career in sugar chemistry, and developed chemical engineering connections later through associations with a water softening company and the management of explosives works. Regarding his explosives expertise, see W. Macnab, 'Chemical engineering in explosives manufacture', *Trans. IChemE 13* (1935), 9–13.

singled out for Macnab's praise.[20] By 1917, such methods had led to new manufacturing efficiencies: the cost of TNT production had fallen from the pre-war commercial cost of 1s 9d per pound to 8½d per pound at Queen's Ferry, a 60% saving.[21] The Superintendent of Gretna trumpeted that his factory was working at a plant efficiency 'within a few per cent. of the theoretically possible maximum' and at a significant profit: 'No longer could it said that a Government undertaking must be more extravagant than a private concern'.[22]

Yet these improvements were attained only with the greatest difficulty. Quinan himself was discouraged early, when Lord Moulton wanted to pass control of the first plant he had designed back to its owners a few months after the construction was complete. He wrote:

> The Oldbury installation was designed by me and its erection was superintended by me, either in person or through my representative, Mr. Pinder, and I know quite well what it is capable of in respect to output, &c. It is not now being operated as it should be, chiefly owing to lack of proper supervision and intelligent study on the part of the Department's managers, Messrs. Chance & Hunt, in fact the present condition of the Concentrators is such as to reflect upon me personally, to say nothing of the Department. Present conditions can easily be remedied by a little skilful attention and adjustment, but this, I regret to state, is not forthcoming from the present overseer, Mr. Calder, nor do I believe an arrangement to be possible whereby he would give his entire time to this Site ... In fact, the whole position, as it has developed, is a source of humiliation and grief to me, and I am, therefore, come to the conclusion that, in justice to myself, I must repudiate further responsibility for the Oldbury installation and regretfully request you to relieve me of further connection therewith ... The plant, as it is being operated, will no doubt continue to muddle along one way or another.[23]

Quinan nevertheless did retain responsibility, while warning other managers not to let William Cullen or others bustle them 'into taking anyone you don't

---

[20] [W. Macnab], Ministry of Munitions, Dept. of Explosives Supply, *Preliminary Studies for H.M. Factory, Gretna* (London: HMSO, 1920), ix; Ministry of Munitions, *Report on the Statistical Work of the Factories Branch* (London: HMSO, April 1919). These powerful methods should be contrasted with production control at firms such as Albright & Wilson, which limited 'statistics' to tabulated figures of batch quantities, purities and costs [BCS MS 1794 Box 81 'Statistical reports 1892–1895'; Box 64, 'Old Red Process Book 1869–1913']. Despite winning praise, however, Quinan's 'systematic management' was not entirely new. Comparative and graphical methods had been adopted by American railroad companies some thirty years earlier, and very similar statistical reporting to that championed by Quinan evolved at the High Explosives Operating Department (HEOD) of Du Pont at Repauno, New Jersey, from 1904 [JoAnne Yates, *Control Through Communication: The Rise of System in American Management* (Baltimore: Johns Hopkins Press, 1989), pp. 235–8, 248–9, 254]. It is intriguing to speculate whether Quinan was aware of these internal company practices before applying them to the traditionally isolationist British chemical industry.

[21] *The Times* 28 June 1917, p. 8. Inflation, which over this period amounted to about 30%, improves the comparison further.

[22] J. C. Burnham, *H. M. Factory, Gretna: Description of Plant and Process* (London, 1918), p. v.

[23] Letter, Quinan to Moulton 2 Sep 1915, PRO SUPP 10/278.

want, or whom you don't regard as first class'.[24] The following year, the new Gretna factory appeared to be even more badly managed, according to the liaison officer between the Ministry and construction contractors, who complained bitterly to Quinan that

> We have both of us been bitterly disappointed about Gretna. The damned place has had a blight over it ... Gretna is the only bit of the Department's work that has attracted 'public' attention. The whole Ministry and, I fear, the whole government, and House of Commons know about it. From the first Addison and L.-G. were doubting, if not hostile, and the whole Finance Branch, from Lever downwards, have been and are now prepared to cry stinking fish.[25]

Quinan re-organised the factory administration and kept close tabs on its managers, warning the most senior, for example, that unless they better delegated tasks

> you stand about as much chance of handling the Gretna Factory as I do of ever reaching Heaven, and as you know in my particular case it is a precious slim one ... You have never had to deal with any Plant bigger than a bunch of hen coops, and you must broaden your mind and learn to look upon large administration problems in the only right and proper way.[26]

'KBQ' also edited numerous technical reports crossing his desk and urged their authors to write clearly and impersonally.[27] According to E. F. Armstrong, 'his experience in training the staff available for his factories, who came to him largely without expert knowledge of the work they had to do, possibly gave him special insight into the ideal education for a technical chemist. He often waxed enthusiastic in private conversation over his schemes for training chemists'.[28] Along with such detailed supervision, Quinan kept copious notes – some 40 notebooks – on plant designs and operational success, both on his own plants and those in America, France and Russia.

But while K. B. Quinan was widely characterised as the ideal technical organiser, William Macnab himself eclipsed him in organising the chemical engineering profession. He disseminated Quinan's methods to a much wider audience than Quinan ever supervised, through the technical studies he edited with the assistance of H. W. Cremer.[29] These had been instigated under

---

[24] Letter, Quinan to Col. R. Waring, HM Queen's Ferry, 1916, PRO SUPP 10/278. Calder became President of the IChemE in 1931, Cullen in 1937.

[25] Letter, 'Willie' [Corbett] to 'Pa' [Quinan], 19 Aug 1916, PRO SUPP 10/278.

[26] Letter, Quinan to J. Burnham, 14 Dec 1917, PRO SUPP 10/278.

[27] Letter, Quinan to H. W. Roberts, HM Queen's Ferry, 4 Aug 1916, PRO SUPP 10/278; notes on J. Riley, Report on the T.N.T. Acid Cycle, HM Queensferry, IChemE library.

[28] Armstrong, op. cit. Despite reference to 'chemists', Armstrong used one-third of the review to develop his view of the chemical engineer as a 'connecting link between chemist and engineer ... [s]peaking the language and, still more, in sympathy with the mentality of both ... akin to the half-back in football.'

[29] Herbert William Cremer (1893–1970) had been in charge of the TNT plant at Queen's Ferry and later an assistant to Quinan in London, being appointed Director of Chemical Warfare Supply in 1918. Regarding Quinan's 'truly immediate influence' on him, see H. W. Cremer, 'Chemical engineering: fact or fiction?', Chem. & Indus. 14 Jan 1950, 31–3.

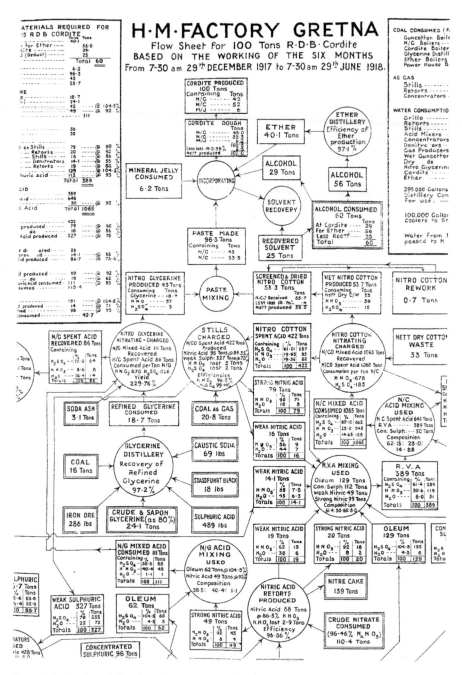

Figure 3-1   Portion of flowsheet for Gretna cordite production. *Source*: Preliminary studies for H.M. Factory, Gretna and study for an installation of phosgene manufacture (London: HMSO, 1920).

Quinan's urging to emulate the example of the National Research Council in the USA, which was planning to disseminate previously classified technical documents at war's end. Quinan argued to Moulton and Christopher Addison, the third Minister of Munitions, that 'there are many manufacturers both large and small who have never even heard of the existence of the data which we have available, and whose practice may still be described as obsolete' and urged that the publication be issued to teaching institutions, manufacturers and technical societies in Britain and its colonies.[30] Consequently, the audience for these *Technical Records of the Department of Explosives Supply* was nicely targeted: industrialists, practising engineers and managers, and the content of the reports was suitably tailored. They incorporated technical aspects of manufacture including flowsheets, blueprints and photographs; actual cost records and the method of determining them; and comparative costs for various raw materials and production processes. Teachers, indeed, supplied them to postwar students of the subject as among the most up-to-date and exemplary chemical engineering works available.[31]

According to Abbott's theory of professionalisation, Macnab's representation of Quinan was important in helping to demonstrate that chemical engineers were competent to handle the problems of rapidly and efficiently scaling up plant. The reports documented that persons such as Quinan employed a rational method of defining and analysing problems, and that their designs were demonstrably efficient. Abbott sees these features of problem classification, reasoning and solution as being crucial in justifying jurisdiction over professional roles.[32] Moreover, Macnab directed his claims to a particularly effective arena: not to the general public or to academics, but to employers and employees in the chemical plant industry.[33]

Yet such jurisdictional claims could not be supported merely by showing that experienced plant designers/managers could do the job. Indeed, while Macnab stressed scientific management, he did not clearly identify chemical engineers as unique specialists. Moreover, the authority of Quinan as a role-model had a limited currency. He shunned publicity and was ambivalent about advancing 'chemical engineering' as a profession.[34] His recognition within the government was limited to a handful of influential movers. The Treasury repeated balked at matching the salary he had received from de Beers, and a

---

[30] Quinan urged the rapid printing of as many as 20 000 copies to be issued 'without restrictions, to all of the Teaching Institutions throughout Great Britain and the colonies, to all of the various manufacturers, and also to the Technical Societies both at home and in the Colonies.' PRO SUPP 10/279, c Dec 1918.

[31] See, for example, Armstrong, *op. cit.* and W. C. Peck, 'Early chemical engineering', *Chem. & Indus.* 2 June 1973, 512–7.

[32] Abbott, *op. cit.*, pp. 39–52.

[33] *Ibid.* p. 68.

[34] Brought to organising meetings by Macnab, Quinan argued that 'very few chemical engineers using the title in its true guise could be found in the whole of Great Britain, or in fact in any other country', and 'doubted whether the number in Great Britain alone was really sufficient to warrant the formation of the Institution', unless combined with 'high moral character' [IChemE Provisional Council Minutes 8 Mar 1922].

£10 000 honorarium had to be wrestled from it with the support of Addison.[35] Quinan's public exposure appears to have been restricted largely to Macnab's efforts, to a few brief mentions in Parliament by Ministers of Munitions, and to being granted a Companion of Honour in August 1917.[36] Quinan received scarcely a mention in the reminiscences of the first Minister of Munitions or in several historical accounts of the Ministry.[37] Moreover, the government diluted its support for this arguably paradigmatic 'chemical engineer' by drawing upon many other chemists and self-described chemical engineers to perform urgent tasks of plant design and management.[38] Quinan also came under criticism in the last months of the war. He and his staff were blamed for delays in constructing new mustard gas plants, and inspections of German chemical factories suggested that the British had been less competent and speedy in scaling up from experimental plant to commercial production than had their opponents.[39] For his part, Quinan was discouraged by the quality of commercial plant supplied by some companies, and by the readiness of 'the old man' (Lord Moulton) to turn over his work on a modified process for fixation of atmospheric nitrogen – pressed by Churchill in the last year of the war – to a consortium of companies.[40] He returned to his South African post with de Beers in February, 1919. Quinan's symbolic role as exemplar faded quickly.

THE MINISTRY OF MUNITIONS AS MODEL OF ORGANISATION

Organisations, and the social context of their activities, can reinforce claims to professional jurisdiction as much as prominent practitioners can. Almost inci-

---

[35]   PRO MUN 7/26. Equivalent to over £200 000 in 1999 currency.

[36]   The citation for the Companion of Honour states vaguely, 'K. B. Quinan, Esq. Special work in connexion with the Explosives Supply Department of the Ministry of Munitions', prompting de Beers to expand the description publicly [*The Times* 25 Aug 1917, p. 8; 30 Aug 1917, p. 9].

[37]   Lloyd George, *Memoirs, op. cit.*, Vol. 2. R. J. Q. Adams, *op. cit.*; Wrigley, *op. cit.* The official history mentions not his competence or thoroughness but the less professional qualities of 'lightning speed' and 'great energy' [*History of the Ministry op. cit.*, p. 45].

[38]   Among them, for example, F. L. Nathan (responsible for propellants and alcohol production), W. B. Davidson (advisor), H. Griffiths (plant designer), S. R. Illingworth (toluol advisor), E. C. B. Wilbraham (superintendent of H.M. Factory Rainham). See IChemE AF nos. 90, 23, 43, 46 and 243 respectively. In late 1916 the Ministry received 'great assistance' from Du Pont in the USA in the planning of a new explosives plant [C. Addison, *The Times* 28 June 1917, p. 8]. Lord Moulton praised E. I. Du Pont de Nemours as a major supplier of war supplies to Britain. Over the war, Du Pont provided some 1.5 billion pounds of explosives to Britain, France and Russia [William S. Dutton, *Du Pont: One Hundred and Forty Years* (New York: Scribners, 1951)].

[39]   Haber, *op. cit.*, 165, 171. The missions are detailed in: ABCM, *Report of the British Chemical Mission on Chemical Factories in the Occupied Areas of Germany* (London: ABCM, 1919), and Ministry of Munitions, *Report of the British Mission Appointed to Visit Enemy Chemical Factories in the Occupied Zone Engaged in the Production of Munitions of War* (London: HMSO, 1919). On Quinan, explosives and poison gas production, see Guy Hartcup, *The War of Invention: Scientific Developments, 1914–18* (London: Pergamon, 1988), pp. 6–10, 44–60, 108–9.

[40]   Letter, Quinan to Turner, 3 Jan 1918; letter, R. Forster McCale, Min. of Munitions, to Quinan, 12 Apr 1919, both PRO SUPP 10/277. Quinan's work at Billingham, Teesside, on a modified Haber process for nitrogen production, which had been praised and encouraged by Winston

dentally but persuasively, the Ministry of Munitions helped to validate the legitimacy of chemical engineers. It did this in several ways: by creating a new technical problem (the need to rapidly scale-up and make more efficient chemical plants); by identifying 'chemical engineers' as being superior for solving this problem compared to academically-based chemists; and, by providing an occupational environment in which problems could be solved and solutions publicly demonstrated.[41] The Ministry provided this disinterested professional legitimation to organising chemical engineers by promoting the wider professionalisation of technical workers, including the formation of representative bodies and encouragement of research and instruction.

In the months approaching the war's end, the government's industrial strategy was flavoured by a mixture of enthusiasm for new organisational relationships and anticipation of heightened international competition. Such concerns were exacerbated by direct comparisons with the evidently still-strong German chemical industry. E. F. Armstrong, for example, who had been speaking to colleagues 'visiting the occupied territories of the Rhine and who had seen the works put up by the great German chemical firms during the war', complained that in Britain the state factories 'were mostly constructed in a way which made them unsuitable for peaceful purposes'.[42] Such criticisms, not surprisingly, opposed the Ministry's own evaluations.[43]

Perhaps accelerated by such practitioners' gossip, the Department of Explosives Supply shortly afterwards began publishing the technical reports of its wartime work. The volumes were therefore not intended – at least officially – to serve the specific interests of chemical engineers. Yet this was one of the effects they had. The Introduction to the first of these publicised higher aims of the Ministry of Munitions, and justified them by portraying its accomplishments positively. Far from having been merely a wartime makeshift creating 'totally unsuitable' factories for ordinary industry as argued by its critics, the Ministry, claimed Macnab, could serve as a model for the postwar chemical industry. The reports were intended to provide much more than simply illustrations of the value of 'chemical engineering' expertise. They included suggestions for revitalising technical education and industry, which had exhibited 'too little mutual trust and appreciation', by promoting 'closer co-operation between the more strictly theoretical and technical workers ... and above all a class of business and financial men'. The principal concern was that 'the old rule-of-thumb practice which prevailed in so many of our chemical factories' be replaced by 'thorough scientific control of every stage of the process'.[44] Thus

Churchill, fourth Minister of Munitions, was transferred to Brunner Mond and Co. in 1919. See *The Times*, 26 Apr 1918, p. 10.

[41] Organisations may be a source of new professional 'tasks' which can then be claimed by a professional group. See Abbott, *op. cit.*, 39.

[42] E. F. Armstrong, *J. SCI* 38 (1919), 100R. Armstrong was SCI President 1922–4.

[43] For example Christopher Addison praised Quinan for producing factories 'which to a very large extent will be of permanent value to peace industries' [*The Times* 28 June 1917, p. 8].

[44] [W. Macnab], Ministry of Munitions, Dept. of Explosives Supply, *Preliminary Studies for H.M. Factory, Gretna* (HMSO, 1920), pp. ix-xii. This was a common theme during the period. See, for example, C. Divall, 'A measure of agreement: employers and engineering studies in the universities of England and Wales 1897–1939', *Soc. Stud. Sci.* 20 (1990), 65–112.

a professional space was being cleared for chemical engineers as plant designers, educators and technical mediators. They should, Macnab effectively argued, colonise existing territory and stake out new ground for their discipline.

Macnab was not the instigator of these sermons, even if he did imbue them with particular relevance to chemical engineering. His overarching themes were drawn from attitudes and rhetoric prevalent in government.[45] During and after the war, Lloyd George described the melding of industry, labour and science by the Ministry as one of its most important accomplishments, claiming that it had demonstrated the value of acting in close co-operation with industry by appointing 'important members of the trade' to official posts, by discussing matters with representative associations of the trade and, where necessary, promoting the formation of such associations.[46] While such discourse may have met widespread indifference in some quarters, there is evidence that it reflected contemporary beliefs in sectors of the chemical community and Whitehall departments. Direct government interest indeed dates from this time. Ninety industrial trade associations were encountered by the Ministry of Munitions; many registered or extended their scope during the last years of the war and the postwar period.[47] Similarly, professional association was on the ascendant for scientists, educators, and civil servants: the war removed their occupational isolation and enabled collective action.[48]

The Technical Reports complemented other government initiatives favourable to chemical engineers. Most of these provided an organisational structure into which new bodies could fit. The state took an active part in postwar organisation of the chemical workforce it had mobilised, principally through the Ministry of Reconstruction.[49] For example the Minister in mid 1917, Sir Keith Price, invited the new Association of British Chemical Manufacturers, one of the largest new bodies, 'to co-operate with him in preparing to deal with the problems which will arise in the chemical industry after the war' and, the following year, to help to organise an industrial council. The ABCM,

---

[45] One reviewer wrote that Macnab was 'fully in sympathy with the ideals of some of the leading teachers and industrial leaders who are striving to elevate British chemical industry to its rightful position. His preface is not the least valuable portion of the book and ... it should receive the widest possible publicity'. E. F. Armstrong, 'The Gretna record', *J. SCI* 39 (1920), 312R–313R.

[46] See Ministry of Munitions, 'Mr Lloyd George's farewell address to the staff of the Ministry' (London, 1916), p. 6, and Lloyd George, *Memoirs, op. cit.*, 566–7. K. B. Quinan's liaison officer, on the other hand, complained that Ministry organisation was complex and inefficient: 'Gretna is the greatest Socialist experiment of our time, it is State Socialism gone mad' [letter, W. Corbett to Quinan, 19 Aug 1916, PRO SUPP 10/278].

[47] *Industrial Trade Associations: Activities and Organisation* (London: Political and Economic Planning, 1957), pp. 3–19.

[48] See R. & K. MacLeod, 'The social relations of science and technology, 1914–1939', in: C. M. Cipolla (ed.), *The Fontana Economic History of Europe: The Twentieth Century – Part One* (Glasgow: Collins, 1976), pp. 301–63, esp. pp. 318–9.

[49] From 1917 the Ministry published pamphlets on issues ranging from education to domestic help to town planning. Among those relevant to the chemical engineers were 'New fields for British engineering', 'Scientific business management' and 'Trusts, combines and trade associations' [*Reconstruction Problems* Nos. 5, 21 and 31 (1918–19)].

initially representing some 120 firms, had been formed the preceding year with the support of the SCI, the Chemical Society and the Society of Dyers & Colourists to promote co-operation between manufacturers, lobby government, organise research, and promote university and technical education. ABCM members variously stressed the role it could play in resolving labour issues and in participating in direction of the chemical industry.[50] To cap its close identification with government initiatives and wartime successes, its first president in 1918 was Lord Moulton. Industry was thus mapping itself directly onto the state structure established during the war. Moreover, Christopher Addison, the former Minister of Munitions, was later to head the Ministry of Reconstruction.

The inter-linking of state, industry and labour along co-operative lines was promoted in other ways.[51] The coalition government organised so-called 'Whitley Councils' of employers' associations and trade unions to improve not only arbitration procedures but also industrial development and competitiveness through industrial training, 'co-operation with the educational authorities' and 'design and research, with a view to perfecting the products of the industry'.[52] Similarly, the Demobilisation and Resettlement Department of the Ministry of Labour organised postwar employment of chemical engineers by asking the SCI and other bodies to fill advisory committees.[53] Such means are an example of what Abbott has described as the use of existing infrastructures to permit state penetration of civil society.[54]

The Ministry of Munitions, disbanded in 1921 after accumulating ample reserves of munitions but in the midst of economic retrenchment and public weariness for war, had thus provided ammunition with which to organise a new profession.[55] It brought together practitioners from disparate backgrounds; allowed them to share knowledge which previously had often been narrowly specialised, secretive or incomplete; dramatically and rapidly increased the scale of British chemical engineering and chemical manufacture; fostered a new relationship between industry, government and labour; and identified the need for more chemical plant specialists trained along academic lines. Perhaps most importantly, it released these semi-legitimated practitioners at the end of the war to fend for themselves.

---

[50] *The Times* 3 Aug 1917, p. 3. On the ABCM, see C. Bedford, 'The organisation of British chemical manufacturers', *J. SCI 35* (1916), 1040–5 and 561–2, 995; *J. SCI 37* (1918), 175R. Because its constitution debarred it from labour and wage issues, the Association restricted itself to technical and commercial questions and declined participation in industrial councils, a role which it relegated to the Chemical Employers' Federation founded in 1917. See also H. Levinstein, 'Chemical industry and the outlook in Europe', *Trans. IChemE 15* (1937), 12–6.
[51] For more on the broader political context and the trend to 'corporate bias', see Middlemas, *op. cit.*
[52] Ministry of Labour, *The Whitley Report* (London: HMSO, 1917); Ministry of Reconstruction, *Industrial Councils: The Whitley Scheme* (London: HMSO, 1919).
[53] SCI *Minutes of Council* (henceforth CM) 18 Dec 1918, p. 135.
[54] Abbott, *op. cit.* 121.
[55] For an enthusiastic contemporary account, see G. A. B. Dewar, *The Great Munitions Feat* (London, 1921).

POSTWAR COHERENCE

The political context

The immediate postwar climate engendered by the successes of the Ministry of Munitions infused the campaigners for new chemical engineering organisations. The experiences of the Ministry had suggested that industry, chemical production, labour and higher education could subsequently be efficiently coupled.[56] A number of these men had received practical training in chemical engineering or occupied their first posts in the highly atypical wartime economy, during which the government had rationalised their firms, controlled their production and allocated their labour.[57] Some of their older colleagues had held executive positions in the Ministry and were keen to further its organisational examples.

The political spectrum of organisers was broad, but there was considerable consensus on the future of a corporate organisation of the economy.[58] John Hinchley, for example, a strong supporter of Fabian socialism,[59] was in tune with the thinking pervading the end-of-war Ministry of Reconstruction, which included a large complement of Fabians and Guild Socialists.[60] The concept of industry-wide collaborations of industry, labour and state was widely supported both among socialists and the Coalition government. In a June 1918 meeting, Hinchley urged the continuation of the provisional British Association of Chemists (BAC) 'in order that it might have a voice in the deliberations of the Industrial Councils as outlined in the Whitley report'.[61] Similarly, William Macnab, as a Council member of the Institute of Chemistry and member of the Chemical Engineering Group Provisional Committee, initiated discussions with the Ministry of Labour for technical representation on such a Council.[62]

Even five years after the war, however, party political perspectives on industrial organisation were not highly distinct. The decision to continue the coali-

---

[56]   For an economic history of the business side of this postwar rationalisation movement, see Leslie Hannah, *The Rise of the Corporate Economy* (London: Methuen, 1976), Chap. 3. On the enforced State corporatism during and following the first world war, see Harold Perkin, *The Rise of Professional Society: England since 1880* (London: Routledge, 1989), 174–86.

[57]   Of the first 300 applicants for IChemE membership (Jul 1922 – Nov 1924), 54 had worked in explosives during the war, 38 of them directly for the Ministry of Munitions. 20% of the first Provisional Council of the Chemical Engineering Group and 21% of the IChemE Provisional Committee had had connections with the Ministry of Munitions or the explosives industry.

[58]   I.e. co-operative decision-making between the state, professional and business bodies rather than by market pressures or government alone.

[59]   E. M. Hinchley, *John William Hinchley, op. cit.*

[60]   M. Cole, *The Story of Fabian Socialism* (London: Heinemann, 1961), 187–8. See also B. Russell, *Principles of Social Reconstruction* (London: George Allen & Unwin, 1916), 141–2. On the tenets of the 'reconstructionists', see Craig R. Littler, *The Development of the Labour Process in Capitalist Societies* (London: Heinemann, 1982), pp. 99–100.

[61]   C. A. F. Hastilow and L. P. Wilson, *Chem. Trade J. 63* (1918), 248 and 273.

[62]   C. A. Russell, N. Coley & G. K. Roberts, *Chemists by Profession* (Milton Keynes: Open University Press, 1977), pp. 239–40. Within two years the chemical industry and 26 others had formed industrial councils. By the economic slump of 1920–1, however, the Building Guilds had collapsed, and no more was heard of other National Guilds.

tion in November 1918 had been a sign of continued national unity, and as late as 1921, conservatives remained allies of the interventionist Coalition Liberals in government.[63] Charles Garland, like Hinchley a campaigner for both the CEG and the IChemE, also had a hand in organising the labour-oriented BAC. His political allegiances were different, however: a long-time officer of the National Union of Manufacturers and a founder of the South London branch of the Junior Imperial League (a right-of-centre political organisation), he was elected Unionist (Conservative) MP for South Islington in 1922–3 during the Conservative government of Andrew Bonar Law which gained power in the midst of the economic slump.[64]

Nearer the political median, W. J. U. Woolcock, General Manager of the Association of British Chemical Manufacturers, was a Coalition Liberal MP for Hackney (Central) in the 1918–22 Parliament and Parliamentary Private Secretary to the Minister of Munitions.[65] Woolcock took an active part in guiding the postwar Dyestuffs Act and Safeguarding of Industries Act through the House of Commons. With his ABCM and government connections, Woolcock was well positioned to translate and apply the experiences of the Prime Minister and former Minister of Munitions to organise postwar chemical engineering, and he did so enthusiastically via the embryonic IChemE.[66] The ABCM was in other respects a close ally of both the IChemE and Ministry of Munitions: besides its first President Lord Moulton, Charles Carpenter was an organiser and became Vice President of both the ABCM and Institution.[67]

Another organisation with a sympathetic corporate outlook was the British Chemical Plant Manufacturers Association (BCPMA) formed in 1920.[68] Woolcock, as General Manager of the ABCM, had initiated its organisation

---

[63] On postwar reconstruction, see K. Grieves, *Sir Eric Geddes: Business and Government in War and Peace* (Manchester: University of Manchester Press, 1989), Chap. 5.

[64] Garland (1887–1960) devoted his few Commons speeches to issues other than the chemical industry [see, for example, *Parliamentary Debates 168*]. He had served an apprenticeship at an iron works and was afterwards manager and director of several firms. From 1925, he was President of the BAC and vice president of the National Union of Manufacturers; he became President of the IChemE for 1941–2. See IChemE AF 153; *Who Was Who 5*, 407; *TCE* Feb, 1961; *Chem. & Indus.* 7 Jan. 1961, 25.

[65] During the War, while acting secretary and registrar to the Pharmaceutical Society, William James Uglow Woolcock (1878–1947) was appointed assistant director of Army Contracts and acted as chairman of the Committee responsible for medical supplies to the Army. He became President of the SCI 1924–6 and later a vice-president of the Federation of British Industries. Woolcock never claimed to be a chemical engineer, and was made the first honorary member of the Institution. See 'Nominations', *The Times*, 5 Dec 1918, p. 14; IChemE AF 1a; 'Founders' biographies', Gayfere archive box VII/1, IChemE; and *Chem. & Ind.* Jubilee Number (1931), 89.

[66] A successful candidate in the 'coupon election' in which Prime Minister David Lloyd George asked voters to support those to whom he gave his 'letter' or 'coupon', Woolcock shared Lloyd George's political aims and experience of the chemical industry.

[67] Also a member of Moulton's Committee for Explosives Supply in 1914 and a Council member of the SCI, Carpenter resigned as VP in March 1924 owing to ill health. Carpenter (1859–1938) had studied science at Birkbeck college, and from the early 1880s had a lifetime association with the South Metropolitan Gas Company. He was President of the SCI 1915–7, and a founder of the ABCM in 1916. See *Chem. & Indus.* 57 (1938), 878–9.

[68] *J. SCI 39* (1920), 229R.

and served as its Secretary. Four other officers and members of the IChemE comprised nearly half of the officers of the BCPMA.

While all these individual and institutional actors demonstrated support for some form of industrial collectivism, it would be misleading to attribute a single philosophy to them as a group.[69] In various ways threads of what we might now label Taylorist and technocratic thought stitched together advocacy for the technical and managerial expertise of chemical engineers with the wider social and political elements of corporatist thinking. Some historians have suggested that Taylorism inspired and permeated the early (North American) academic discipline of chemical engineering, influencing how technical tasks were identified and solved: Macnab's reports did appear to place the American Quinan's methods of chemical engineering in the context of Taylorist principles, and something like Taylor's notion of 'scientific' management of factory production was certainly extended to management in the contemporary chemical industry.[70] But we should not make too much of these appearances: the influence of Taylorism on the intellectual roots of chemical engineering (discussed in the next chapter) is doubtful, and there is no obvious connection between the new postwar bodies in the British chemical industries and Taylor's ideas of industrial organisation and efficiency, which cannot be mapped convincingly onto larger political structures. As John Wilson has argued, managerial techniques successful in the USA were not necessarily viable in the different market environment of Britain.[71] The same is true of technocracy, understood here as a system of thought which sees technical experts employing 'scientific' principles of economics, sociology and management in the pursuit of social harmony and mutual improvement among all interest groups. Like Taylorism, technocratic thought was first envisaged as a means of optimising industrial production and relations, and only later extended to the wider social context.[72] While Macnab's writings can be interpreted in this vein, little vocal support is evident among his political contemporaries. Lloyd George, for example, preferred problem solving and management to be undertaken by businessmen, for their adaptability and vigour, rather than by technically adept administrators.

Despite these mixed philosophical bases for the collective and organising movement in the chemical industry, it is clear that wartime organisations

---

[69]   Peter Williamson *Varieties of Corporatism: A Conceptual Discussion* (Cambridge: Cambridge University Press, 1985); *idem.*, *Corporatism in Perspective: An Introductory Guide to Corporatist Theory* (London: Sage Publications, 1989), in attempting to clarify the definition, distinguishes at least three varieties of corporatism, ranging from authoritarian regimes to liberal democracies. See also O. Newman, *The Challenge of Corporatism* (London: MacMillan, 1981), Chap. 1.

[70]   See, for example, N. Swindin, 'Bonus systems for chemical works', *Chem. Age* 16 Aug 1919, 240–3.

[71]   Divall and Johnston, *op. cit.*; John F. Wilson, *British Business History 1720–1994* (Manchester: Manchester University Press, 1995), pp. 162–5.

[72]   C. S. Maier, 'Between Taylorism and Technocracy: European ideologies and the vision of industrial productivity in the 1920s', *J. Contemp. Hist.* 5 (1970), 27–61. See also J. Meynaud, *Technocracy* (London: Faber & Faber, 1968), passim.

provided both the immediate incentive and an example for continued efforts. As Roy and Kay MacLeod have discussed for the optical industry, the war showed that 'an economic and scientific alliance between government and industry was, whether immediately or in the long term, of vital interest to both'.[73]

Organisers of chemical engineering

Translating these various strands of collectivist thought into a new professional organisation, the IChemE, required the building of an alliance across traditional occupational categories. Six persons stand out for the range and enthusiasm of their organisational activities: Hinchley, Macnab, Woolcock, Garland, J. A. Reavell and Harold Talbot. Significantly, each was prominent in at least three organisations linking government, industry and professional engineers.[74]

The representation of organisers with direct wartime experience was particularly strong among eventual officers of the Institution: ten of the first sixteen Presidents of the IChemE had Ministry experience. In approximately chronological order, they were: F. L. Nathan, Advisor to the Admiralty 1914–5 on cordite supplies (essentially K. B. Quinan's counterpart supplying the sea-borne forces), who designed and erected the Royal Naval Cordite Factory at Holton Heath, Dorset;[75] W. A. S. Calder, Director of Chance & Hunt and manager of the first national TNT plant at Oldbury;[76] Macnab and Cremer, the chief promoters of Quinan;[77] W. Cullen, in the Technical

---

[73] R. & K. MacLeod, 'Government and the optical industry in Britain, 1914–1918', in: J. M. Winter (ed.), *War and Economic Development* (Cambridge: Cambridge University Press, 1975), pp. 165–204.

[74] See Table A-I, Appendix.

[75] Sir Frederic Lewis Nathan (1861–1933) later criticised Quinan's Gretna factory which 'had been put up quickly during the war' as opposed to Holton Heath, intended as a 'permanent establishment with the best labour saving plant', and which was better organised and 'cleaner than Gretna', and hence better suited for postwar production ['Evidence given by Sir Frederic Nathan before the committee on production of cordite', typescript, 15 Mar 1920, 'F.L.N. 4' folder, Nathan papers, IChemE]. The government description of Gretna was 'a well-designed factory built comparatively slowly', and a Parliamentary Committee proposed that Gretna 'being a modern and up-to-date factory, should therefore be kept in preference to the older factory at Waltham Abbey' [*History of the Ministry, op. cit.*, 67; *Parliamentary Debates 163* (1923), 2523–31].

[76] William Alexander Skeen Calder (1874–1940) was IChemE President in 1931. Calder had studied chemistry at the Royal College of Science and Germany in the early 1890s, then as chemist at F. C. Hill & Co. before joining Chance & Hunt. CRO DIC/BM 20/191; W. C. Peck, 'Early Chemical Engineering', *Chem. & Indus.*, June 2, 1973, 512–7; 'Lord Moulton', *Trans. IChemE 17* (1939), 184–5; R. E. Threlfall, *The Story of 100 Years of Phosphorus Making 1851–1951* (Oldbury: Albright & Wilson, 1951), pp. 178–9.

[77] Macnab was sixth IChemE President, 1934, Vice President of the Institute of Chemistry 1921–4, and a Council member for 12 years. See IChemE AF 123; *Who Was Who 4*, 743–4. Chairman and Honorary Secretary of the Chemical Engineering Group, he was also Honorary Secretary of the IChemE 1931–7, and President 1947–8. Cremer was also a Vice President of the SCI, and Secretary, Vice President and Treasurer of the RIC. See F. Warner, 'Famous men remembered', *TCE* (Dec. 1985); *Who Was Who 6*, 255–6.

Investigation (Explosives and Chemicals) group;[78] F. H. Rogers, working on Explosives and Food installations during the first world war (including designing H. M. Guncotton Plant Colnbrook);[79] F. A. Greene, constructing shell-filling plants;[80] H. Griffiths, responsible for the design and erection of various explosives plants;[81] D. M. Newitt, working with Griffiths as members of research team under Nathan at Nobel's Explosives Co;[82] and H. Hartley, Controller of Chemical Warfare Dept.[83] Among the Councils and first members of the CEG and of the IChemE, about one in five had had direct experience of the Ministry.[84]

But while participants in war-related 'chemical engineering' formed an influential minority of early members of the IChemE, they were not the only interest group. The age and career stage of most Presidents during its first decade generally precluded their having had hands-on experience. For example, the first President, Sir Arthur Duckham, had been closely associated with the gas industry and industrial management.[85] The third (Sir Alexander Gibb) was

---

[78] Working as a consultant after the war, William Cullen (1867–1948) was elected President of Institution of Mining and Metallurgy 1929–30; for the SCI he was Council member, Vice President 1934–7 and President 1941–3. He became IChemE President 1937–8. See Ministry of Munitions, *List of Staff and Distribution of Duties* (1918), 190; *Who Was Who 4*, 272; AF 345.

[79] Francis Heron Rogers (1877–1955) was Hon. Treasurer, 1923–9; Vice President; President 1939–40. A patent agent and designer of metallurgical plants, Rogers sat on the Boards of several firms. AF 74. Guncotton (or nitrated cotton, or nitrocellulose) could be produced in powdered form as a high explosive.

[80] Frank Arnold Greene (1876–1962) later consulted for the paper and gas industries. Among his numerous affiliations he was Hon. Treasurer of the CEG, IChemE and Diesel Engine Users' Association, and IChemE President 1943–5. See IChemE AF 53; *Chem. & Indus.* 12 May 1962, 862–3.

[81] Following the war Hugh Griffiths (1891–1954) ran his own consulting business, and took over chemical engineering lectures started by John Hinchley at Battersea Polytechnic between 1917–34. He was President of the IChemE 1945–6. AF 37 and DNB.

[82] Dudley Maurice Newitt (1894–1980), a former student of Hinchley and member of the IChemE Provisional Committee, was elected President for 1949–50. See 'Founders' Biographies', box VII/1, Gayfere archives, IChemE; P. Danckwerts, 'Famous men remembered', *TCE* (Oct. 1984); Obituary, *TCE* (July 1982), 295; *Who Was Who 7*, 580; and 'Memoirs of the early days', *TCE* (April 1972), 137.

[83] Sir Harold Hartley (1878–1972) was active in fuel research, railways and scientific management, and twice became President of the IChemE, in 1951 and 1954. See *International Who's Who 1948*, 382.

[84] Even John Hinchley, who had concentrated on a scheme for peat-drying for the War Office, had been asked in 1916 to submit a design for a State TNT plant [E. M. Hinchley, *op. cit.*].

[85] Duckham (1879–1932) had held supervisory and engineering appointments since 1899 in Gas and Water companies. His invention of an improved retort for coal carbonisation led to formation of the 'Woodall-Duckham' group of companies of which he was sole Director from 1923. During the first world war he had been Director-General of Aircraft Production and Deputy-Comptroller of the Munitions Inventions Dept, one of those industrialists Lloyd George sought in State administration as 'men of push and go'. Duckham was also President of the Society of British Gas Industries and Vice President of the British Commercial Gas Association. See obituary in *Chem & Indus. 51* (1932), 167. Company histories and business archives of Woodhall-Duckham Ltd are at the University of Glasgow, UGD/309/154.

mainly a civil engineer;[86] the fourth (J. Arthur Reavell) had trained in electrical engineering and founded a chemical engineering contracting company. Reavell, like Woolcock, was active in promoting the interests of both chemical engineers and manufacturers in several organisations.[87]

Managers or plant owners, in fact, made up over a third of the first members of the IChemE. The remainder were divided about equally between self-described engineers and chemists employed by others; independent chemical engineering consultants comprising only 6%. And, emphasising its interstitial position between science and engineering, half the members of the IChemE held degrees, typically in chemistry or engineering.[88] Thus the supporters transcended traditional class and occupational boundaries, empowered by the ferment of wartime change.

New industries, new imagery

The first years of the IChemE were buoyed by an enthusiasm from several sources. The political and organisational components of this ebullience were matched by a comparable commercial fervour. The identification of chemical engineering as a newly viable activity adopted within at least some parts of the chemical industry is suggested by the promotions of publishers and advertisers eager to exploit a potential market. For example, a remarkable series of some three dozen small books, the *Chemical Engineering Library*, published by Benn Brothers, blanketed the market between 1922 and 1926.[89] These publicised methods of design and plant operation for practising chemical engineers, most of whom, during this period, were without formal training in the subject.

Among new periodicals, the *Proceedings of the Chemical Engineering Group* (1919–), *The Chemical Age* (also published by Benn Bros., 1919–), *Chemistry and Industry* (1923–) and the *Transactions of the Institution of Chemical Engineers* (1923–) created a public image of the subject for an increasingly wide readership. Articles often emphasised the technical details of particular processes or materials rather than general methods – information that could inspire or directly alter the working practices in small firms. Periodicals did more than books, however, to promote an identity. Advertisements, focusing almost exclusively on ready-made components such as grinders, boilers and pumps before the war, now increasingly promoted firms' less tangible (and previously less

---

[86] Alexander Gibb (1872–1958) was owner of an engineering consultancy from 1921 and President of the Institution of Civil Engineers in 1936. See G. Harrison, *Alexander Gibb: The Story of an Engineer* (London: Geoffrey Bles, 1950), in which chemical engineering receives two sentences.

[87] James Arthur Reavell (1872–1973), owner of the Kestner Evaporator and Engineering Company from 1908, was a council member of SCI, chairman of the CEG 1920–4 and 1930–2, chairman of the BCPMA 1927–30, and IChemE President for 1929–30. See *The Kestner Golden Jubilee Book 1908–1958* (London: Kestner Evaporator and Engineering Co., 1958); *Who Was Who* (1996). For each of the above, see also *Presidents' Book*, IChemE.

[88] See Chapter 4.

[89] Some of the books' authors were to join the IChemE. Another more limited series was published by Lockwood & Son.

marketable) design skills as a feature of their new chemical engineering perspective.[90] Silica Gel Ltd announced 'Our chemical engineers are always at your service'.[91] Morris Ltd. warned that 'to lay down a complete chemical plant practical engineering experience coupled with a chemist's knowledge of the problems of the laboratory are essential'.[92] Advertisements in *The Chemical Age* and *Proceedings of the CEG* began to portray manufacturers such as Scotts ('The Chemical Engineers'), The Chemical Engineering & Wilton's Patent Furnace Company, John F. Carmichael ('Chemical Engineers & Contractors') and Guthrie & Company Chemical Engineers, as firms applying a distinctly effective approach to design and installation.[93] These were newly adopted identities. Significantly, in an earlier, undated catalogue, Scotts had advertised themselves simply as 'Engineers since 1834'; by 1934, however, their new description had proven so commercially fertile that they extended their newly clarified identity backwards by a century, titling their anniversary brochure *One Hundred Years of Chemical Engineering*.[94] Such retroactive claims and current public portrayals did more than assert their firms' competence: they staked the claim of chemical engineers to new jurisdictions previously served by other, less self-coherent, groups. Advertisements, articles and books raised the perception of the public and industry – and chemical engineers themselves – of a new professional group ready to compete for territory.

New technologies, too, signalled new opportunities. During the 1920s, the most prominent of these was high-pressure plant, pioneered in Britain by the synthetic ammonia facility begun during the first world war, and taken over and developed by ICI at Billingham.[95] The plant required expertise in compressors, pumps, reacting vessels, control devices and moving machinery. At Billingham, work 'of the chemical engineering type' included 'heat transfer, lagging, heat exchanger design and pressure drop in catalysts', and engineers were responsible for testing in engineering and instrument design, planning, consulting, and the design and maintenance of plant. In conjunction with

---

[90] Indeed, advertisers seeking to promote themselves on the back of the new organisations proved to be an early problem for both the CEG and the IChemE. The CEG was criticised by the SCI council for accepting advertisement revenue without their approval, and the early IChemE quickly dissociated itself from a 'chemical engineering catalogue' of plant and machinery after the publisher, J. L. Hill, used the Institution's name without permission in its promotions [CM 26 Sep 1923].

[91] *The "Chemical Age" Year Book, 1928*, p. 74. Some forty firms were listed as specialising in chemical engineering that year.

[92] Advertisement, E. Ford Morris Ltd, *The "Chemical Age" Year Book, 1923*, p. 192.

[93] Advertisements, *The "Chemical Age" Year Book*, pp. 107–10, 142 (1923), 44 (1924), 57 (1928).

[94] The Balfour Group of Chemical Engineers, Leven, SRO GD410.

[95] S. G. M. Ure, 'Progress of chemical engineering during the last twenty-five years', *Chem. & Indus. 54* (1935), 432–4. Boiler pressures after the war rose considerably. See C. O. Brown, 'High pressure synthesis – basis of new chemical engineering industries', in: Sidney D. Kirkpatrick (ed.), *Twenty-Five Years of Chemical Engineering Progress* (New York: AIChE, 1933). Kirkpatrick, chemical industry journalist and President of the AIChE in 1942, later praised British and German research in the field, noting the commendable British safety record of eleven years without a fatality. Letter, Kirkpatrick to M. A. Williamson, 18 Aug 1936; Brennan archive.

Figure 3-2 Advertisement for Scotts. For smaller-scale equipment, a short employee was sometimes photographed to emphasise size. *Source*: '*The Chemical Age*' *Year Book 1923*; interview, W. E. Bryden with Johnston, 5 Dec 1997.

chemists, they did flowsheet designing and other chemical engineering tasks.[96] The German Haber process, the inspiration for all subsequent work, was, according to Reuben and Burstall, 'the first modern chemical process: the first to use gases at high temperatures and pressures, the first to require specialized plant, the first to be genuinely capital-intensive'.[97] These demanding technologies, widely reported in the press, were emblematic of a new scale and approach to process engineering.

New materials for chemical plant were also a frequent source of enthusiasm and optimistic forecasts by the emerging chemical engineer. Novel metal alloys and corrosion-resistant materials had been sinks for speculative capital for decades, and continued to proliferate with somewhat more success in the interwar period. For example, IChemE founding member Norman Swindin had been employed before the war in the commercial development of 'Tantiron', an acid-resistant high-silicon iron alloy promoted by metallurgist Robert Lennox. Swindin subsequently built his own successful business based on rubber-coated plant components, pumps and submerged burners.[98] The Tantiron business was bought in the early 1920s by another founder member, J. Arthur Reavell, to expand his Kestner Evaporator and Engineering Company. Indeed, new materials for chemical plant were the basis of the company's continued success, creating a business dynasty for Reavell, his sons and grandson, all to become members of the IChemE.[99] 'Revolutionary' developments in tantalum, tellurium lead, and copper and nickel alloys were just as exciting during the interwar years.[100] New metals, in fact, transformed the environment of postwar chemical plants. Stainless steel, first used in German and Austrian nitric acid and munitions plants after 1916, became increasingly applied from the 1920s in an industry which had, pre-war, relied on cast and wrought iron, mild steel, lead, copper, tin, wood and bitumen.[101] Chemical stoneware, vitreosil and inert resins similarly multiplied the options available to designers, and improved welding techniques allowed stronger, more reliable vessels than could be produced by forging or rivetting. The welding and annealing of aluminium, developed by Richard Seligman's Aluminium Plant and Vessel Company (later known as APV) from 1910 and applied to vats and fermenting tanks before the war, allowed the firm to enter the more promising

[96] Victor E. Parke, *Billingham: The First Ten Years* (Durham: Imperial Chemical Industries, 1957), esp. pp. 64–5.

[97] B. G. Reuben and M. L. Burstall, *The Chemical Economy* (London: Longman, 1973), p. 17.

[98] Norman Swindin, *Engineering Without Wheels: A Personal History* (London: Weidenfeld and Nicolson, 1962), an excellent illustration of the wide variety of tasks that might be expected of a chemical engineer in the smaller firm.

[99] *The Kestner Golden Jubilee Book, op. cit.*, pp. 1–16. See AF 71, 932.

[100] W. M. Cumming and F. Rumford, 'The trend of chemical engineering', *Chem. & Indus.* 57 (1938), 905–10.

[101] Yet studies of commercially available stainless steels in 1927 by the Chromium Alloys Sub-Committee and the Joint Research Department of the BCPMA nevertheless found most of them 'disappointing'. *Reports from DSIR and Research Associations 1925–8*, PRO WO 188/371. The first major use of stainless steel in America was at Du Pont's Repauno, NJ, nitric acid plant in 1926 and shortly after by Allied at Hopewell, VA.

heat exchanger market in 1923 – changing it from 'one of metallurgical engineers primarily involved with vessels, to one of process engineers supplying complete plant lines'.[102] Other singular technological innovations became just as pervasive. Electric motors, more compact and easily maintained than their alternatives, helped promote a transformation in the practice of pumping fluids, which had frequently been based on pneumatically driven lifts and 'eggs'. Chemical engineers saw themselves as particularly well adapted for exploiting connections with such new technologies.

AGENDA FOR ORGANISATION

Not surprisingly, this diversity of enthusiasms was equalled by that of professional aims for new chemical engineering bodies, both in terms of scale and in execution. For a small but active minority, the prevailing theme was 'organisation to boost national fortunes'. They equated the professional development of chemical engineers as a crucial step in making British industry efficient and internationally competitive.[103]

By contrast, most of the 'supporters of formation' of the IChemE clearly had more limited aspirations. When prompted to supply a reason for their support, they typically listed the importance of professional qualification to recognise experience, or the need to improve standards of training.[104]

The first group was thus sympathetic to the postwar government aims and sensitive to external economic threats; the second, to the more local problems of professional rivalry over jurisdiction. The leaders wanted a strong national industry; the majority wanted jobs within it.

FORMATION OF THE CHEMICAL ENGINEERING GROUP OF THE SOCIETY OF CHEMICAL INDUSTRY

From the last months of the war, chemical engineers were meeting to consider new organisations.[105] A group devoted to chemical engineering as a specialism,

---

[102] G. A. Dummett, *From Little Acorns: A History of the A.P.V. Company Limited* (London: Hutchison Benham, 1981), p. 52. Dr. R. J. S. Seligman (1878–1972), a founder member of the ABCM in 1916, the BCPMA in 1920 and IChemE in 1922, passed control of APV to his son in 1958.

[103] Of 206 registered 'supporters of formation' several, particularly the pre-1922 campaigners, stand out as having a proselytising aim and employing distinct rhetoric. Remarks added to their forms registering them as supporters of the movement include 'An institution which would ensure the proper training of the Chemical Engineer and a recognition of his status is an urgent National necessity' [H. J. Pooley]; 'the science of Chemical Engineering which vindicated its existence so worthily during the war will do *more* to restore industry in the future in this land, than any other branch of the great engineering profession' [F. H. Rogers]; 'the objects with which it is proposed to form the Institution are of the greatest national importance to the development of a great national chemical industry' [W. J. U. Woolcock]; and, 'of vital importance to the industrial development of Great Britain and the Empire' [K. B. Quinan]. See 'Supporters of formation of the Institution', box VII/1, 'Formation Papers', Gayfere archive, IChemE.

[104] Such a difference in objectives, largely dividing the organisers and officials from the members, has arguably persisted through the history of the Institution.

[105] 'History of the formation of the group', *Proc. CEG 1* (1919), 3–6.

rather than to chemical engineers as professionals, proved easier to negotiate. The role of chemical engineers in industry was still contentious; the need for chemical engineering, now, was not.

The most promising route to such a new body was via the Society of Chemical Industry. The SCI, at least, was not an accrediting organisation, and so posed little threat from other claimants to the same occupational territory.

At the end of July 1918, some 70 persons attended a meeting chaired by Prof. G. T. Morgan.[106] John Hinchley, who had previously circulated a petition for the inauguration of a chemical engineering Section of the SCI, subsequently put forward a formal request to the SCI with signatures of the supporters and, with Harold Talbot, who had also championed its creation, provided a detailed plan.[107] A mood for separatism was evident. About a third of the SCI members replied to a poll, three-quarters of them favouring 'a separate Institution of Chemical Engineers if any difficulty were experienced in securing inclusion in, or affiliation with, the Society of Chemical Industry'.

For a body which included only a handful of educators, it is significant that the stated purpose of the Chemical Engineering Group was to promote chemical engineering research and the education of chemical engineers. At the organising meeting, Hinchley sketched the proposed work as arranging frequent conferences at various centres, and to mount a campaign for original investigation in technical colleges, schools and works. As one attendee noted, 'such a programme could be obtained if a responsible and fully accredited body were charged with its supervision and general organisation. Most of the speakers deplored the backward state of research work of this kind in the United Kingdom, and the meeting was very strongly in favour of energetic measures being taken to establish chemical engineering education and research on a satisfactory and progressive basis'.[108] A related aim was the development of 'data sheets and curves useful in chemical engineering design'.[109]

This emphasis on promoting research and communication of knowledge is noteworthy. Such provision of academic knowledge was another important means of protecting the nascent profession from outside interference by competitors such as chemists or mechanical engineers.[110] It allowed them to claim authority over such data and its means of determination. Such a function would, for a stronger profession, have been provided by academic institutions. For chemical engineers, however, with only a handful of mainly part-time, non-degree courses available, three 'academic' functions were effectively shared by three entities: (1) the Ministry of Munitions had demonstrated a disciplinary distinction for chemical engineering; (2) the CEG supported research and

---

[106] Morgan, Professor of General and Industrial Chemistry at the University of Birmingham, had been an associate member of Chemical Warfare Committee, Ministry of Munitions.
[107] SCI CM, 19 Apr 1919, pp. 90–1. Talbot (b. 1885), then a company director, had held posts as Chief Chemist and Manager at lighting, dye and other companies, notably General Manager of the Welsbach Light Co. See IChemE AF 159 and *Chem. & Indus. 51* (1932), 860.
[108] *J. SCI 37* (1918), 296R.
[109] *Chem. Age 1* (1919), 442.
[110] See Abbott, *op. cit.*, 56.

publicity for the subject; and (3) a few relatively weak academic departments such as Imperial College provided instruction. Although rather vulnerable to attack in themselves, these three activities together protected the jurisdiction of chemical engineers from competitors.

But the new Group had other, less inward-looking, aims, notably to act as 'broker' between users and suppliers of chemical plant to improve efficiency while preserving commercial secrecy.[111] In all these objectives the CEG was mirroring at least some of the aims being published in the Ministry technical reports, and of government policy at large. Implicit in them was co-operation with both industry and government.

Professional activities were an unstated aim of CEG activities. Unlike the Association of Chemical Technologists and British Association of Chemists, also set up with the support of John Hinchley, the Chemical Engineering Group sought a membership of professional engineers. Where the Association of Chemical Technologists had set an annual subscription of 10s. 6d. 'to bring membership within the reach of all technologists in the country', the Group decided to set its subscription at a more professionally respectable one guinea.[112] Hinchley showed his hand by alluding to ambitions to 'even form an Institute which could confer degrees or diplomas, for there are comparatively few chemical engineers in this country, and in the future we must "grow" them'.[113] According to a contemporary, Hinchley in particular had wanted to form a separate Institution rather than the Chemical Engineering Group, but 'in consultation with friends he gradually came to the conclusion that the time was not ripe for such an organisation'.[114]

Three weeks before the Armistice, the Group officially formed with Hinchley as Chairman, and by the end of 1919 the CEG claimed 510 members. The organisation began to achieve results, too: the first conference, on power plant in chemical works, was held that July. The initial optimism was not sustained, however. The postwar trade slump in 1920–1 led to 'pronounced diminution of membership': by 5% in 1921 and 20% in 1922.[115]

FORMATION OF THE INSTITUTION OF CHEMICAL ENGINEERS

The disappointing membership numbers were likely not only a consequence of the worsening commercial climate but also of the limited mandate of the Chemical Engineering Group. While the new Group was successful in focusing interest and promoting the technical aspects of the subject, it was impotent in developing a professional presence for chemical engineers in a difficult postwar economy. Prevented by its constitution (as a sub-group of the SCI and hence

---

[111] *Chem. Age 1* (1919), 537.

[112] *Chem. Eng. & Works Chemist 1* (1911), 3–7; *J. SCI 37* (1918), 296R.

[113] *J. SCI 37* (1918), 413R.

[114] *J. SCI 38* (1919), 97R; 'The history of the formation of the Institution of Chemical Engineers', *Trans. IChemE 1* (1923), vii–x, esp. vii, and J. A. Reavell, '[J. W. Hinchley:] His work for science', *Trans. IChemE 9* (1931), 28.

[115] Anon., *Proc. CEG 3&4* (1922), 1–3.

constrained by its conditions of formation) from accrediting engineers, the CEG had no means of tying and anchoring the occupational tasks it studied to a discrete type of professional.

Through 1920 a number of the members continued to agitate for a separate Institution specifically to recognise chemical engineers as distinct professionals. During the group's Fourth Conference held that July in Newcastle, a meeting was held to promote the idea, but no immediate action was taken. Lord Moulton, later in the year, provided ammunition for the cause in a speech at University College, London, urging the provision of *laboratory* facilities in which 'truly *manufacturing* processes' could be carried on.[116] Coupled with his wartime declarations of the need for both chemical engineers and adequate training, Moulton was a vocal sympathetic outsider in what was increasingly becoming a process of jurisdictional negotiation between engineers and chemists.[117]

By the following year, nationalistic motivations were giving way to heightened demands for professional qualification. Suggestions were made to form a north of England Society separate from the CEG.[118] Finally Hinchley, still a Council member, called a meeting in November 1921 presided over by Sir Arthur Duckham. During the 80 minute meeting motions to agree the desirability of an Institution, to form a provisional Institution and to elect a Provisional Committee were supported by at least ten speakers from a variety of backgrounds.[119] No dissenting voices were heard, and the approximately one hundred persons attending unanimously carried the motions.[120] Among the members of the Provisional Committee elected from those moving and seconding resolutions were the Chairman and Honorary Secretary of the Chemical Engineering Group, Reavell and Talbot. Indeed, nearly half the members of the new Provisional Committee of the IChemE were also

---

[116] Our emphasis. For the complete text, see *Trans. IChemE 17* (1939), 186–91.

[117] Sir Alex. Gibb, presiding at the 1929 Annual General Meeting, observed that Moulton would have been the Institution's first president had he not died in 1921.

[118] Some one hundred persons were distributed between the industrial centres of the north-west (around Manchester), north-east (near Teesside) and Yorkshire. An early compilation by the CEG listed 32% Northern, 36% London, 18% Midlands, 9% Scottish and 4% foreign locations for its first 395 members [*J. SCI 38* (1919), 100R]. The corporate membership of the early IChemE to 1929 (598 members) was similarly 29% Northern, 32% London and home counties, but 8% Midlands, 5% Scottish and 20% foreign [IChemE AF].

[119] These included H. M. Ridge (b. 1873), the managing director of a furnace and engineering company; consulting chemical engineers F. H. Rogers, F. A. Greene and J. A. Reavell (who also were members of the Institution of Mechanical Engineers); chemical engineers C. J. Goodwin, D. Brownlie and J. MacGregor; and W. J. U. Woolcock. Camillo J. Goodwin (b. 1884) took over his father's chemical engineering practice in 1910 (which had for a time employed J. W. Hinchley) and consulted for Nobel's Explosives and the Ministry of Munitions during the war. David Brownlie (1880–1951), after part-time technical school training, served an apprenticeship as a technical chemist and then became a company Director. James MacGregor (b. 1878), also with part-time chemistry training and wide industrial chemistry experience, became a company manager and owner. See IChemE AF 1a, 33, 36, 53, 71, 74, 120, 271.

[120] 'Proposed Institution of Chemical Engineers', *IChemE Minute Book, 1921–3* (hereafter IChemE *Minutes*); 9 Nov 1921.

Council officers of the CEG.[121] Hinchley had, from the outset, argued that the Institution would 'quicken the efforts' of the CEG and other societies with cognate aims, and others stressed the 'quite different functions' of the sister organisations.[122] In practice, the two bodies gravitated towards the complementary roles of staking a claim for a specialism and an appropriating class of professionals.

But there were nagging concerns over contesting this 'professional space'. Apparently still uncertain of their mandate, the first meeting of the Provisional Committee two weeks later set out to reiterate the justification for their activities. Hinchley, the Honorary Secretary, claimed that inter-professional rivalries were absent, noting 'the astonishing success of the project, and the support it had received from all important industrial centres of the United Kingdom'. But Reavell and Macnab suggested investigating whether any existing institution could meet the need of chemical engineering. Could not chemical engineers be seen as a variant of some existing type of professional? 'After considerable [but unrecorded] discussion it was agreed that no established institution could do the work, and that the formation of an Institution of Chemical Engineers be proceeded with, and that affiliation or co-operation with other bodies be left for future consideration'.[123] Nevertheless, employers and employees implicitly melded; the Provisional Committee met over the next five months to draft a constitution and set up finances, usually in the boardroom of the Association of British Chemical Manufacturers provided by William Woolcock.[124]

As with the CEG, the precise aims of the new organisation were variously interpreted. The vague calls for co-operation to promote national interests were soon replaced by more specific aims. Both the 'hall marking' and training of chemical engineers were clear objectives, with Hinchley stressing the latter. A strong academic component would undoubtedly have helped to validate the vulnerable profession at that time. But as with the publication of a 'Chemical Engineering Hand Book' he had planned for the CEG, Hinchley's notion of summer schools of chemical engineering never materialised.[125]

---

[121] The 16 IChemE Provisional Committee Members (1922–3) and 21 CEG (1923) council members included seven in common: Hinchley, Talbot, Reavell, Macnab, Rogers, Goodwin and S. G. M. Ure.

[122] J. W. Hinchley and Editor, *Chem. Trade J. & Chem. Engineer*, 9 Jul, 1921.

[123] IChemE *Minutes*, 'First meeting of the Provisional Committee 7 Dec 1921'.

[124] Duckham missed all but one of the first eight meetings, apparently because of illness. The first four meetings were held at his offices, but the following ten – one of which was on Boxing Day – were held at the ABCM, leading Woolcock to be made chairman [*J. SCI* 40, 268R]. The best attendees were Hinchley, Macnab, Talbot, Garland, Nathan and Gibb. Others (Baker, Brownlie, Drew, Smith) attended only one of the first nine meetings. Their enthusiasm is to some extent reflected in their subsequent popularity as Council members: in the vote to elect the first Council, the decreasing order of preference for sixteen candidates was Talbot, Reavell, Garland, Davidson, Gibb, Macnab and Nathan. See Gayfere archive, box III/3, 'Foundation Documents'.

[125] *Ibid.*, informal meeting of IChemE Provisional Committee 14 March 1922. His widow noted that Hinchley personally had three publishers' contracts for text books on his death in 1933, for which he delivered nothing [E. M. Hinchley, *op. cit.*, pp. 119–20].

The Institution's promise of legitimating the new profession nevertheless generated moderate support. Following the Inaugural Meeting in May 1922, seventy-three guarantors of its costs came forward, including most of the Provisional Committee and several manufacturers.[126] Over the following months some 200 completed 'Supporters of Formation' forms were collected, and by July a raft of application forms for membership was being accepted.[127] The Provisional Committee met another six times to settle the final Articles of Association and by-laws, which were approved by the Board of Trade in October 1922; the Institution was formally incorporated in December, and the first Corporate Meeting was held the following June.

Early relations with other bodies

Importantly, the competition by chemical engineers for a place among existing occupations did not take place at work sites as much as within organisations. Abbott has noted that boundaries between professions tend to disappear at places of work, and the pre-war experiences bear this out. Instead, a professional niche was found through negotiation with existing bodies. There, functions were relatively separated and idealised.

But translating the enthusiasm for co-operative postwar bodies to the realities of an accrediting institution proved awkward. The goal of close co-operation and 'active support' of the CEG, SCI and other organisations was scaled down almost as rapidly as the constitution of the IChemE was solidified. The important task was soon identified as fitting into the existing professional system. Hinchley initially contacted the Mechanicals, Civils, Gas and Electricals, and eventually consulted eleven scientific and engineering societies, including the Institute of Chemistry and the American Institute of Chemical Engineers regarding a proposed constitution. The Institute of Chemistry alone requested a meeting, and the AIChE merely cabled its suggestion for reciprocal privileges for members of the two organisations. Despite Hinchley's illustration of the American counterpart as a model of success (and thus an important legitimation for the British profession) and Garland's little-known attempts to found a British Division of an 'internationalised' American Institute of Chemical Engineers, contact between the two bodies remained limited, if amicable.[128] In

---

[126] Gayfere archives, box III/3, 'Formation documents', folder 'Guarantee fund'.

[127] Applications were completed by all prospective members, although one Provisional Council member (A. C. Flint, Assistant Secretary) died in 1924 without applying for membership. CEG members were more dilatory in joining, with some of their Provisional Council applying as late as 1930, and others not at all.

[128] For example: '... the corresponding American Body has been in existence for over 15 years, and constitutes an influential organisation which is extremely anxious to co-operate with the British Institution'; and, five months later: '... it is intended that a British Institution, if and when formed, shall work in the closest harmony with the Transatlantic organisation'. Garland, in later years, mentioned an abortive 1921 proposal he and others made to the AIChE in New York, and voted that year at the General Meeting in Baltimore, that 'the American Institute of Chemical Engineers should change its title so as to become an international body, of which any institution formed in this country should become the British branch'. [See Hinchley, 'Re Proposed Institution of Chemical Engineers', October 1921; open letter, Hinchley to Managing Directors, March 1922; Founders' Biographies, June 1938, respectively, box VII/1, Gayfere

any case, the American organisation was hardly a powerful ally. It had fewer than 300 members at the end of the war, although a spurt of growth doubled its size by 1923.[129]

The standoffishness of the other professional bodies is not surprising. British chemical engineers still had little visibility and threatened either to perturb, or overturn, existing and developing disciplinary and professional arrangements.[130] Nor did chemical engineers fit in with prevailing agendas. The Institution of Civil Engineers, the largest and most senior technical organisation, was seeking to monopolise status through legal restrictions on practice. They organised an Engineering Joint Council (EJC) in 1922–3 with the Electricals, Mechanicals and Institution of Naval Architects. These 'big four' sought to oppose further splintering of technical professions by blocking the issuance of further Royal Charters. They also obstructed collective action by other institutions, particularly a private bill in Parliament in 1926–7 for legal registration of the profession.[131] Besides being on the 'wrong side' against these older scientific and engineering organisations, the IChemE was seeking an explicit accommodation both with manufacturers and government. This relationship was cemented in the very foundations of the ABCM, BCPMA and IChemE by their overlapping administrations and by the constitutions they adopted. Nevertheless Woolcock, and other IChemE Council members associated with the manufacturers' organisations, attempted few formal alignments between them. This was probably unnecessary given the multiple personal interconnections between these individuals. They may have been content, too, for each to fulfil its role in a system of co-operating complementary bodies satisfying the needs of plant manufacturers, chemical manufacturers and professional chemical engineers. In practice their separate spheres of activity rapidly became well defined and distinct, even if their officers were largely shared.

Relations with the SCI were cooler. The ubiquitous Woolcock, in his role as SCI Council member and IChemE Provisional Committee member, soothed his associates with the intimation that 'there was no intention on the part of the members of the Provisional Committee that the proposed Institution should interfere in any way with the activities of the Society of Chemical Industry. On the contrary, it was generally hoped that the new Institution would form a mutually helpful unit to other chemical and engineering societies in existence'.[132]

---

archive, IChemE; and C. S. Garland, 'The chemical engineer in reconstruction', *Trans. IChemE* 21 (1943), xvi]. The Council Minutes, however, record no evidence of any more direct connection between the AIChE and Provisional IChemE at this time other than joint offers of papers in their journals, the sharing of membership lists and the exchanging of publications.

[129] See Appendix. AIChE membership exceeded that of the CEG only in 1921. The slow start of the American organisation is suggested by the fact that the IChemE membership until the second world war lagged behind its American counterpart's by only five years, despite the fifteen year AIChE head-start and a national population three times as large.

[130] The Institution of Mining and Metallurgy, for instance, had obtained a Royal Charter in 1915, and the Institution of Gas Engineers was seeking one.

[131] Hamish B. Watson, *Organizational Bases of Professional Status* (PhD thesis, University of London, 1976), pp. 80–5.

[132] SCI CM 11 Apr 1922, p. 324–5.

Nevertheless, the SCI Council discussed IChemE matters only five times during the Institution's first six years.[133] Indeed, the SCI intersected few of the professionalising aims of the IChemE.

But the CEG, as a new specialist body still tied to its parent, the SCI, had a troubled existence. CEG activities were the most frequent topic of discussion at SCI Council meetings, where fears of organisational schism caused agitation.[134] From its beginnings, the Group occupied separate offices and charged separate dues. Besides its vexing independence, relations were further strained by the Group's chronically difficult financial position, with financing and organisation quickly coming to dominate its original aims of education and technical promotion.[135]

The relationship between the new Institution and its progenitor, the Chemical Engineering Group of the SCI, was amicable. The first topic of discussion was the mundane matter of office accommodation.[136] But even with shared occupation of premises and interests, the alliance between the Group and IChemE did not link their fortunes. The CEG membership, initially higher even than that of the American Institute of Chemical Engineers, was overtaken by that of the IChemE in 1926, and rose by scarcely 200 members over the following 57 years of its life.[137] Clearly the accreditation offered by the Institution offered chemical engineers considerably more than did the Group: it vaunted their members' competence and claims to authority over the practice of chemical engineering, rather than signalling a mere technical interest or economic involvement with it.

INTO A NEW LANDSCAPE

Competition for professional 'space' in industry and academia demanded the weapons of public recognition and certified status. But while recognition seemed assured immediately after the war, the support of government in fostering educational programmes, and industry in employing chemical engineers, proved difficult to retain into the postwar years. The period of full employment, stable markets for firms and state intervention came quite rapidly to an end. Herbert Cremer recalled that the early IChemE was received with 'suspicious glances of other members of the engineering fraternity ... feeble jokes of some industrialists and ... almost complete indifference of our universities'.[138] Without the legitimation of influential sponsors, chemical engineers could not effectively press their jurisdictional claims.

As long as the parochial interests of chemical engineers chimed with the

---

[133] SCI CM 1916–22, 1922–28.
[134] SCI CM 1916–22; the CEG was discussed fifty times over four years.
[135] No other subject group was formed until 1931, nor charged separate annual dues. See *Proc. CEG 2* (1920), 1; *3&4* (1921/2), 1.
[136] IChemE CM, 7 Mar 1923; *Proc. CEG 5&6A* (1923/4), 1.
[137] The Chemical Engineering Group ceased publication of separate *Proceedings* in 1972 and was re-named the Process Engineering Group in 1976.
[138] Cremer, *op. cit.*

state's needs at a time of direst emergency, all was well. But once the military crisis passed, chemical engineers were faced with the huge task of retaining even the limited ground they had gained during the war. For a brief period, the emergence of a political consensus concerning the desirability of industrial structures involving – by British standards – a fairly high degree of state intervention and centralisation ('state corporatism') seemed to suggest that the chemical engineers could continue to employ ideologies forged during the war to lay siege to the industrial fortresses occupied by chemists and engineers. But the mood for collective action by the state and industrial bodies was short-lived. The IChemE was founded just as the train of political thinking to which it was coupled took off in a new direction: co-operation in industry soon involved 'societal corporatism' – the establishment of relatively autonomous, representative bodies coming together in voluntary association. In this trans-formed environment, the Institution and its specialists struggled for a place.

CHAPTER 4

DESIGNING PROFESSIONALS

The existence of new organisations for chemical engineering in Britain had mixed benefits in the early interwar years. They unquestionably promoted professional self-identity, increasing the sense of common purpose among members of the CEG and, even more, those of the IChemE. On the other hand, membership as such conferred few immediate advantages. Employers remained largely indifferent to the specialism. Nor was the state to continue its active measures to bolster the nascent profession.[1] Declaring a 'stronghold' for the subject was not enough; its borders had to be constructed, its territory occupied by suitable recruits and defended from invaders. Organisers of the Institution consequently devoted much of their energy to publicising the occupation, justifying the profession and solidifying the discipline.

Because the academic differentiation of chemical engineering was already more marked and easily asserted than its occupational presence, education assumed an early importance in the IChemE. The aspirations of the nascent profession soon became closely linked with the creation of a distinct intellectual specialism. The identity of the chemical engineer was nevertheless woven and refined by the organisers of the Institution by seeking to exploit social and political trends, by drawing upon aspects of an emerging American disciplinary model, and by seeking to set standards of training.

DEFINING THE CHEMICAL ENGINEER

The task of solidifying the identity of this new breed of technical worker advanced along several fronts between the wars. Explicit definitions of the professional skills were tailored to a variety of audiences. For technologists generally, the promoters of chemical engineering stressed the general territory they hoped to claim. For the state and industrialists, chemical engineers justified themselves by accentuating how they could improve production economics. For cognate occupations such as industrial chemistry and chemical technology,

---

[1]   Industrial and state support are the subject of Chapter 5.

promoters distinguished chemical engineers by how they differed functionally, that is, by their particular know-how, skills and working methods. For practitioners and would-be IChemE members, professional definition reduced to the set of educational qualifications and experience necessary for corporate membership in the Institution.[2] And beyond these lay a disciplinary definition aimed at rival academics and potential sponsors of university programmes.

The image most convincingly portrayed, and having the widest currency, was of chemical engineering as being a hybrid profession emerging from, and reliant upon, others. This was represented explicitly in the first seal of the IChemE, designed by John Hinchley's wife Edith in 1926. It equated the profession to a weak sapling supported by a firm stake and nourished by streams from chemistry and mechanical engineering, represented by glassware and a gear wheel, respectively, on solid stone pedestals and accompanied by doves.[3] Thus the perceived weakness, need for external support and reliance on established and peacefully co-existing professions was graphically acknowl-

Figure 4-1 Original seal of the IChemE.

---

[2] 'Corporate' members (those permitted to vote) were those of 'Associate Member' or 'Member' status, as opposed to 'Graduate' or 'Student' grade. Membership grades were revised in 1971, with 'Associate' becoming 'Member' and 'Member' renamed 'Fellow'.

[3] The seal was replaced with an heraldic emblem (with its own implicit connotations of long and noble lineage) in 1965.

edged. For an audience of engineers, Hinchley was explicit about the engineering connections:

> The functions of the chemical engineer were formerly performed by engineers with slight knowledge of chemistry, by industrial chemists with a little knowledge of engineering, and by the co-operation of both. In all cases, however, those particular functions which pertain to chemical engineering were inadequately carried out and the scrap heaps of the recent past bear testimony to this point. Chemical engineering is that branch of engineering which is concerned with processes and plant in which chemical or physical changes are the principal features. A chemical engineer is therefore essentially an engineer with a knowledge of fundamental science, and a special knowledge and experience of those processes which are carried out on a large scale in the manufacture of chemical and technical products. Since these processes cannot be carried out unless a commercial profit is obtained, the chemical engineer must possess a knowledge of industrial economics and factory management.[4]

Such iconography and rhetoric sought to convince potential adversaries that chemical engineers would inhabit uncontested territory beyond the borders of established professions.

Yet a definition of the profession still needed relevance for employers. The most urgent professional credential was that of occupational justification. Lord Moulton, in a lecture at University College, London, in 1920, picked up a theme he had developed during the wartime Ministry, arguing that chemical engineers demonstrated their importance by improving profitability and competitiveness. 'The pursuit of cheapness', he said, would determine whether 'chemical engineering is to live'. This attribute, claimed Moulton, 'transforms the subject and makes it stand by itself'.[5]

John Hinchley also stressed production efficiency in an article published four months after Moulton's death, claiming that a doubling of plant output would be possible 'if there were complete control by competent chemical engineers'. The chemical engineer was thus 'the man whose job it is to raise the efficiency of chemical plants'.[6] The new connection between chemical engineering and the self-publicity of plant manufacturers was mentioned in the last chapter. Typical of this piggy-backing of commercial and professional interests, a 1923 advertisement stated that 'Chemical Engineering is the science and art of using power to manufacture economically chemical and allied products'.[7] Through the 1920s campaigners continued to advertise economy and profit as the prime

---

[4]   J. W. Hinchley, 'The training of the chemical engineer', *World Engineering Congress, Tokio, October-November 1929* (unpublished, Nathan papers).

[5]   Lord Moulton, 'The training and functions of the chemical engineer', *Trans. IChemE 17* (1939), 186–91. Based on the content of this invited address, the audience appears to have been a general one.

[6]   J. W. Hinchley, 'The need for an Institution of Chemical Engineers', *Chem. Trade J. & Chem. Eng*, Jul 9 1921.

[7]   Advertisement, E. Morris Ltd, *Chemical Age Yearbook 1923* (London: Benn Bros., 1923), p. 192. See also Fig. 3–1.

justification for chemical engineers.[8] Sir William Alexander, for example, observed that 'the final court of appeal for the chemical engineer and the employer is the costs sheet', and Sir Alexander Gibb described the business of the chemical engineer as being 'to take the best that exists and to improve on it, and only by doing so shall we justify our existence'.[9] Such themes retained currency through the 1920s in the face of continued international competition and industrial reorganisations.

For contemporary chemical workers, on the other hand, the definition of the chemical engineer initially was based on what people who thought of themselves as chemical engineers *did*. It was not entirely a historical accident that chemical engineering had gained a foothold through munitions. Some of the key players in the early IChemE, such as William Macnab and Sir Frederic Nathan, had worked in explosives well before the war. Trained as chemists but later identifying themselves as chemical engineers, they recognised that knowledge of physical characteristics (e.g. the size and density of gunpowder granules, or the method of pulping cotton fibre for guncotton production) was at least as important as the chemical details (e.g. the alkaline post-rinse of nitrocellulose). Two technical cultures merged in explosives research, and promoted – indeed, required – a new species of technical specialist.[10] Lord Moulton nevertheless based his functional definition on other aspects of what he had seen at the Ministry of Munitions. Planning for change of scale, he contended, 'is the secret of successful chemical engineering'.[11] Similarly, in 1919 Charles Garland described the chemical engineer as 'a chemist who transferred results obtained in the laboratory to operations conducted on an industrial scale.[12] The first definition adopted officially by the IChemE, nevertheless, stressed J. W. Hinchley's preference for engineering associations:

the branch of engineering which relates to the "design", construction, erection and

---

[8]   The AIChE, too, tried to facilitate the promotion of chemical engineers into positions of executive responsibility by emphasising the goal of production for profit in its policy on training. 'The training and work of the chemical engineer' *Trans. Faraday Soc. 13* (1917–18): 61–118; C. L. Reese, 'Presidential address', *Trans. IChemE 3* (1925): 13–15; T. S. Reynolds, *75 Years of Progress: A History of the American Institute of Chemical Engineers, 1908–1983* (New York, 1983); Reynolds, 'Defining professional boundaries: chemical engineering in the early 20th century', *Technol. & Culture, 27* (1986), 694–716.

[9]   Sir William Alexander, 'Annual dinner', *Trans. IChemE 6* (1928), 17–9; Sir Alex. Gibb, 'Presidential address: The economics of power as applied to chemical engineering', *Trans. IChemE 6* (1928), 12–6.

[10]   Seymour Mauskopf, 'From an instrument of war to an instrument of the laboratory, the affinities do not change: chemistry and munitions', Seminar, Science Museum, London, 4 Nov 1998, for example, mentions the combination of the expertise of academic chemists with that of École Polytechnique graduates in 19th century French explosives research. Development of explosives by that time relied increasingly on the instantiation of this chemical and engineering background in a single individual. An analogous need for both chemical and physical knowledge was particularly obvious in the other two major occupations of the first IChemE members, coal-gas/coal tar/coke-oven technology and mining/extractive metallurgy/metal founding.

[11]   Moulton, *op. cit.*

[12]   C. S. Garland, *J. Soc. Chem. Indus. 38* (1919), 160R.

operation of plant and works in which matter undergoes a change of state or composition.

This was expanded in one of the earliest public claims to professional territory, an article written by Hinchley for the 13th edition of the Encyclopaedia Britannica.[13]

Thus an identity shaped by occupational tasks was problematic. It forced a comparison with chemists and mechanical engineers – both established competitors. Nevertheless, the issue of professional uniqueness was not, for some, pushed strongly enough. Indeed, such working definitions had not evolved much by 1933, when W. E. Gibbs, professor of chemical engineering at UCL, described the chemical engineer as 'essentially a chemist who has been trained to be industrially effective'.[14] This identification as a sort of industrially-savvy chemist is significant in a context in which the alternative was alliance with the undifferentiated and probably practically-trained 'engineer'. Both a social and class dimension were significant. To distinguish chemical engineers from 'chemists with a smattering of knowledge of engineering' and 'engineers with a smattering of knowledge of chemistry', an editorial that year in *The Chemical Age* called for 'an authoritative ruling from the Institution of Chemical Engineers as to the status, functions and legitimate scope of the chemical engineer'.[15]

Such contentious boundary-setting was no easier than defining the scope of the academic discipline. Here, descriptions became contentious and increasingly painstaking. The function of such definitions was to circumscribe the intellectual content and occupational competence of chemical engineering and to distinguish it from cognate disciplines. The audience, consequently, was composed of competing professionals – heavily weighted with academics – rather than chemical engineering practitioners. During the first decade of the IChemE, such comparative definitions were widely discussed as the Institution attempted to set admissions standards and a syllabus for training. Little consensus could be found among employers, few of whom were convinced that the financial benefits of the 'efficiency' argument justified the organisational changes required. Moreover, the largest employers were dismissive of the need for an occupational hybrid, and criticised moves to institute chemical engineering courses; F. Rogers of Nobel Industries observed that 'it would be very undesirable in large chemical works to endeavour in any way to combine in one class of person the qualifications and experience of the chemist and engineer'.[16] Some, by contrast, wanted

---

[13]   Sir Alexander Gibb, the current president, had recommended an article on chemical engineering after being approached by the Editor to write other articles for it. Hinchley wrote the article during the winter of 1925–6. CM 9 Sep 1925; 11 Nov 1925; 10 Feb 1926.

[14]   W. E. Gibbs, 'The functions and training of the chemical engineer', *Chem. Age 28* (1933), 67–9. Gibbs's Inaugural Lecture claim that the Chemical Engineer was 70% chemist and 30% engineer had irked Hinchley, according to his wife. [E. M. Hinchley, *John William Hinchley*, *op. cit.*].

[15]   'The chemical engineer in industry', *Chem. Age 28* Feb 11 1933, 1.

[16]   IChemE, 'Extracts from replies received to the memorandum on "The training of the chemical engineer", circa late 1924.

a thorough grounding in physical chemistry, physics or metallurgy. Yet others thought that any four-year course would be inadequate to train a chemical engineer with adequate chemistry, physics and engineering knowledge, as opposed to those who stressed the necessity of long periods of works experience before, during or after college.[17]

CONSTRUCTING MEMBERSHIP

In this morass of contentious demarcations, the leaders of the new Institution saw themselves literally in defining roles. Hinchley argued that, if every practising chemical engineer were expected to become a member of the IChemE, the institution would become 'a powerful means of raising the standard of chemical engineering practice in this country'.[18] Indeed, the power most directly under the control of the IChemE leadership was the selection of suitable members. Claims to authority were partly dependent on the social standing of those sharing the identity of 'chemical engineer'. The Chemical Engineering Group had never applied membership controls, and so had been unable to clearly define the attributes of specialists associated with chemical engineering. The Institution could, however, through its membership policy, filter and prune the available pool of ostensible 'chemical engineers' into a membership and profession of its own making. More positively, it could actively mould the skills of its younger membership by an explicit educational policy for training the next generation of chemical engineers. Distrusting its prospects in a competition for the 'survival of the fittest profession', the IChemE sought to actively breed a more competitive strain of engineer. Such 'occupational eugenics' was in contrast to the majority of contemporary would-be professions, which were content merely to corral their members within easily-traversed disciplinary fences. Indeed, multiple 'technical identities' were common during this period, when individuals commonly held memberships simultaneously in a handful of professional bodies.

Despite such potential power in shaping both their contemporary and future membership – and thereby the identity of the 'chemical engineer' – the IChemE Admissions Committee had to tread a narrow line. Standards of qualification that were too harsh would create an elite profession, but one that was too small in numbers to become viable and effectively populate industry. Too lax a policy would swell the ranks of professional chemical engineers at the expense of the 'genetic purity' of its expertise, and hence the reputation of the profession. Either extreme would result in failure of the profession to thrive.

Admissions standards

As the Institution was taking shape in 1921, the plan for admissions was that 'men only who have by study reached a certain standard of scientific attainment,

---

[17]   *Ibid.*, e.g. R. Seligman of The Aluminium Plant & Vessel Co, Colonel O. C. Armstrong of Greenwood & Batley Ltd, and member Neils Rambush.

[18]   Hinchley, *op. cit.*

and by their practice have justified the title of chemical engineer, are to be admitted, but eventually it will itself set the standard of knowledge found desirable'.[19] Such a laudatory but vague standard, however, could be – and was – interpreted along a wide spectrum of accomplishment. In any case, at least 60 applications arrived before the officers of the Institution had discussed an admissions procedure or explicit standards.[20] In January 1923, the Provisional Council elected 61 Members and twenty Associate Members, some weeks before a Nominations Committee was formed. At its first meeting, twenty applicants were made Members or Associate Members, five were rejected, and six were deferred. The committee left no surviving records concerning the standards or procedures employed. As with any aspiring profession having little academic presence or external justification, admissions standards in the IChemE had to be mapped from the qualifications of the organisers. Such cloning was inherently sensitive to these 'seed' individuals, but was extended by their aspirations for the profession. The decision was taken to have few 'Honorary Members', for instance, thus maintaining the purity of the stock.[21] For the first few meetings, at least, those applicants accepted immediately were judged on the basis of letters of reference or personal acquaintance. A minority was interviewed.[22] A good example of the use of reference letters to refine membership standards is the case of James Carlyle, of whom H. J. Pooley wrote:

> I take it at the present stage there are very few members of the Institution who have real training on the lines that we hope chemical engineers in the future will be trained. From what I have seen and know of many of the existing members they were originally either chemists or engineers with a useful working knowledge of the companion profession in either case. That being so I think I have said enough to show that Mr Carlyle is at least as well qualified as most of them to be admitted as a full member.[23]

Even two years later, the general council of the Institution felt it necessary to emphasise the threshold between corporate members and less-skilled Graduates:

> A GRADUATE is a man who has had a good general and scientific education (and has taken a scientific degree or some equivalent thereof) but who has not had sufficient practical experience or adequate Chemical engineering training.[24]

For an organisation later to vaunt its international identity, the admissibility

[19] 'An Institution of Chemical Engineers', *Chem. Trade J. & Chem. Eng.* 9 Jul 1921, 1.
[20] The application records for IChemE membership contain details of training, job history, letters of reference, correspondence with referees and other supporting material. Except for the sixty applications of Volume 1, all are bound in order of receipt.
[21] This had been praised by K. B. Quinan, who noted 'the absence of intention to court support for the Institution by granting Honorary Membership, or other recognition to titled persons, or prominent individuals in the commercial or scientific world, who have no personal knowledge of Chemical Engineering in the full sense.' CM 8 Mar 1922.
[22] Minutes of Nomination Committee, 25 Jan 1923, 1 Feb 1923.
[23] Letter, Pooley to Garland, 2 Aug 1923 (AF 184).
[24] CM 11 Feb 1925.

of foreign (i.e. non-Commonwealth) members was initially questioned. In 1927, when a Berliner submitted an application, Council was forced to consider its policy for Corporate membership explicitly. It quickly decided, however, that 'there was no nationality bar in the Articles of Association', so 'members of foreign, and especially enemy nationality' were allowed.[25]

The Associate Membership examination

Qualifying examinations were becoming an established part of several engineering organisations (see Table 4-1), and the IChemE saw them as fulfilling several roles. Examinations would select and distil the prospective membership by a more objective process than mere recommendation, serve as an advertisement of entry standards, and promote the reputation of the Institution as a qualifying body.

Table 4-1   Introduction of qualifying examinations for engineering institutions[26]

Institution	Date of formation	First examinations
Institution of Civil Engineers	1818	1897
Institution of Mechanical Engineers	1847	1913
Institution of Gas Engineers	1863	1926
Institution of Electrical Engineers	1871	1913
Institution of Mining Engineers	1889	1947
Institution of Mining and Metallurgy	1892	1950
Institution of Production Engineers	1921	1932
Institution of Chemical Engineers	1922	1926

John Hinchley, as Honorary Secretary and one of the few academics teaching the subject, was the motive force behind such exams. While the early Institution was eager to influence educational standards for chemical engineers, its council initially wanted teaching institutions to have ultimate responsibility for setting and testing them. At the first meeting of 1923, Hinchley reported that he had been negotiating with the London City & Guilds Institute to found an annual examination under their auspices.[27] He proposed that this would serve as the qualifying examination for Graduate membership. Whether a similar examination could be devised as a qualifying test for Associate membership, he thought, 'was a matter of doubt'. The City & Guilds Institute asked the Institution to draft an examination syllabus, which Hinchley suggested could be divided into several independent sections.[28] The first meeting of the Education Committee that year ruled out liaison with the Institute, because its highest examination omitted tests for general and scientific education. Sir Frederic Nathan recom-

---

[25]   CM 27 Apr 1927, 18 May 1927.
[26]   Adapted from H. B. Watson, *Organizational Bases of Professional Status* (PhD Dissertation, U. London, 1976), p. 140.
[27]   The City & Guilds had been holding examinations in salt and alkali manufacture which were not well attended, and proposed to replace them with chemical engineering.
[28]   CM 4 Jul 1923.

mended an exam administered by the Institution itself, and Hinchley proposed something along the lines of the Unit Studies in Chemical Engineering adopted by the Massachusetts Institute of Technology (MIT). He also suggested an examination for which 'the student was compelled to read up all the literature he could obtain upon the subjects of examination if he wished to answer the questions satisfactorily, so that the examination became a test of the student's own wits and training, rather than a test of information and knowledge'.[29] Such a 'take-home' examination was an unusual format for its time, and the origin for the Institution's long-running 'Home Paper'.

Within three years, the Institution had prepared such an examination designed for Associate membership qualification. The paper explicitly set out, for the first time, the cognitive abilities expected of a chemical engineer. The first, which set the format for decades to come, had six sections. The 'Home Paper', to be completed within about one month by the candidate, had two sections: the first (A), to design a specified plant and the second (B), to write an essay or speculative design proposal. The remaining four sections were to be completed in supervised sessions of three hours each over two successive days. Section (C) dealt with physical and engineering principles, with questions on thermodynamics, flow, friction, conveying and grinding, and calculating the requirements for both heating and cooling (what was to become known as 'process design'). Section (D) specialised in physical chemistry-related problems such as concentration and distillation. Section (E) concerned materials for chemical engineering plant such as appropriate materials, construction method, precautions and lifetime. Finally, Section (F) concentrated on the selection of components and economics of plant design. Within each section there was a range of problems from descriptive to calculation-intensive. There was virtually no 'chemistry' as such, nor mechanical or civil engineering (chemical equations, calculation of loads and site selection were all absent, for example). Instead, the candidate for Associate Membership of 1926 was required to be adept at physical chemistry, engineering physics and design economics.[30] The first examinations, taken by six entrants that summer, yielded only one successful candidate. By 1939, 55 of the 403 Associate Members had qualified this way.[31]

Profile of the early membership

How well did the early members mirror the organisers' expectations, and to what extent did their identity devolve from a uniform stock? The details of the membership and admissions policy of the Institution can be probed by analysis of application forms. These records, the most carefully preserved over the

[29] MEC 25 Sep 1923.
[30] IChemE, 'Examination for Associate Membership 1926'.
[31] Nomination Committee Minutes 1926–39. For the subsequent evolution of the examination, see G. U. Hopton, 'The Associate Membership examination', *Trans. IChemE 28* (1950), 173–6 and *idem.*, 'The examination of the Institution of Chemical Engineers', *Quart. Bull. IChemE* Jun 1955, xliv-xlviii.

Institution's history, witness the central importance accorded to entrance qualifications.

Occupation

The first members of the IChemE had particularly disparate training and occupational experience. The term chemical engineer was not widely employed by applicants – scarcely one in twenty claimed the title through the 1920s, and one in five during the Thirties. On the other hand, some letters of reference from employers indicated that the description had long been used by some die-hards; one, for example, from Douglas A. MacCallum, a Glasgow 'consulting chemist & metallurgist, chemical engineer', dated from July, 1908. Such mixed designations were among the most common: a quarter of members between the wars described themselves primarily as chemists. Over a third were managers, supervisors or directors of companies; most of the remainder were engineers of various descriptions. A few percent described themselves as consultants or teachers. Thus members were largely split, at least nominally if not occupationally, into chemists and engineers, with many of them in supervisory roles. Nevertheless, such categories masked the reality of working life for many members. Many 'managers' were general managers of small firms, figuratively wearing a variety of hats from trouble-shooter to accountant to purchaser. The job site was, as often as not, a physically dangerous environment in which innovation meant adapting and making do.

Other affiliations

This breadth of members was reflected, also, in their other professional affiliations. Through the 1920s, nearly three-quarters of the new corporate members belonged to at least one other professional or trade organisation. Over half of

Table 4-2 New corporate membership by occupation, 1923–39[32]

Self description	1923–9	1930–9
Chemical engineer	5%	17%
Chemist	24%	16%
Engineer	14%	9%
Manager	35%	39%
Educator	6%	6%
Consultant	7%	3%
Technologist/assistant	8%	4%
Metallurgist	1%	1%
Sales representative	0%	4%

---

[32] Sample: 98 of a population of 474 new corporate members (1923–9) and 69 of 258 new corporate members (1930–9). Self-described 'engineers' among new members fell from 30% for the first 60 (1923–4) to average 23% over the entire interwar period. Statistics on members after 1924 are derived from data sampled by R. Mackie from the IChemE application forms. For later periods, see Tables 5-1, 6-1, 7-1 and 8-1.

these members belonged to a chemical or scientific organisation such as the Chemical Society or Institute of Chemistry; over two-thirds to another engineering or trade body such as the Institution of Mechanical Engineers or the ABCM; and over a third to another chemical engineering-related body such as the Society of Chemical Industry, Chemical Engineering Group or the American Institute of Chemical Engineers. Through the 1930s, such affiliations fell, but maintained roughly the same proportions. New corporate members having American affiliations fell, however, from a high of about 10% through the 1920s to some 1% to 3% over the following three decades.[33] The significant transferral of allegiance to the home-grown institution suggests that tangible professional benefits were of more importance to members than were learned society functions.

Qualifications

In accordance with the attempts of the IChemE leadership to breed a more competitive professional, the academic qualifications of the new members were impressively higher than those of most other contemporary engineers (although not necessarily of members of scientific bodies such as the Institute of Chemistry). In the first decade of the Institution, nearly half of the new corporate members held at least one degree, and 18% a master's or doctorate. The fraction increased through the 1930s, with over two-thirds holding at least one degree, and over a quarter of the membership having more than a bachelor's degree, by the second world war.[34] Over half of the bachelor's degrees were awarded by colleges of the University of London, and nearly half of the doctorates from German teaching institutions. The degree subjects in the interwar period were most frequently chemistry, mechanical or civil engineering.

Location

The largest fraction of corporate members joining between the wars worked in London and the Home Counties of South-East England. Foreign members comprised about one in five new admissions.[35] Council officers, too, were mainly London-based. Even the first Council, however, had eight 'corresponding members'. K. B. Quinan, the first Joint Vice-President, for example, seems never to have attended a council meeting.[36]

Membership in the remainder of Britain was dominated by four contingents that together shared over a third of all new members: North-West England (especially Manchester, Widnes, Port Sunlight and Liverpool), North-East England (especially Stockton-on-Tees and Durham), Yorkshire (especially

---

[33] See Appendix.

[34] 1923–32: 29% BSc; 8.5% MSc; 10% PhD. 1933–42: 39% BSc, 19% MSc, 10% PhD.

[35] Both these figures have changed little over the life of the Institution.

[36] Annual Report, July 1924. Other corresponding members of council were James MacLeod, J. H. Young and A. Cottrell (Scotland), W. A. Fraymouth (Bhopal, India), W. G. Weaver (Cape Town), J. A. Wilkinson (Johannesburg) and Neils E. Rambush (Stockton on Tees).

Table 4-3   New corporate membership by region, 1923–39

Region	1923–9	1930–9
London and South-East	32%	43%
Overseas	22%	19%
North-West	11%	10%
North-East	13%	10%
Midlands	8%	6%
Yorkshire	6%	8%
Scotland	5%	3%
Wales	2%	0%
Southwest	1%	1%

Huddersfield, Sheffield and Hull) and the Midlands (especially Birmingham and Derby).

DESIGNING A DISCIPLINE

The meticulous care taken by the Nominations Committee in checking the references of candidates was both time-consuming and still open to criticism from outsiders and other professionals as being subjective and self-serving. Such disadvantages were counteracted by the Institution's strong public support for educational programmes. But what cognitive foundation was claimed as the basis of a model syllabus?

The American connection

The evolution of the *discipline* of chemical engineering – that is, the body of knowledge on which the academic component was based – illustrates the importance of intellectual apparatus in claiming and maintaining jurisdiction over occupational practices.[37] Practitioners and educators developed a new way of conceptualising chemical manufacturing; an intellectual framework which, through the interwar period, gained growing acceptance as being founded on more than just a synthesis of earlier technical disciplines.

The distinct occupational origins of British and American chemical engineering (i.e. in explosives and coal-related industries on this side of the Atlantic, and petroleum on the other) have been mentioned. Nevertheless, most historical studies of chemical engineering have focused on the emergence of the profession in the United States, and have tended to assume that the intellectual substance of chemical engineering in Britain was almost entirely the result of American

---

[37]   Andrew Abbott, *The System of the Professions* (Chicago: University of Chicago Press, 1988), pp. 52–8.

influences.[38] Other scholars, however, have highlighted certain national differences in the cognitive scope of the discipline in European countries other than Britain.[39] As we shall stress, the historical evidence forces the acknowledgement of the indigenous roots and importance of local context of British chemical engineering, and the particular circumstances under which chemical engineering emerged as a profession in Britain, shaped the associated intellectual discipline.

The early academic connections of chemical engineering were important because they gave a plausible sense of identity to the would-be profession, made explicit its cognitive scope, and, by providing practitioners with a unique intellectual foundation for the changing techniques of chemical manufacturing and the design of chemical plant, gave them the means to exclude competing occupations from these and certain other related kinds of industrial work. But if we are to understand the evolution of the intellectual apparatus of chemical engineering in Britain, we need to be clear about the peculiarities of other countries, and particularly the American case.

By comparison with many of the countries of continental Europe, both Britain and the USA were characterised before the second world war by political systems that laid great emphasis on the voluntary, 'professional' regulation of those scientific or technical occupations that commanded social and economic prestige. Similarly, the universities were not subject to any significant measure of control by the state. Thus in both countries intellectual developments relating to the chemical industries were largely a matter of tacit negotiation between academics and certain groups of business officials, consultants and other would-be professionals. But the British chemical industry was by no means identical to the American in terms of the markets it served or the typical structure of firms, and advocacy of educational programmes of chemical engineering in the two countries took place under rather different circumstances.

Generally speaking, British chemists and engineers were more firmly entrenched in the industrial division of labour, and enjoyed older and more effective forms of professional organisation than did their counterparts in the United States. Chemistry, in particular, was well established as a university discipline. It was therefore more difficult for chemical engineers in Britain than in America to press their claim to professional status, and to establish a novel

---

[38]   Many of these accounts were written by former practitioners; the two principal collections of such material being *History of Chemical Engineering* (Washington: American Chemical Society, 1980), and *A Century of Chemical Engineering* (New York: Plenum Press, 1982), both William F. Furter (ed.). On the US experience see, for example, F. J. Van Antwerpen, 'The origins of chemical engineering', *History of Chemical Engineering*, pp. 1–14; David F. Noble, *America by Design: Science, Technology and the Rise of Corporate Capitalism* (New York, 1977), pp. 26–7, 38, 79, 192–5; J. W. Servos, 'The industrial relations of science: chemical engineering at MIT, 1900–1939', *Isis*, 71 (1980), 531–49; T. S. Reynolds, *75 Years of Progress: A History of the American Institute of Chemical Engineers, 1908–1983* (New York, 1983), *passim*; Reynolds, 'Defining professional boundaries', *op. cit.*, 694–716.

[39]   E.g. K. Buchholz, 'Verfahrenstechnik (Chemical Engineering) – its development, present state and structures', *Soc. Stud. Sci.* 9 (1979), 33–62 and, less convincingly, K. Schoenemann, 'The separate development of chemical engineering in Germany', in *History of Chemical Engineering*, *op. cit.*, 249–72.

intellectual underpinning for their expertise. But while the scale of their achievement was more limited than that in the United States, British chemical engineers
nevertheless proved remarkably inventive when it came to defining the intellectual content of their discipline.

Terry Reynolds has shown that by the first decade of the twentieth century,
groups of senior chemists employed in American chemical manufacturing were
seeking to distinguish themselves from analytical chemists, who had a low
status, by claiming as their own the task of scaling up chemical processes from
the laboratory to the industrial level. They were joined by some consultants to
the chemical manufacturers, and some businessmen.[40] These nascent chemical
engineers were opposed by the existing professional society for chemists, the
American Chemical Society, the leaders of which argued that the development
of new manufacturing processes could be understood within the conceptual
framework of industrial chemistry. Mindful of the need to establish its claim
to a distinctive specialism, the American Institute of Chemical Engineers
(AIChE), founded in 1908, chose to restrict full membership to persons over
30 years of age who possessed several years' practical experience in both
chemistry and another branch of engineering. Would-be members also had be
actively involved with the application of chemical principles to manufacturing.[41]
Thus early members of the AIChE used their knowledge of physical chemistry
as a way of distinguishing themselves both from self-trained 'factory hands'
and from mechanical engineers who had become involved with chemical
plant design.

These measures had some limited success, but according to Reynolds the
key to legitimating American chemical engineering as a profession was the
evolution of a new cognitive base – the unit operation. This conceptual framework was not formally employed in the struggle for professional status until
1922, when the AIChE's Education Committee chaired by Arthur D. Little,
head of one of the largest American consulting engineering firms, recommended
that it be employed to define the scientific core of academic educational
programmes. However the unit operations had already been evolving over a
period of 20 years or more in several American universities, chief among which
was the Massachusetts Institute of Technology (MIT).[42] Hence Reynolds sees
the American profession as being shaped by at least three factors: first, the
desire of certain groups of technical workers to differentiate themselves from
the more heterogeneous and less prestigious category of 'chemists'; second,
their appropriation of a cognitive realm (first physical chemistry, and then unit
operations) as a way of underpinning claims to technical expertise; and third,

[40] Reynolds, 'Defining professional boundaries', *op. cit.*, 697–708.
[41] *Ibid.*, pp. 697, 706. The corresponding age threshold adopted by the IChemE was 25.
[42] By 1905 the concept of unit operations implicitly informed the curriculum of a new chemical
     engineering course at what was to become MIT. From 1900, William K. Walker, the course
     instructor, had a close association with the Arthur D. Little engineering consultancy. Servos
     (*op. cit.*) has discussed the currency of the notion of unit operations in the US in this early
     period, showing that Little was linked closely with such concepts in the curricula of American
     institutions before 1915.

a gain in legitimacy in academic, engineering and industrial circles following the successful application of this intellectual apparatus to the design and operation of large chemical plants.

Tracing the origins of concepts is difficult when the inchoate professions that are trying to appropriate them are in a state of flux, and the intellectual features of American chemical engineering up to 1939 have been described in various and inconsistent ways. Olaf Hougen and Franklin J. Van Antwerpen, for example, identify the unit operations as being crucial from 1916 to 1925, although they also note the increasing importance of material and energy balances to 1935, and of applied thermodynamics and process control from then until the end of the second world war, and beyond.[43] On the other hand, Gianni Astarita's work on Italy distinguishes unit operations as the sole organising concept.[44] Clive Cohen largely concurs with this simple genealogy for both the United States and Britain, while moving the influence of unit operations back to about 1910, and suggesting that the profession in both countries could have developed on the basis of other – but largely unspecified – intellectual foundations.[45]

All sources agree, however, that the unit operations were first described as such in 1915 by Arthur D. Little. He defined them as discrete physical processes employed in chemical manufacturing, such as distillation, roasting, filtering and condensation. Each described a particular way in which material could be transformed physically – for example, by the reduction in size of solid matter, or by the mixing or separation of solids, liquids or gases. These basic processes would be performed in sequence to obtain a final product; while the number and order might vary from chemical to chemical, any manufacturing process could be understood in terms of the same set of building blocks. The steps could either be carried out on batches of material, or performed as a continuous process in which the material is transformed at different locations in a plant.

Accounts of the origins of the unit operations before 1915 diverge considerably, however. Martha Trescott identifies them as almost exclusively an American innovation. While acknowledging that George E. Davis may have described unit operations in his 1901 *Handbook of Chemical Engineering*, she claims that the concept was popularised and integrated with physical chemistry exclusively by American practitioners. Trescott suggests that three national characteristics explain this American 'lead': first, the American predilection for rational mechanical design, leading to the careful design of equipment intended for specific uses; secondly, the influence of Frederick W. Taylor's system of 'scientific management' in American engineering, which stressed the technically and financially efficient division of manufacturing tasks into unitary operations arranged sequentially; and finally, an American willingness to innovate, to scale

---

[43] Van Antwerpen, *op. cit.*; Olaf A. Hougen, 'Seven decades of chemical engineering', *Chemical Engineering Progress*, 73 (1977), 89–104.

[44] G. Astarita, 'L'Evoluzione dei fondamenti teorici dell'ingegneria chimica', *Chimica e Industria (Quaderni dell' Ingegnere Chimico Italiano* (supplement), 8 (1972), 112–4.

[45] C. Cohen, 'The early history of chemical engineering: a reassessment', *Brit. J. Hist. Sci.* 29 (1996), 171–94, esp. 175–86.

up industrial processes, and to engage in the production of goods for mass markets.[46]

Chemical engineering in the United States was certainly associated with the manufacture of chemicals on a large scale. But beyond this, none of Trescott's observations is particularly convincing. In the first place, the distinctive American system of manufacturing long runs of relatively small items of machinery and consumer goods of a mechanical or electrical nature was of limited relevance to the design of large chemical manufacturing plants or the bulk production of chemicals, many of which were intermediates destined for further processing.[47] Indeed, Philip Scranton has provided a more nuanced interpretation of this second industrial revolution in which he highlights the importance of specialty production in the American context, but contrasts it with oil refining which he describes as 'routinised' and 'the domain for engineering as a species of laboratory science: professionals carefully testing multiple factors bearing on a relatively stable problem set'.[48] Moreover, chemical manufacturing on a considerable scale arose from the scaling up of laboratory procedures long before unit operations were conceived, and efforts to rationalise chemical production – particularly by the introduction of labour-saving processes – pre-dated Taylor's activities.[49]

The close connection between unit operations and mass production is, however, remarked upon by other advocates of the pioneering role of the United States. Klaus Buchholz postulates that the practical problems of exploiting the country's vast oil fields played a substantial part in the American genesis of chemical engineering. He puts forward a particular direction of evolution: from market forces and existing chemical technologies, to intellectual concepts and an emergent profession. Attributing the occupation's growth to the 'need for mass-production prompted by the expanding market' in petrochemicals from the turn of the century, Buchholz asserts that this pressure promoted the adoption of economical production methods and encouraged the rational design of industrial plants.[50] More importantly, he claims that the particular technical requirements of petrochemical processing – namely physical operations such as distillation – explain the readiness of US manufacturers to accept the conceptual framework of the unit operations.[51] By contrast to the American

---

[46]   M. M. Trescott, 'Unit operations in the chemical industry: an American innovation in modern chemical engineering', *A Century of Chemical Engineering*, pp. 1–18.

[47]   David A. Hounshell, *From the American System to Mass Production, 1800–1932: The Development of Manufacturing Technology in the United States* (Baltimore: Johns Hopkins University Press, 1984); John K. Brown, *The Baldwin Locomotive Works, 1831–1915: A Study in American Industrial Practice* (Baltimore: Johns Hopkins University Press, 1995). Brown finds no unique features in the American context of the capital goods sector.

[48]   Philip Scranton, *Endless Variety: Specialty Production and American Industrialisation, 1865–1925* (Princeton: Princeton University Press, 1997), pp. 66–67.

[49]   Taylor's first paper on scientific management appeared in 1895; he died in 1915, but Taylorist analysis and management were applied most widely in American industry through the interwar period.

[50]   Buchholz, *op. cit.*, 42.

[51]   *Ibid.*, 37.

case, a unique occupation combining mechanical and chemical expertise failed to coalesce in Germany; the petrochemical industry there was negligible, and much more complex chemical syntheses dominated the dyestuffs and pharmaceuticals industries. The occupational specialism of *Verfahrenstechnik* was organised around specific industrial products and their manufacturing processes, and chemical plants were designed and maintained by a combination of chemists and mechanical engineers with a strict division of labour.[52] In a similar vein, Jean-Claude Guédon argues that American manufacturers, particularly in the petrochemical industry, were much more welcoming towards unit operations than were the British, and that the American system of higher education was more receptive to the new conceptual framework.[53]

Yet Buchholz says that neither the American experience nor the concept of unit operations were essential for the solution of the problems of expanding industrial production in petrochemicals.[54] The point may be made the other way around: the establishment in the United States of the profession of chemical engineering and its associated intellectual apparatus was not the inevitable outcome of the growth of the manufacture of bulk petrochemicals. This particular market provided fertile ground for the emerging profession in the United States, but gaining a measure of acceptance for a novel conceptual framework also required agreement between academics, chemists and engineers working in the chemical industries, and their business leaders that this was to their mutual advantage.

The British context: professional organisation, plant design and the unit operations

Consultants dominated the leadership of the IChemE in its early days, and they were particularly influential in defining the intellectual basis of chemical engineering. Their views were shaped by their experience of designing chemical plants and in many cases, as we have discussed, that of constructing and operating manufacturing facilities for explosives, chemical weapons, dyestuffs and intermediate products that had been sponsored by the state during the first world war.[55] This shared background was important for securing agreement within the IChemE over the details of the intellectual framework

---

[52] *Ibid.*, 38, 44–51.

[53] J.-C. Guédon, 'Conceptual and institutional obstacles to the emergence of unit operations in Europe', *History of Chemical Engineering, op. cit.,* 45, 56.

[54] Buchholz, *op. cit.*, 37. Other historians reverse Buchholz's arrow of causality. Van Antwerpen, for example, concludes that the explosive development of large-scale chemical plants had to await the development of chemical engineering as a distinct engineering discipline. Van Antwerpen, *op. cit.*, 11.

[55] See Chapter 3 and R. MacLeod, 'Chemistry for King and Kaiser: revisiting chemical enterprise and the European war', in: Anthony S. Travis, Harm G. Schröter and Ernst Homberg (eds.), *Determinants in the Evolution of the European Chemical Industry, 1900–1939* (Dordrecht: Kluwer, 1997) and *idem.,* 'The chemists go to war: the mobilisation of civilian chemists and the British war effort, 1914–1918', *Ann. Sci. 50* (1993), 455–81.

In the Chair at the Annual Dinner of the Institution of Chemical Engineers, 1930.

Figure 4-2   Key members of the early IChemE. *Source*: *The Industrial Chemist* (Apr. 1930), 169.

for chemical engineering, and it proved critical as well for defining the links between the Institution and academe.

The IChemE settled on a set of intellectual principles by 1925; in its fundamentals, this served the profession until well after the second world war. By 1939, accredited chemical engineers were proportionally more numerous in

At the Annual Dinner of the Society of Chemical Industry, 1930.

Figure 4-3 Key figures in the SCI, 1930. *Source: The Industrial Chemist* (1930), reprinted in *The Kestner Golden Jubilee Book 1908–1930* (London: Kestner, 1958).

Britain than America,[56] but it is arguable whether the idea of the unit operations and associated concepts was as convincingly accepted in Britain as in the USA. The links between British educational programmes, professional organisation and industrial practice were, on the whole, fairly weak. While increasingly uniting academic curricula, unit operations had yet to make convincing inroads at the workplace.

The influence of North American ideas cannot be denied. The highly successful enlargement and reconstruction of the explosives industry during the war had indeed been guided by the design techniques of K. B. Quinan, who had spent at least his early career in the USA and kept abreast of developments there.[57] Links thereafter between the IChemE and the AIChE ensured a ready awareness in Britain of the American technical literature, particularly the AIChE periodicals available to IChemE members, and American textbooks occupying a less extensive British market.[58] More particularly, the successful programme of chemical engineering at MIT was well enough known in Britain to be presented to the IChemE in 1922 as an ideal to emulate.[59] Not surprisingly, there were striking parallels between the list of subjects judged to be distinctive of chemical engineering that the IChemE published in 1925, and the recommendations of Arthur D. Little's report of 1922 for the AIChE. Both lists placed the unit operations at the core of the new academic discipline.

There were, however, marked differences in emphasis. Reflecting the dominance of plant manufacturers and consultants among its leadership, the IChemE made a capacity to design manufacturing plant on a commercial basis central to its definition of the intellectual foundations of chemical engineering. This embodiment of separable chemical stages in discrete equipment was conceptually and practically distinct from the American version of unit operations, and the British also proved quicker to adopt new technologies such as control equipment.[60] The continued focus on cost-effective chemical staging of processes also helped to maintain educational programmes for chemical engineers in a more intimate connection with business ideals than did courses for chemists

---

[56] By 1940, there were 761 accredited chemical engineers in Britain (IChemE 'corporate' members) compared with 1349 in the US (AIChE 'active' members). However, the number of student members in the two institutions (for example, 50 in the IChemE versus 1400 in the AIChE in 1932) implies a significantly higher rate of production in America. This is supported by academic records: by 1940 some 40 American colleges offered degrees in chemical engineering, MIT alone averaging 58 graduates per year between 1920 and 1934. In Britain, the combined output of diploma, certificate and undergraduate degree programmes was some 15–20 per year. For PhDs in chemical engineering, the US rate by 1940 was over fifty per year; the British output was less than one-tenth this figure.

[57] Quinan, although 28 years younger than G. E. Davis had, like him, been educated and trained in an environment devoid of recognised 'chemical engineers' and related practices.

[58] British 'textbooks' were nearly absent, although there was a variety of specialist guides and monographs available. See Appendix, 'Literature'.

[59] R. E. Johnstone, 'The dark days', *TCE* Jan 1987, 38–9. M. B. Donald, an IChemE officer and later an important educator had, in fact, undertaken his chemical engineering studies there in 1921–2.

[60] S. Bennett, 'The use of measuring and controlling instruments in the chemical industry in the UK and the USA during the period 1900–1939', in: Travis, Schröter and Homberg, *op. cit.*

and engineers.[61] J. M. Coulson has argued that early plants were mainly batch processes, in which the main task was to 'maintain the correct chemical conditions between the various stages'; thus plant was 'not too dissimilar from that of the laboratory ... and the thread woven through this kind of process was essentially that of chemistry; it was essential that the chemistry was right'.[62] Economic considerations took second place. From the late 1930s came the beginnings of a change both in the nature of plant and the method of assessing its performance. There was increasing demand for intermediate products (especially in the oil industries) encouraging the development of larger units and continuous processes. Larger plant also offered lower cost per unit although with higher capital requirements, and the financial scrutiny of plant performance consequently tightened.

There was one other important respect in which British conceptual definitions were more extensive than those current in the United States: not until the mid-1930s was there any significant attempt in the latter country to define a conceptual framework for chemical engineers when dealing with the *chemistry* of processes that would function as unit operations did for the *physical* transformations of chemical manufacturing. In Britain, by contrast, the 1925 IChemE enumeration listed various 'reaction treatments of materials', which were to become more commonly known as unit *processes*, and later still as 'chemical reaction engineering'.[63]

Thus the defining concept for the emerging profession, at least in Britain and the USA, was the set of largely empirical categorisations collectively known as 'unit operations', but the origins of the concept appear to have had distinct national origins. In the USA, unit operations have been postulated as an outcome of the Taylorist programme in manufacturing, or as a consequence of American enthusiasm for mechanical innovation and large-scale manufacturing. The unit operation, as first coherently articulated by A. D. Little, was defined as an isolable physico-chemical stage in a manufacturing process. In Britain, by contrast, unit operations initially were popularised as plant-based chemical operations rather than the manufacturing processes per se: to some extent, the notion of chemical stages was reified in mechanical equipment. The adoption of this intellectual principle was initiated by professional consultants such as G. E. Davis from the 1880s and eventually taken up by academics and professional societies, particularly after the first world war. The reasons for the adoption are various: for consultants, the concept provided a technical vocabulary for discussing plant design in operational terms without divulging company secrets; for academics, unit operations generalised and organised the particularities of chemical plant design into a manageable curriculum.

---

[61]   IChemE, *The Training of the Chemical Engineer* (London, 1925), 5–6; Colin A. Russell, Noel G. Coley & Gerrylynn K. Roberts, *Chemists by Profession* (Milton Keynes: Open University Press, 1977), pp. 158–85, 264–83.

[62]   J. M. Coulson, 'Chemical engineering as a science and an art', *TCE* Apr 1962, A69-A74.

[63]   The list of treatments included calcining, electrolysis, hydrolysis, oxidation and fermentation. N. A. Peppas and R. S. Harland, 'Unit processes against unit operations: the educational fights of the thirties' in: Peppas, *op. cit.*, pp. 125–42.

The appropriation of the unit operation and other intellectual constructs as unique features of chemical engineering helped to consolidate the profession's identity and to explicate its works to its clients, to the wider public, and to practising engineers themselves.

TEACHING CHEMICAL ENGINEERING BETWEEN THE WARS

Unit operations were increasingly the basis for explicit training in the new professional specialism. This unique training set out the boundaries of knowledge and defined the standard of achievement required of chemical engineers. Indeed, these newly-identified specialists were distinct from their colleagues in other engineering specialisms in that they devoted much more attention to educational issues.[64]

Immediately after the first world war, the close association of professional goals with management responsibilities encouraged, in courses of academic training, an emphasis on the economic tasks of personnel involved in manufacturing. The syllabus was shaped also by the needs of consultant engineers and of teachers. Consultants were more concerned with the design of chemical plant than with its operation and maintenance, and so they tended to attach significance to the cultivation of design skills. Because the early teachers were, for the most part, also consultants, they readily adopted this point of view; it also helped them distinguish the new discipline from that of industrial chemistry.

Between the world wars, the needs of consultants and teachers on the one hand, and of chemists on the other, were reflected in the dispute over the balance that should be struck over the teaching of 'engineering' and of 'chemistry' in the earlier part of an academic training. The IChemE council was dominated by consultants and teachers, but it did not enjoy complete control over the development of courses in the universities and a variety of curricula was developed. In colleges where the teachers of chemistry were sympathetic, courses of chemical engineering tended to pay greater attention to the teaching of the chemistry of manufacturing processes. But through the interwar period, the syllabi of the new courses in chemical engineering – particularly the undergraduate courses – marked a growing distinction between the disciplines of chemical engineering and of industrial chemistry.

Although a few courses on chemical engineering existed before the foundation of the IChemE, few chemical engineers received their education and training there. The interwar battles of the Institution were fought largely in university departments and councils, where the profession could vie for disciplinary

---

[64] See, for example, R. A. Buchanan, *The Engineers: A History of the Engineering Profession in Britain 1750–1914* (London: Jessica Kingsley, 1989), pp. 161–79; M. Davis, 'Technology, institutions and status: technological education, debate and policy, 1944–56' in: P. Summerfield and E. J. Evans (eds.) *Technical Education and the State Since 1850* (Manchester, 1990), pp. 120–44; I. Glover and M. Kelly, *Engineers in Britain: A Sociological Study of the Engineering Dimension* (London: Allen & Unwin, 1987), pp. 93–115; C. Divall, 'Professional organisation, employers and the education of engineers for management: a comparison of mechanical, electrical and chemical engineers in Britain, 1897–1977', *Minerva 32* (1994), 241–66.

authority. Education therefore attained considerable importance. The teaching of the discipline in institutions of higher education differed significantly from that of civil, mechanical and electrical engineering, although in all cases teaching may be understood as an accommodation between the demands of employers and of the professional needs of teachers.[65]

A number of factors might help explain the distinctive characteristics of the contemporary curriculum. The early teachers of chemical engineering had to define the academic boundaries of their discipline against those of the well established subject of chemistry. It was also necessary to increase the employability of graduates, despite the indifference or opposition of many industrialists. The presence in the works of chemists, already engaged in the development and operation of chemical manufacturing processes, and that of engineers, employed in the design and construction of chemical plant, were additional barriers to enhancing opportunities for employment.[66] Teachers of chemical engineering thus came to share with practising chemical engineers an interest in developing a very distinctive kind of academic training as a means of distinguishing and establishing their professions. The situation was quite otherwise in the established branches of engineering, where the professional communities were split over the wisdom of encouraging instruction in the universities.

As will be discussed in the next chapter, firms were not particularly mindful of the need for distinct specialists. However, leading industrialists who supported the efforts of the IChemE between the wars included the chemical plant manufacturer Sir Arthur Duckham, the detergent and food manufacturer William Hulme (Viscount Leverhulme) of Lever Brothers, the dyestuffs manufacturer Herbert Levinstein, and Sir Josiah (later Lord) Stamp. The early support given by W. J. U. Woolcock, the former parliamentary private secretary to the minister of munitions, was particularly significant. As manager from 1920 of the Association of British Chemical Manufacturers (ABCM), which represented particularly the interests of the medium and smaller manufacturing firms, he favoured employing chemical engineers because they could undertake the tasks of both the industrial chemist and the mechanical engineer.[67] The ABCM and its sister organisation, the BCPMA, co-operated with the IChemE on matters of training to a much greater extent than did the trade associations in the engineering industries with the principal professional societies of engineers.[68]

By the late Twenties, with the government forgetting its wartime dependence on the industry, the Institution brandished education as a multiple weapon: it

---

[65] C. Divall, 'A measure of agreement', *Soc. Stud. Sci. 20* (1990), 65–112; *idem.*, 'Fundamental science versus design: employers and engineering studies in British universities, 1935–1976', *Minerva 29* (1991): 167–94.

[66] J. F. Donnelly, 'Chemical engineering in England, 1880–1922', *Annals of Science 45* (1988): 555–90. See also his 'Industrial recruitment of chemistry students from English universities: a reevaluation of Its early importance', *Brit. J. Hist. Sci. 24* (1991): 3–20; Reynolds, 'Defining professional boundaries' *op. cit.*, 697–716.

[67] 'A character sketch of the new President', *Chem. Age 11*, 12 July 1924, 32.

[68] Divall, 'A measure of agreement' *op. cit.*, pp. 76–77.

could promote recognition of the discipline of chemical engineering in academia by making space within, or even converting existing, departments; it could establish standards of training that might improve the competitiveness of industry and the availability of jobs; and it could nurture the profession by helping to create a demand in industry for accredited chemical engineers and thereby displace lower-qualified workers. In doing so, the Institution could draw upon existing educational programmes in American universities as well as the pioneering courses at colleges such as Battersea and Imperial.[69]

Training chemical engineers

The tiny Institution differed in at least one important respect from the principal societies of professional engineers in Britain, namely the Institution of Civil Engineers (ICE), the Institution of Mechanical Engineers (IMechE), and the Institution of Electrical Engineers (IEE).[70] These institutions had all been founded before it was generally accepted that an academic training was desirable for engineers. Although by 1914 they had made provisions for those young men who wished to undertake a period of study in a university or a technical college, their requirements for membership still strongly reflected the tradition of pupillage. By contrast, the potential membership of the IChemE lay principally among men who would receive an academic training in chemistry.[71] It was therefore necessary to devise a course of training that would distinguish the chemical engineer from the chemist.

Proponents of an academic training in chemical engineering had to establish a need in the face of two disciplines that were well established in the universities. Two groups were most concerned with the provision of a novel form of training. The first was the consultants responsible for the design of plant and, in conjunction with the plant manufacturers, for its construction and commissioning. They were in a weak position with regard to the contracting companies, for the latter often designed plant directly to the specifications of their clients. Consultants hoped to improve their position by basing their expertise on a new form of knowledge. The second group was the chemists involved in the control of manufacturing as plant managers or superintendents. They feared that their standing was under threat from their association with analytical chemists, who were not held in high regard by industrialists.

Interwar education: establishing authority

The IChemE's policies on training were, as we have remarked, influenced by those of the American Institute of Chemical Engineers and, more generally, by

---

[69] F. Rumford, 'Early chemical engineering teaching: chemical engineering at the Battersea Polytechnic in the period 1924–1936', *TCE* Dec 1967, CE263, CE267.

[70] 'Lord Moulton', *Trans. IChemE 17* (1939), 184–85; Moulton, 'The training and functions of the chemical engineer', *Trans. IChemE 17*, 186–91; 'Annual corporate luncheon', *Trans. IChemE 18* (1940), 21; William C. Peck, 'Early chemical engineering', *Chem. & Indus.* 2 Jun 1973, 511–517; Donnelly, 'Chemical engineering', *op. cit.*, pp. 581–587.

[71] Donnelly, 'Industrial recruitment of chemistry students', *op. cit.*

educational developments in the USA. In both countries, the profession of chemical engineering appealed to chemists dissatisfied with their low standing as analysts and research workers in industry. The first president of the IChemE, Sir Arthur Duckham, characterised the chemical engineer as an individual with considerable managerial authority – a situation more true of the leadership of the Institution than of its general membership. From the first statement of policy, nevertheless, there was a strong emphasis within the institution on training in the economic function of chemical engineering. By 1926 the education committee, upon which teachers and consultants were in the majority, had agreed a policy on education and training which set the institution apart from the other professional societies of engineers.[72]

Many employers who expressed an opinion on the matter agreed that an academic training was desirable. However, their opinions were characterised largely by their brevity; by default, the 'special training' which was urged by all parties might simply have combined instruction in elements of chemistry with that in mechanical, civil and electrical engineering.[73]

In 1925, the institution had agreed the details of a syllabus of academic training. As with the membership examinations, this was largely drafted by J. W. Hinchley and Sir Frederic Nathan, assisted by Hugh Griffiths and E. C. Williams. A consultant himself since 1908, Hinchley employed the concept of unit operations in his Battersea and Imperial College courses. Since consultants were often expected to advise on all aspects of the civil, mechanical and electrical engineering of a chemical plant, Hinchley regarded a training in these subjects as a necessary complement to that in unit operations.[74] Nathan, a military man trained at the Woolwich Military Academy with long experience of both the government ordnance and the private explosives industry, did not have Hinchley's experience with teaching. He was, however, acquainted with contemporary ideas on the training of chemical engineers through his position as Director of Propellant Supplies during the first world war, and through his subsequent appointment as head of the Intelligence Section of the Fuel Research Board of the Department of

---

[72] There were 17 members active in the period 1923 to 1925. Three (J. W. Hinchley, F. G. Donnan, the professor of physical chemistry at University College, London, and E. C. Williams, the first professor of chemical engineering at the same college) were engaged principally in teaching. Of the remaining 14 members, at least seven (W. G. Adam, E. A. Alliott, W. Newton Drew, C. S. Garland, Sir Frederic Nathan, E. W. Smith and F. R. Tunks) were men with considerable experience of industry or the state's technical services, and the remainder (R. G. Browning, W. J. Gee, Sir Alexander Gibb, F. A. Greene, H. Griffiths, W. Macnab and F. H. Rogers) were consultants. Gee was distinctly unusual in admitting to having attended 'elementary school only', being 'self taught' in scientific education and, following some nine years' practical experience with two chemicals firms, gaining his income through the firms working his various inventions. See AFs 2, 13, 27, 44, 64, 74, 92, 123, 153, 170, 185, 218.

[73] E.g., 'The training and work', *op. cit.*, pp. 61–118; Duckham, 'Presidential address' (1924), *op. cit.*, 14–6; 'The chemical engineer "arrives"', *Chem. Age 14* (1926), 91–2; Donnelly, 'Chemical engineering in England', *op. cit.*, pp. 584–5. An exception was the M.P. and former soldier, Sir William Alexander. 'Annual dinner', *Trans. IChemE 6* (1928), 17–9.

[74] D. M. Newitt, 'J. W. Hinchley 1871–1931', *TCE* Feb, 1971, 49.

Scientific and Industrial Research (DSIR).[75] Griffiths had become a consultant in 1917 after extensive experience during the war of the design and erection of plant for the explosives industry. In the same year, he took over from Hinchley the course at Battersea. Evan Clifford Williams was the first Ramsay professor of chemical engineering at University College, London.[76] Appointed in August, 1923, he had been trained as a chemist and had considerable experience of the manufacture of intermediate products. Williams had been trained initially as a chemist at the University of Manchester and, at the time of his appointment, was in charge of the manufacture of intermediate products at the British Dyestuffs Corporation. It is of no little interest that the position had attracted applications mostly from persons in responsible posts in works management, rather than the career academics who by this time tended to make up the majority of those applying for chairs in the major engineering subjects.

As suggested above, although he shared with Hinchley a basic commitment to the conceptual basis for an autonomous discipline of chemical engineering, Williams was more inclined to regard favourably theories drawn from chemistry, particularly physical chemistry, when it came to more detailed conceptualisation of the subject.

The final version of the memorandum on 'The training of a chemical engineer', published in 1925, bore the hallmarks of Hinchley's catholic philosophy, also expressed strikingly in the IChemE's 'Home Paper', intended from the start as a guide to what was expected from the universities.[77] With minor alterations made through the 1930s, the 'Training' document remained the institution's official view until 1944. Four years' study at university or technical college was felt to be desirable; three years of chemistry, physics and engineering (for which no further recommendations were made) were followed by a year in the study of 'those subjects specially within the province of a chemical engineer'. This course was presented by Hinchley as a version of that at the Massachusetts Institute of Technology, modified and reduced in length to suit British conditions. The most obvious omission was of 'practice schools', periods of industrial experience taken under the tutelage of academics. Two further sets of recommendations were added to the MIT syllabus, covering the mechanical and civil engineering skills thought necessary for 'Chemical plant construction' and for 'Factory design and construction'.[78] Finally, the IChemE

---

75   'Colonel Sir Frederic Lewis Nathan, KBE', *Trans. IChemE 3* (1925): 23; S. R. Tailby, 'Famous men remembered', *TCE* Jun 1986, 92–3; Nathan papers, IChemE, *passim*.

76   The first meeting of the IChemE council in 1923 agreed to invite Williams, who was not a supporter or member, to join and 'help in the organisation of chemical engineering training' [CM 4 Jul 1923].

77   MEC (1927–37); A. J. V. Underwood, 'The chemical engineer and his training for industry', *Chem. & Indus. 54* (1935), 235; H. W. Cremer and A. J. V. Underwood, 'The education and training of the chemical engineer' in G. H. Ford (ed.), *The Transactions of the Chemical Engineering Congress* Vol. 4 (London, 1937), pp. 32–47; Underwood, 'Chemical engineering – reflections and recollections', *Trans. IChemE 43* (1965), T302–16; Newitt, 'Memoirs', *op. cit.*, pp. 137–8.

78   A. *Chemical Engineering Processes*: conveyance and storage of materials; production, transference and conservation of heat; treatment of materials (including unit operations). B. *Chemical Plant Construction*: properties of materials; resistance of materials to corrosion; principles of

committee recommended the study of 'General principles of factory organization and management'; no more details were given.[79]

Although industrialists comprised a minority on the committee, the opinions
of chemical manufacturers and of manufacturers of plant were canvassed. Of
the 'many firms' drawn from the membership of the ABCM which were invited
to comment in 1924, 138 replied; 72 approved of the committee's proposals
without further comment, while the remainder voiced only minor reservations.[80]

The education committee regarded the ability to design chemical plant and
manufacturing processes on the basis of the standardised unit operations as a
sufficient academic training for employment in either the construction or the
operation of plant. Industrialists did not dissent from this arrangement, even
though the evidence is that the majority of chemical engineers (whether graduates or not) had some responsibility for the operation of plant and that few
were engaged solely in design. The philosophy of the examination also received
considerable support in the trade press.[81]

Many employers were pleased to welcome an extension of these measures.[82]
The attitude of the IChemE towards the training of potential managers had
repercussions for its policy towards courses below the university level; that is,
in the technical colleges run by the municipal authorities. There were clear
advantages to the institution if this sector were to expand. The cost of attending
a technical college meant that young men from the lower social classes, who
were often denied a university education, could receive a training which might
fit them for membership of the IChemE. On the other hand, firms seeking
technically qualified personnel for management positions often preferred to
recruit graduates because of their higher social standing. The admission to the
IChemE of large numbers of men without degrees might have diluted the
regard with which employers held membership of the institution. This conflict
of interests was not resolved formally by the IChemE. It encouraged the
development of facilities for training in the technical colleges but, in contrast
to the policies of the IMechE and the IEE, it made clear that these courses
were not as well thought of as those in the universities. It did not grant

---

design and construction. C. *Factory Design and Layout*: layout and construction; power, heat
and light.

[79] IChemE, 'Training of the Chemical Engineer' *op. cit.*, 5–7.

[80] 'Extracts from replies received on the memorandum on "The training of a chemical engineer"',
circa late 1924; MEC (1923–5); F. Nathan, 'Memo on the education work of the Institution'.

[81] MEC (1927–37); 'Chemical engineering associateship', *Chem. Age 14* (1926), 431; 'The first
chemical engineering exams', *ibid. 15* (1926): 153; 'Extract from the Report of the Board of
Examiners ... 1926', *Trans. IChemE 4* (1926): 197–203.

[82] E.g. 'The Education Committee', *Trans. IChemE 2* (1924): 16–22; A. Duckham, 'Presidential
address, Sir Arthur Duckham, K.C.B.', *ibid.*: 14–6; Duckham, 'Presidential address by Sir
Arthur Duckham', *ibid. 3* (1925): 16–8; Duckham, 'Annual dinner', *ibid.*, 5 (1927): 18–20;
'Training the chemical engineer', *Chem. Age 10* 16 Feb 1924: 157; G. Brearley, 'Economics of
production as exemplified in process industries', in H. W. Cremer and T. Davies (eds.), *Chemical
Engineering Practice* Vol. 1 (London: Butterworths Scientific, 1956), pp. 53–4; C. S. Robinson,
'Kenneth Bingham Quinan, C.H., M.I.Chem.E. 1878–1948', *TCE*, Nov. 1966, CE290-CE297.

diplomates from the technical colleges the same degree of exemption from the institution's examinations as graduates.[83]

Interwar academic programmes

Few academic institutions developed or modified courses between the world wars to meet the educational requirements of the IChemE. Teachers in the universities and the technical colleges were keen to receive the institution's endorsement if it involved no restructuring of the syllabus; in 1925, the education committee rejected about 45 applications for the recognition of existing courses. However, by 1939, there were still only five colleges of university rank offering full time courses in chemical engineering: Imperial College, University College, and King's College, all in London; the Royal Technical College, Glasgow (affiliated to the University of Glasgow); and the Manchester Municipal College of Technology (affiliated to the University of Manchester).

In 1926 the Institution granted exemption from all of its examinations to those who passed 'creditably' either the taught portion of Hinchley's postgraduate course at Imperial College or that instigated, in 1924, by Williams at University College; both of these included a practical training in design.[84] In 1929, the course at King's College, London, organised by another active member of the IChemE, Herbert W. Cremer, was accepted, after the teaching in design had been strengthened at the insistence of the institution. However, the exemption granted in 1933 to graduates from the honours degree in Oil Engineering at the University of Birmingham did not extend to the Home Paper because the course paid insufficient attention to design. The only other course granted full recognition prior to 1939, that at the undergraduate level at the Manchester Municipal College of Technology, met the institution's requirements in this respect.[85]

The education committee's recommendations of 1925 and its policy on the recognition of courses together provided a reasonably clear perspective on what was considered intellectually distinctive about the training of a chemical engineer. But there was still considerable scope for disagreement over what should be taught in the first three years of study. The theoretical articulation of chemical engineering phenomena was in its infancy, and the debate therefore tended to focus on whether subjects drawn from 'engineering' or from 'chemistry' should form the greater part of the first stage of an academic training.

Teachers of chemical engineering and the consultants prominent within the IChemE made up the greater part of those favouring the first option. They

[83] MEC (1926–39); Divall, 'A measure of agreement', *op. cit.*, pp. 68–74. This two-tier policy continued for Home Paper exemption after the war, with graduates of institutions such as Battersea and Bath, for example, not gaining the exemption accorded their university counterparts.

[84] Hinchley's course in the 1920s was attended principally by graduates with a degree in chemistry and several with experience as industrial chemists. Letter, K. L. Emler to Johnston, 3 Mar 1997.

[85] MEC (1926–39); Minutes of the Chemical Engineering Committee, King's College, London (MCEC) (1928–66), KDCH/M1.

wished to weaken the occupational as well as the intellectual links between chemical engineering and chemistry; they regarded it important for graduates to possess sufficient knowledge in engineering to allow them to design a complete chemical plant. The supporters of the teaching of chemistry came from a variety of backgrounds. They were concerned principally with the training of the majority of chemistry graduates, who were finding employment in the operation of plant. For this task, an understanding of mechanical and civil engineering was less important than that of industrial chemistry. However, the views of the former group were the more influential within the IChemE. By the late 1920s, some 60 percent of the questions in the examination for membership were said to require a knowledge of 'engineering'. There is some evidence that, in the 1930s, many employers agreed with this policy.[86] The slump of 1930–2 brought an influx of graduates made redundant and wanting to retrain via a post-graduate diploma in chemical engineering.

The issue was thrown into relief in 1931, when J. Arthur Reavell, then president of the institution and a senior figure in the chemical plant industry, raised the possibility of an undergraduate course in chemical engineering. As an equipment and plant manufacturer, Reavell thought that the first two years of the course should emphasise 'engineering'. However, he had few suggestions to make as to the details of the syllabus. Of ten commentators, only the consultant F. H. Rogers, and the industrialist C. S. Garland (both of whom had been involved the preparation of the IChemE's recommended syllabus) expressed a clear preference for a separate course at the undergraduate level. They thought that a training in both 'engineering' and 'chemistry' was desirable. By way of contrast, William Rintoul, of ICI, and Sir Robert Robertson, Government Chief Chemist (and formerly of the Woolwich Arsenal Laboratory), were opposed to an undergraduate course; they, like two other commentators, thought that an initial training in chemistry was more important.[87]

The proponents of undergraduate teaching were in a sufficiently strong position within the IChemE to advocate the introduction of a course within the University of London. Throughout the early to mid 1930s, private discussions were held between teachers (Hinchley and S. G. M. Ure of Imperial College; W. E. Gibbs, M. B. Donald and H. E. Watson of University College, and H. W. Cremer of King's), consultants (Greene), and industrialists (Nathan, Reavell and Alliott) over the best way of securing support for their ideas within the university. Although they differed over the importance of a training in chemistry, these men wished to see chemical engineering develop as an autonomous discipline. They were therefore prepared to characterise chemical engineering as primarily an engineering discipline, and they were not greatly

---

[86] E.g., 'The education committee', *Trans. IChemE 2* (1924): 18–22; 'The education and training of the chemical engineer', *ibid. 9* (1931): 15; J. H. West, 'The functions and training of a chemical engineer', *Chem. Age* 28 (1933): 94–5; 'Training the chemical engineer', *ibid.*, 89.

[87] 'The education and training', *op. cit.*, pp. 14–20. Both men supported the courses at the postgraduate level in the University of London.

disturbed by the opposition within the faculty of science of the university to the introduction of a course which might draw students away from established courses in chemistry. However, the need for the utmost financial economy meant that the syllabus had to be developed largely from existing courses in chemistry and in engineering. Both of the two undergraduate courses, introduced in October 1937 – one at Imperial College, the other at University College – led to the award of a degree in the faculty of engineering.[88]

Evolution of university departments

The development of courses in the universities was influenced by factors other than the policies of the IChemE. Variations arose from the institution's recommended syllabus, partly because of the different backgrounds of the teachers. They were also related to differences in the strength of ties between teachers and employers, and by the degree of support for the teaching of chemical engineering given by the professors of chemistry and of engineering. Responding to disparate pressures, chemical engineering programmes evolved distinctly in each case. And the first generation of teachers influenced their students significantly by the courses they taught and often by the industrial contacts they utilised for job placements.

Imperial College, London

The history of chemical engineering at Imperial College, the second teaching institution at which John Hinchley taught, is important for at least two reasons. Via Hinchley, the subject was established there despite intense opposition; it is a particularly early and explicit case of what were to become common battles for academic jurisdiction. Secondly, Hinchley was active not only in an academic sphere but several professional ones. Imperial tied these domains together.

The development of chemical engineering at Imperial College was most obviously marked by the antipathy between Hinchley and his head of department, W. A. Bone, the professor of chemical technology. The latter strongly opposed the teaching of chemical engineering other than as a subject subordinate to chemical technology. The support of industrialists within the college's governing body, including that of C. S. Garland, proved significant between the wars in sustaining the contrary view. However, there was no permanent forum in which industrialists could intervene directly in disputes over the organisation of teaching.[89]

The teaching at Imperial College was strongly shaped by its prior history. Chemical technology was not initially taught as the professor of the period (W. A. Tilden) thought there was no practical difference between 'pure' and

---

[88]   MEC (1930–7); Minutes of the Senate of the University of London (MSUL) (1935–6), University of London Archives (ULA) ST 2/2/52, Minutes 3140, 3268.

[89]   Annual Report of the Delegacy of the City and Guilds College (1937–8), Imperial College Archives (ICA) ICA/FG1, p. 57; M. de Reuck, 'History of the Department of Chemical Engineering and Chemical Technology 1912–1939'.

'applied' chemistry.[90] The initiative came from industrial representatives on the Pure and Applied Sciences Sub-Committee in 1908. They suggested a four year course in Chemical Technology, but only an 'advanced' course in Gas Manufacture was initiated that year.

Tilden was replaced by T. E. Thorpe in 1909. Thorpe was more positive, recommending, the following year, the establishment of a lectureship in Chemical Engineering to offer fourth year courses. Vague proposals from the Governors for a Department of Chemical Technology were defeated by financial restrictions.

In November 1910, Hinchley was appointed to teach an *ad hoc* course on 'Design of Plant Required for Chemical Manufacturing'.

Thorpe prepared a memorandum on an independent Department of Applied Chemistry and Chemical Technology in 1911, at the instigation of Edward Divers of the Society of Chemical Industry: a 'composite of many specialists teaching particular industrial fields'. This was to be a full time course, preceded by a three year course in Science. There were, however, problems over finding a Director, owing to the 'doubtful unity and independent status' of the department. An appointment was offered to Threlfall (a consulting engineer), who declined owing to pressure of work.

Problems of identity included Hinchley's suggestion that a study of chemical engineering was a sufficient study in itself. Eventually, Bone, teaching Fuel Technology, was offered a Chair notionally under Thorpe. The 'Department' consisted of Bone and Hinchley's existing courses, plus a new one in electro-chemistry.

Hinchley's initial course, of 1910, was aimed at men who had already graduated in chemistry, and this facilitated an emphasis on the engineering aspects of plant design. He wished to press the division of labour in industry to its fullest extent. Because the managerial structures of works now embodied a framework of knowledge beyond that of the individual, Hinchley's lectures consequently tended to involve a brief outline of theory but to concentrate on 'empirical problem-solving'. Students undertook about 300 hours of formal study over one year; they had to prepare the design of a commercial plant based upon an appreciation of unit operations and the economics of manufacturing plant.[91] The post-graduate diplomates found employment largely through Bone's personal contacts and through the Chemistry department.

Increased war-time interest encouraged Bone to propose expanding the activity of the Department into a 'broader basis of work and study'. A course on 'Explosives' was given by G. I. Finch, one of the two part-time lecturers who comprised the department along with Hinchley and Bone. In 1917, it became apparent that University College, London was considering the estab-

---

[90] See Robert Bud and Gerrylynn Roberts, *Science vs Practice: Chemistry in Victorian Britain* (Manchester: Manchester University Press, 1984).

[91] *Calendar*, Imperial College of Science and Technology (1912–3), p. 26; J. W. Hinchley, 'The work of the Imperial College in the training of engineers' in 'The training and work', *op. cit.*, 87–8.

lishment of teaching in chemical engineering. Largely to meet this threat, and to enhance the standing of the Imperial courses, Bone secured both Hinchley's appointment to a permanent teaching post, at a level below that of full professor, and the expansion of facilities for chemical engineering.[92] Bone also manoeuvred the following year to block the establishment of an independent chemical engineering department at UCL, partly on the grounds that chemical engineering should be taught only in conjunction with chemical technology. Hinchley, still allowed two days per week consultancy, was able to avoid explicit conflict with his nominal superior.

During the formative years of the CEG and IChemE, the deadlock in the Imperial College Chemical Technology Department changed only slowly. A lecturer, S. G. M. Ure, was appointed in 1921 to teach chemical engineering.[93] The college authorities were usually more sympathetic to Hinchley than Bone, because Fuel had continued to receive greater financial support. Relations between Hinchley and Bone remained turbulent. A series of disputes between the two men culminated in 1925 with a proposal from the college authorities to form a separate department of chemical engineering. Bone's objections and a lack of finance combined to force a compromise whereby Hinchley was made a professor the following year, but still under Bone. This, however, was sufficient to enable Hinchley to introduce, in 1928, a taught course of up to two years' duration, leading for the first time to the college diploma specifically in chemical engineering.[94] The new course naturally followed the recommendations of the IChemE.[95]

The new postwar high-pressure processes also introduced new concerns not covered by Hinchley's and other syllabi, such as the failure of metals by creep stress.[96] High-pressure boilers, new metal alloys and protective coatings became widely applied, as did techniques of recovering heat and particulates from flue gases.[97] Such innovations demanded links between chemistry, physics and engineering for the first time by their demands for knowledge of equilibrium reactions and catalysis, high pressures, properties of engineering materials and large-scale design. Materials and techniques, too, were improving noticeably.

[92] Approximately £20000 was spent upon the recommendation of a small committee which included Rudolph Messel, the representative of the Society of Chemical Industry; Richard Parsons, the civil engineer; Sir Robert Mond; Sir Robert Hadfield, the steel industrialist; A. Hutchinson, professor of mineralogy at the University of Cambridge; and F. H. Carr, chairman of British Drug Houses.
[93] Sebastian Greig Monteith Ure (1881–1941) had a BA Eng from Glasgow and MA Maths and Physics, and five years' experience in shipbuilding. He was a member, when appointed, of the IChemE provisional committee along with Hinchley. AF 93.
[94] de Reuck, 'History of the Department of Chemical Engineering', op. cit., 8–72.
[95] Nevertheless, the college refused quinquennial inspection of the course by the IChemE.
[96] Letter, K. L. Emler to Johnston, 3 Mar 1997. Emler afterwards found employment with the Vitalite Co, manufacturers of composite billiard balls, of which Hinchley was a director. See AF 1373.
[97] 'Developments in chemical plant during 1930', Chem. & Indus. 59 (1931), 79, 102; S. G. M. Ure, 'Progress in chemical engineering during the last twenty-five years', Chem. & Indus. 54 (1935), 432–4.

Yet the rate of advance of the discipline seemed discouragingly slow for some. In a yearly review of chemical engineering for *Chemistry & Industry*, A. J. V. Underwood wrote that, judged on research, 'the year 1928 can show but little achievement or cause for satisfaction':

> Admittedly, there is increased interest in chemical engineering education ... which is all to the good for future progress. But, for the present, significant contributions to the science of chemical engineering are rare.

Both chemical engineering principles, and knowledge of construction materials had to be developed, argued Underwood, if the subject were to thrive.[98] Avoiding such criticism, Hinchley strengthened the teaching of the scientific aspects of chemical engineering – for example, by studying problems of corrosion – and strove to expand the curriculum to include subjects such as the planning and design of factories, and industrial administration and bookkeeping. The result was a very full curriculum.[99]

Owing to Bone's antipathy to the teaching of chemical engineering, no attempt was made to replace Hinchley as professor after his early death in 1931; the existing courses continued under Hinchley's assistant, S. G. M. Ure, now promoted to reader, and a new lecturer the following year.[100] With Bone's retirement imminent, in 1935 a temporary advisory committee of ten (including three individuals who were prominent in the IChemE) recommended as his replacement A. C. G. (later Sir Alfred) Egerton, reader in applied thermodynamics at the University of Oxford.[101] Egerton had worked under K. B. Quinan during the first world war and was sympathetic towards the committee's desire to establish an undergraduate course in chemical engineering.[102]

Under Egerton, and without the rivalry of Hinchley and Bone, old barriers disappeared. A draft syllabus for a course of four years' duration (three for those who had reached the intermediate level through study at school) had already been prepared by a small group of teachers, including the professor of mechanical engineering, Cecil H. Lander, the professor of physical chemistry, J. C. Philip, and two of the staff in chemical engineering who were heavily

---

[98] A. J. V. Underwood, 'Chemical engineering in 1928', *Chem. & Indus. 48* (1929), 33–5.

[99] *Calendar*, Imperial College (1930–1), pp. 182–4, 189–90a; Newitt, 'J. W. Hinchley', *op. cit.*, 49. E. H. T. Hoblyn, whose father had been taught by Hinchley, noted that the syllabus had changed little by his time as a student. Kenneth Emler, another graduate of Hinchley's diploma course in the mid-Twenties, noted nevertheless that practical design was largely empirical, based 'on current practice and experience, either personal or in the literature', and using a factor of safety seldom less than five. *TCE* 29 Nov 1990, 50; K. L. Emler to Johnston, 3 Mar 1997.

[100] Francis Bailey, appointed in 1932, had trained as a chemist at Armstrong College, then in Chemical Engineering at Imperial (1921–3). He had experience in industry and research associations from 1923–32.

[101] The three were Garland, Gibb and Sir Harold Hartley, chairman of the Fuel Research Board of the DSIR. The committee also included Rintoul, who was a member of the advisory committee at University College, and another employee of ICI, Dr E. F. Armstrong.

[102] Egerton (1886–1959) had graduated in Chemistry from UCL (1908), then studied at Nancy and Berlin and worked under Quinan during the war. See D. M. Newitt, 'Alfred Charles Glyn Egerton', *Biographical Memoirs of Fellows of the Royal Society 6* (1960), 39–64.

involved with the IChemE, Ure and D. M. Newitt. The committee saw chemical engineering as 'a synthesis of Chemistry, Mechanical and Electrical Engineering, with a top dressing of borderline subjects such as fuel technology, plant design and works organisation' – hardly a firm foundation for a distinctive identity for the subject. For the reasons already outlined, the teaching in the first three years of the course drew upon existing courses in chemistry and engineering; the fourth year of study was substantially the same as that given to postgraduate students.[103]

University College, London

At UCL, the very different personal and institutional arrangements resulted in a course with a different emphasis to that at Imperial. The first professor of chemical engineering, E. C. Williams, had been trained as a chemist. He agreed with Hinchley that it was the physical operations of chemical engineering which 'most vitally concern' the designers of chemical plant, and the postgraduate course that he introduced in 1924 shared with that at Imperial College a concern for inculcating design skills through a thorough training in the unit operations. However, Williams and the professor of physical chemistry, F. G. Donnan, demonstrated none of the antipathy found between Hinchley and Bone. Moreover, Donnan's sympathy for the academic claims of chemical engineering was marked by his support for the establishment of a separate department.[104] Hence the two professors could agree that the teaching of physical chemistry was more important than Hinchley allowed, without running the risk of a dispute over the precise sphere of influence of each.[105]

Although Williams hoped to devote time to research, teaching developed as the first priority of his staff; lack of money prevented much research until the late 1930s. Teaching was entirely postgraduate in nature, although the desirability of an undergraduate course was implied in Williams' inaugural lecture. His argument for the distinct identity of the chemical engineer, particularly with respect to the academic chemist, echoed that of Moulton; the chemical engineer dealt with chemical processes on a increased scale of operation, and had to work for profit. Moreover, while chemical processes were obviously involved, it was the 'physical aspect' which 'mostly vitally concern' those responsible for the design of plant. For Williams, chemical engineering was, in its academic dimension, 'another term for applied physical chemistry', and this formed the

---

[103] 'Report of Chemical Technology Committee of Imperial College of Science and Technology', MSUL (1935–6), *op. cit.*, Minute 3268, Appendix AC4; *Calendar*, Imperial College of Science and Technology (1937–8): 226; G. S. Bainbridge 'The Department of Chemical Engineering and Chemical Technology', *Journal of the Imperial College Chemical Engineering Society 14* (1962), 16–7.

[104] Donnan (1870–1956) had worked at UCL on the synthesis of ammonia during the war, and was a founder member of the CEG and IChemE, having provided K. B. Quinan's reference for membership.

[105] MEC (25 September 1923); E. C. Williams, *The aims and future work of the Ramsay Memorial Laboratory of Chemical Engineering* (London: University of London Press, 1924), p. 10; *Calendar*, University College, London (1924–5), 91–2.

basis for the distinct identity of the subject. Mere instruction in elements of chemistry, physics and engineering would not lead to the 'thorough grasp of the fundamentals of the various sciences' required by chemical engineer. Moreover, works experience was essential and could only be partially supplanted by practical training in the college laboratory. Williams wished to see students spend six months in industry as part of their course, a desire echoed by Sir Robert Robertson, Chairman of the Chemical Engineering Sub-Committee. Donnan shared these ideas. Indeed, he had become a good friend of K. B. Quinan during the war, who recommended to him a programme for the all-round training of chemical engineers:

> Train a physical chemist (because the field of physical chemistry to my mind comprises practically the whole field of chemistry, in so far as it bears on industrial work), giving him as complementary instruction a thorough grounding in the principles of mechanical and electrical engineering. Teach him to design and cause him to study each of what I call the fundamental processes such as distillation, furnace work, catalysis, sulphonation, nitration etc. Then pull wires and obtain for the partly finished product permission to work as a labourer – and see to it that you underline the "labourer" in the arrangement – and above all teach the young scallywag to get along with his workmen and not regard them as the dirt under his feet.[106]

The course which resulted from UCL's deliberations was given for the first time in 1924–5. It was designed to take two years, with the first year being largely taught and the second session given over to research, either in the college laboratory or in works. Although advertised as open to both graduates in chemistry and in engineering, it was in fact designed primarily for the former.[107] The taught portion of the course focused on 'The Design and Operation of Unit Types of Chemical Plant', although there were other courses in Industrial Chemical Calculations, Heat Transmission and the Dynamics of Fluids; Principles of Mechanical and Structural Engineering; Production and Distribution of Energy in Works, and Materials of Construction. Practical work embraced basic instruction in drawing and workshop techniques, elements familiar on undergraduate courses in engineering, and what would later be called project work, in the chemical engineering laboratory. The latter was intended to consist 'of the investigation of processes with a view to large scale development and the investigation of the theory of mechanism and efficiency of operation of specific unit types of operation', all with the express aim of improving techniques of design and operation industrial plant. This was quite unlike the kind of activity found in other departments of engineering. By the early 1930s, the relevance of theory to industrial practice was also underlined by special lectures from speakers from industry (19 in 1931–2); visits, once a week, to works; and by the use of weekly problem classes requiring students

---

[106] Letter, Quinan to Donnan, 3 Jun 1920, Donnan Papers, f.2, UCL archives; quoted in G. K. Roberts, 'The making of the chemist at University College London, 1914–1939', *Centaurus* 39 (1997), 291–310.

[107] Although complete records are unavailable, it is certain that in 1926 all six entrants were chemists.

to derive data applicable for the design of a component of a chemical plant. Students were also encouraged to spend their long vacation in works.

This first year of study was formalised as a discrete course in 1931, when a college diploma in chemical engineering was inaugurated; students staying the full two years were eligible for the MSc. The one year course had been accepted by the IChemE in 1929, and was subsequently altered little by Williams' replacements.

Williams' successor, William Edward Gibbs (1928–34) had also trained as a chemist.[108] In the early 1930s, he tried to introduce an undergraduate course in order to maximise the use of the new Ramsay Memorial laboratories, completed in 1931.[109] The views of the industrial benefactors of this building project were significant. The advisory committee, established in 1931, had no formal power to comment on the curriculum, but the department relied very heavily on grants from firms for its recurrent and capital spending. This must have made it difficult to ignore the opinions of men on the committee such as Robertson and Rintoul, who tended to favour an emphasis on a training in chemistry. Gibbs shared some of their views.

An address by Gibbs to the Liverpool Section of the SCI in 1933 illustrates the fundamental continuity of his views with those of Williams. The address was in part a plea for a four year course in chemical engineering leading to a first degree. Gibbs said that academic chemistry at the undergraduate level failed adequately to provide for the 70% or so of chemistry graduates who entered industry; a chemical engineer, by contrast, 'is essentially a chemist who has been trained to be industrially effective'. In general terms, he thought that a training to the level of an ordinary degree in chemistry should continue to be a prerequisite for a course in chemical engineering. However, his detailed proposals for the first three years of the undergraduate course included a substantial measure of teaching in engineering as well as that in chemistry.[110]

Gibbs was clear in his belief that the multiplicity of industrially relevant chemical processes could be understood in terms of both physical and chemical operations. In turn, the design and operation of a plant intended to carry out unit operations relied on a proper 'application of the principles of physics, chemistry, mathematics, and economics'. But Gibbs favoured a greater emphasis than his colleagues at Imperial on the analysis of manufacturing processes in terms of the chemical component of unit operations – the so-called 'unit processes' mentioned earlier.[111] These processes, such as oxidisation, combus-

---

[108] Gibbs, like Williams, was a man with considerable industrial experience, having been Chief Chemist to the Salt Union Ltd. Similarly, when Gibbs died in 1934, he was eventually replaced by H. E. Watson, who came from the Indian Institute of Science, an institution dedicated in part to the design and operation of industrial chemical plant on a pilot scale.

[109] H. W. Thorp, '1916–1939: the early years ', *TCE* Jun 1974, 384–5.

[110] W. E. Gibbs, 'Chemical Engineering Education: Professor W. E. Gibbs' Inaugural Lecture', *Chem. Age 19* (1928), 538; 'The functions and training of a chemical engineer', *ibid. 28* (1933): 67–9; 'The education and training', *op. cit.*, 16–17; W. Rintoul, 'The training of industrial chemists', *Chem. Age 31* (1934), 235.

[111] 'Department of Chemical Engineering' (c. 1931) UCL Archives (UCLA) BD 5; Gibbs, 'The functions and training', *op. cit.*, 68.

tion, reduction and electrolysis, had received only a little attention in the IChemE's recommendations of 1925, perhaps because of the need to draw a sharp distinction between chemical engineering on the one hand, and applied and industrial chemistry on the other. Gibbs' successor, H. E. Watson (1934–51) finally introduced the teaching of unit processes at the undergraduate level at University College.[112]

These sciences, applied to the unit operations, formed the basis of the existing course at University College, and could form the basis of the new degree. Yet Gibbs was ambivalent about how far chemical engineering as an academic discipline should be regarded as fully independent of chemistry. Although in terms of industrial *practice*, the chemical engineer could be distinguished from the works chemist by his appreciation of economic realities, the engineering dimension involved 'the skilful use of certain principles and must be learnt by experience'. By contrast, the grounding of chemistry that was required could only be acquired adequately at college, and thus should constitute the vast majority of any new undergraduate course.

By the 1936–7 session, potential chemical engineers had to learn about 'Electricity in Chemical Works' and 'Fuels and Their Utilisation in Industry'; the influence of the Education Committee of the IChemE could just be detected in the altered rubric for work in the drawing office, which now talked of the 'opportunity of carrying out the design and layout of a plant for a complete chemical process'.

Apart from the professors, the teaching staff at UCL between the wars was small: a total of three full time lecturers, and one honorary lecturer. Those in the former group shared little, apart from a relative lack of experience of industry. Burrows Moore (lecturer, 1923–8) was a physical chemist educated originally in engineering at King's College, London and then in Applied Chemistry at the Manchester Municipal College of Technology; his only experience of life outside the college laboratory was three years' experience of research in industry and the DSIR. Similarly J. P. Mullen (lecturer, 1928–58) had originally trained as a mechanical engineer at the University of Liverpool; after graduation in 1920, he undertook research at the University of Illinois until returning to Britain (and Persia) in 1925 to work for Anglo-Persian Oil. M. B. Donald (1897–1978), appointed in 1931, graduated from the Royal College of Science, Imperial College in chemistry and then took his master's degree in chemical engineering from MIT in 1921–2.[113] He had about six years' experience of industry; three in the research division of the Chilean Nitrate Producers' Association, and three as Patent and Technical Adviser to Shell. By contrast, H. W. Thorp (Demonstrator from the 1930s, Lecturer from the

---

[112] *Calendar*, University College, London (1936–7): 113–4; *ibid.* (1937–8): 112–3; 'Department of Chemical Engineering', *op. cit.*, pp. 7–11. The syllabus was drawn up in consultation with the professors of engineering and of chemistry. Letter, Watson to Secretary, UCL 11 March, 1937, UCLA 32/3/7.

[113] Donald's experience of the MIT practice schools helped to introduce the concept of unit operation to UK; see *TCE* Apr 1972, 136; *Who Was Who*.

1950s), originally trained in chemistry at UCL and had nine years' experience of industry.

Institutions, as well as individuals, defined the department. Unlike Imperial, the role of industrial interests, at least in the first fifteen years, was crucial. Indeed, the department owed its very existence wholly to the generosity of industrialists. Immediate responsibility for the foundation of the Ramsay Memorial Chair of Chemical Engineering in 1923 lay partly with Professor F. G. Donnan. At his suggestion, £27 000 was raised largely from industrialists for the establishment of a chemical engineering laboratory. In 1927, a further appeal to secure the financial future of the department called upon the services of eight leading industrialists, including Sir Hugh Bell (Dorman Long & Co.), Sir Robert Whaley Cohen (Shell), F. D'Arcy Cooper (Lever Brothers), Sir Arthur Duckham, Sir Alexander Gibb (Dunlop Tyre); Sir Alfred Mond (ICI) (Chair), Sir Robert Robertson and Sir David Milne (Gas Light & Coke Co). A further £24 000 was raised by this appeal, £10 000 from ICI, and £5000 from Shell. Subscriptions to cover the running costs of the department amounted to £3491 per annum, a figure which increased to £3840 by 1933–4. It was only in the late 1930s that industry showed signs of unwillingness with regard to further funding.[114]

The influence of industrialists, and other outside interests, on the development of the curriculum is less clear. A large Advisory Committee was established in 1931. Along with the professor of chemical engineering, of civil and mechanical engineering, and the two professors of chemistry, there were eleven members appointed in a personal capacity, of whom six, including the Chairman, Sir Robert Whaley Cohen, may be positively identified as industrialists. In addition, provision was made for a further ten representatives of trade associations and professional societies, including the IChemE, the ABCM and the BCPMA.[115] The internal squabbles which had to be endured at Imperial College were absent because there was no tradition of teaching 'industrial chemistry'. The support of Donnan meant that there was no objection to the department being made fully independent from the department of chemistry from the outset.

Michael Sanderson has criticised the universities for failing to provide sufficiently for the industrial demand for chemical engineers in the period between the wars. This appears difficult to justify, given the close analysis of employment prospects below. Moreover, the criticism cannot be sustained as far as University College is concerned. The courses were a success, but only in a minor way. In the 1920s, the numbers each year rarely exceeded half a dozen although there was room for more; prior to the 1930–1 session, a total of 35

---

[114] This was partly a reflection of a more general feeling among industrialists that funding for educational expansion should come from general taxation; but there was also a specific disappointment among those in the gas and oil industries – which had been among the largest donors – that individuals trained as chemical engineers had not proved willing to take employment overseas. The quoted figures should be multiplied by 20 to 25 for their 1999 equivalents.

[115] Unfortunately, formal records of the committee's deliberations do not appear to have survived, although correspondence extant makes clear that not all the senior members of the committee were of 'much use' in furthering the fortunes of the department.

students passed through the department (of whom two-thirds qualified for membership of the IChemE). By 1938, a further 142 students had taken one or other of the courses available; students 'from the colonies and abroad' were said to form an 'appreciable proportion' of the total. Even so, the maximum of 27 new students, recruited in 1932–3, was still inadequate to fill the total number of 50 places (for all years) available once a permanent laboratory, complete with pilot plant, had been finished in 1931. The problem was twofold: first, students were deterred by the cost of postgraduate study. In 1924–5, the first full session of the course, one student was reported as leaving before the end of his course because he had found employment; the problem was still evident in 1937–8. Second, among those who could bear the cost and who had just graduated, the temptation to study in the subject of the first degree was very strong and often encouraged by the members of staff in the parent department; those who did transfer to chemical engineering usually stayed only for the one session.[116]

Nevertheless, some industrial firms, including a few which later gained a reputation for not having employed chemical engineers, eagerly took on students from the UCL courses. Of the 132 students in the period 1931–8, no fewer than 117 obtained employment in industrial firms; a further five went into research or consultancy, five into teaching and two into technical journalism. Not all these men went into posts where their technical expertise was used directly; echoing the aspiration of Sir Arthur Duckham, there was also a 'demand for qualified men to train for higher administrative appointment in industry'. In 1934, however, Watson claimed that only *one* student was engaged in design – all others were employed in some form of process control.[117]

King's College, London

The institutional arrangements at the third college in London to offer teaching for full-time students in chemical engineering, King's College, also encouraged the study of unit processes. King's was the last and the least successful of the three colleges of the University of London to offer chemical engineering as a subject of postgraduate study between the wars.

Early students of engineering at King's were offered a course in 'Chemical Manipulation', which included a grounding in chemical analysis. But chemical engineering as a distinct course of study did not appear until 1928 – just postdating the formation of the IChemE. Although the course was located within the Department of Chemistry, the syllabus reflected the 'force for unification' that characterised the course at UCL and, to a lesser extent, that at Imperial.

The precise origins of the course are unclear, although the principal architect was H. W. Cremer, since 1920 a lecturer in inorganic chemistry at the college. Educated in chemistry at King's (graduating BSc in 1910), Cremer had his major experience of design, construction, management and technical writing

---

[116] *The Chemical Age*, noting the problem of cost, recommended the institution of an external degree in chemical engineering at the University of London.

[117] F. R. Whitt, 'Early teachers and teaching of chemical engineering', *TCE* Oct 1971, 370–4.

under Quinan in the first world war; he had also been involved with the IChemE from its inception, and maintained an interest in consulting. He had offered 'short courses' of twenty lectures in chemical engineering to undergraduates since 1923, and it was this that was to be expanded into 'something further' in the session of 1928–9. 'Applied Physical Chemistry' stood at the heart of the syllabus, a subject which the Chemical Engineering Committee (consisting initially only of professors) took to be the study of the 'general principles' of the 'large scale operation of a given chemical reaction and the manner of their qualitative application so as to secure maximum technical efficiency'. Although both chemists and engineers were to be catered for, the new course was clearly aimed at those 70% of chemistry students who entered industry; moreover, it was thought likely to appeal to those judged not 'particularly fitted' for research in the parent disciplines.[118]

The establishment of a structured course in chemical engineering seems to have been something of a defensive move to safeguard the college's position with regard to developments at Imperial and UCL. Certainly, the college did not intend investing any large sums in the development. Hence the form of the new course was dictated as much by the necessity for economy as any consideration of academic probity or industrial demand. Thus even in the longer term, most of the courses in chemistry were to be undertaken in the main department of chemistry, and 'close co-operation' was intended with the department of engineering. Apart from Cremer, the staff should thus consist only of a laboratory attendant; the annual budget would be about £300, plus the salary of a reader.

The initial syllabus for a college certificate in chemical engineering assumed one year of study, and was worked out in greater detail for the student with a background in chemistry than for one in engineering. The very first student to be admitted, in 1928, took classes in 'Principles of Chemical Engineering', plus those in hydraulics, elementary heat engines, theory of heat engines, theory of machines, electrical technology and strength of materials, as well as attending the drawing office. As finally approved by the Delegacy of the College in 1929, the syllabus had slightly fewer hours devoted to the of teaching subjects in mechanical engineering, and more in 'Pure Chemical Engineering'. Examinable subjects, for chemistry students, were materials of construction and basic principles of structural design, mechanics of fluids and theory of machines, transmission of heat and heat engines, electrical technology; in addition, there was compulsory practical work in engineering drawing and workshop practice. Engineering students took general theoretical chemistry, principles of organic chemistry and chemistry of metals. Both sets of students took courses in physical and reaction treatment of materials (a course based on unit operations), and general principles of organisation and management in chemical works. The IChemE approved the course in June 1929, having apparently convinced the college to accept the desirability of a design paper along the lines of the

118   King's College Chemical Engineering Committee, KCL archive KDCM/M1.

IChemE's Home Paper. This syllabus was to remain substantially unchanged until after the second world war.[119]

The early popularity of the course was modest. One student attended in 1928–9, and two (a chemist and an engineer) the next year. By 1932, a total of only seven students had taken the course full-time, as well as another two, who had undertaken research, and a handful of occasional students. Perhaps because of this poor showing in comparison with the other London colleges, the Chemical Engineering Committee took on another four representatives. These included W. A. S. Calder of ICI and J. Davidson Pratt; the latter, who also served on the UCL advisory board, was chairman of the ABCM and remained associated with the course until its demise in 1966.[120] Within 18 months, the possibility of yet further representation from industry was canvassed, although a decision was deferred until 1935, when two more places for industrialists were made available and the IChemE invited to appoint a representative. It is not clear whether the suggestion, in 1934, to bring the course more into line with that at UCL, by including works visits, vacation experience in firms and a programme of visiting speakers, emanated from the industrialists; but within a year, all three of these elements were in place, partly owing to the influence of Davidson Pratt. In 1936, a Leverhulme scholarship for research or further advanced study had been instigated.

These measures did not greatly enhance the popularity of the course. Although it was reported in 1938 that 'demand for present students far exceeded the supply', by 1938, the last year for which figures are available, only 24 students had taken the course, of whom 16 had passed. Neither the college nor industrial firms were prepared to finance the improvement of facilities.

Other courses

Courses elsewhere in Britain may be treated more briefly. The universities of the Midlands and of the north of England, and of Scotland, were closer than the London colleges to the major centres of the chemical industries. They were local provision for local need. By the early 1920s, several of the provincial universities had already developed specialised courses in chemical technology suited to the needs of their regional industries. For example, the City of Cardiff Technical College was probably the earliest provincial course. Started in 1919 by Harry W. Webb, its curriculum was influenced by the teaching of Hinchley at Battersea. Webb had worked at Oldbury during the first world war, developing continuous nitration processes in explosives. His Cardiff colleague, E. A. Rudge, also had 'some industrial experience'. Webb left the college in the mid-1920s, after which the course developed with IChemE requirements in mind. Rudge left for West Ham, 1937, when residual teaching was transferred

---

[119] Recognition was deemed by the IChemE to have 'lapsed' in 1948, but was subsequently renewed in 1950 after inspection by Prof F. H. Garner (Birmingham) and F. E. Warner.

[120] 'perhaps the best informed man in England on economic and commercial aspects of chemical industry', according to S. D. Kirkpatrick, later President of the AIChE. Letter to M. A. Williamson, 17 Aug 1936, Brennan archive.

to the School of Mines, Trefforest. The Trefforest course, in Fuel and Mining Engineering, had begun in 1920.[121]

Similarly, Huddersfield Technical College launched a part-time course in 1919 intended principally as an internal training programme for chemists and design engineers at the Delton Works, Huddersfield, of British Dyes Ltd., although it was open to other college students. Although having no laboratory, the course dealt with the properties and use of materials of construction and the design of plant, emphasising operations such as heat exchange and distillation. Academic aspects were not stressed until its revision late in the second world war.

Local firms often partly financed the departments. This was the case, for example, with the Department of Oil Engineering and Refining at the University of Birmingham; with the Department of Fuel Technology at the University of Sheffield; and with the Department of Fuel (and Metallurgy) at the University of Leeds.[122]

By the second world war, over a dozen such courses were being offered.[123] Teaching in chemical engineering, however, was developed extensively only in Glasgow and Manchester.

Royal Technical College (University of Glasgow)

At Glasgow, the courses were at the undergraduate level, growing from, and then running in conjunction with, the work of departments of industrial chemistry. A degree in applied chemistry in the faculty of science was introduced at the Royal Technical College shortly after it affiliated with the university in 1913; the local sugar manufacturers presented much of the laboratory equipment.[124] In their third and fourth years, students could take courses in 'manufacturing operations' based on the unit operations, but they did not undertake a

---

[121] Whitt, *op. cit.*

[122] A. J. Biddlestone and J. Bridgewater, 'From mining to chemical engineering at the University of Birmingham' in Peppas, *op. cit.*, 237–44; A. W. Chapman, *The Story of a Modern University* (London, 1955), pp. 305–9; A. L. Roberts, *The Houldsworth School of Applied Science: The Fuel Department after 60 Years* (London, 1966), pp. 1–6.

[123] Others included the South Wales and Monmouthshire School of Mines (1913); University of Edinburgh (1922); Royal Arsenal, Woolwich, Apprenticeship scheme (1925); CIGB and Aldermaston Court Centre for Adult Education (1929); Hackney Technical Institute (1929); London Technical Institute (1931); Leicester College of Technology (1931); Widnes Technical College (1933–5); Liverpool College of Technology (1936); West Ham College of Technology (1937); Sir John Cass Technical Institute (1937). See, for example, W. M. Cumming, 'Early chemical engineering teaching', *TCE*, Mar 1968, CE33, and S. R. Tailby, 'Early chemical engineering education in London and Scotland', in: W. F. Furter, *A Century of Chemical Engineering, op. cit.*, 65–126

[124] The first lecturer, W. M. Cumming (1891–1976) teaching applied chemistry and chemical engineering at RTC from 1920 after five years at British Dyestuffs Corp. manufacturing explosives, claimed that the Glasgow sugar plant manufacturing industry had an important influence on the development of chemical engineering in Glasgow. Day and evening courses were instituted from 1911 (in the School of Sugar Manufacture, a constituent part of the Dept of Technical Chemistry). From 1924 onwards, a course on beet sugar production was introduced.

design problem. From 1923 these courses became the basis for a new degree in chemical engineering run in parallel with the existing one in applied chemistry. The former came under the aegis of the faculty of engineering, although the teaching was undertaken with the department of applied chemistry. Although it included some training in design, the degree in chemical engineering was not recognised by the IChemE, probably because the college failed to submit examination papers for scrutiny. No degree was awarded until 1928, and there were only three graduates, plus five associates of the college, prior to 1939.[125]

Manchester Municipal College of Technology (University of Manchester)

The undergraduate course at the Manchester Municipal College of Technology was considerably more successful. The city could claim one of the longest traditions of teaching chemical engineering, Manchester Technical School having hosted the part-time course by George E. Davis in 1887. This was not continued, however, and its successor, the Manchester Municipal College of Technology, had no specific course of study again until the early 1930s.[126]

In 1918, suggestions from the Applied Chemistry department were made for a post in 'Engineering Chemistry' at the University of Manchester, but these came to nothing, probably owing to lack of finance. Similarly, a part-time course of seventeen lectures on 'various aspects' of chemical engineering was recommended in 1918 and approved in 1919. This was later revised to 20 lectures, with the syllabus drafted by Norman Swindin. Records are unclear as to whether the course was ever offered.

Applied Chemistry, however, was a major activity at the College. As of 1919, studies in Fermentation, Fuel, Dyestuffs and Baking were active.[127] This meshed poorly, however, with the standards later set by the IChemE: a degree course in Applied Chemistry was refused accreditation in 1926, and no opinion was expressed on the desirability of a course in chemical engineering.[128] There is no record of further developments until 1932, when the professor of applied chemistry, James Kenner, proposed new undergraduate and postgraduate courses based on those at MIT. An expert in organic chemistry, he had a strong interest in reforming the teaching of industrial chemistry. However, shortly before his arrival in Manchester in 1927, the college authorities agreed to the removal of textile chemistry, one of the most important sections of Kenner's department, from his control. Since Kenner had gained experience during the first world war in the manufacture of chemical warfare agents under Quinan, it was not surprising that he should meet the challenge to his position

---

[125] MEC (1923–7); Cumming, *op. cit.*; Tailby, *op. cit.*

[126] J. J. Walsh, 'Higher technological education in Britain: the case of the Manchester Municipal College of Technology', *Minerva 34* (1996), 219–57.

[127] In 1919 there were a professor, lecturer and associate lecturer in applied chemistry.

[128] A subcommittee of the Board of Faculty concluded that none of its courses met the IChemE's outline, but that *some* students should be recommended for exemption.

by proposing the new course in chemical engineering.[129] Kenner intended to produce a new type of graduate who could compete with the university product.

Local employers had an unusually large degree of control over the curriculum. They were heavily represented on the 'sectional committee' for chemistry, which held formal control of the syllabus. The committee was generally supportive of the suggestion of a new course. Like Gibbs at University College, Kenner wished to provide a training principally for 'process chemists'. His first proposal, in 1932, drew inspiration from the syllabus and practice schools at MIT, although the three postgraduate courses offered in the University of London, as well as continental courses in chemical technology, were later considered.[130] Financial considerations played as large a part in shaping the curriculum as they had in London.[131] As finally developed, and first offered in 1934, a course of roughly one year's duration in chemical engineering – described as 'most essentially engineering' – was superimposed onto a training in applied chemistry carried to the standard of an ordinary degree.[132] The requirements of the IChemE in matters of design were met by a practical examination taken over four days. By 1939, the course was sufficiently well established for Kenner, perhaps mindful of the unhappy situation at Imperial, to persuade the sectional committee and the college authorities to refuse an offer of £25 000, made by the petroleum industry, to establish a chair in Fuel Technology.[133]

PREWAR OUTPUT OF ENGINEERS

By the late 1930s, the combined efforts of teachers and of consultants, working within and without the IChemE, and supported by sympathetic employers, had secured a modest provision for the teaching of chemical engineering in the universities. The completion, or even the partial completion, of a course in the subject was thought by some employers to provide a suitable academic training for those who sought employment in the manufacture of chemicals or in the design and construction of chemical plant. Some 80 to 90 students, at least 90

---

[129] Lord Todd of Trumpington, 'James Kenner', *Biographical Memoirs of Fellows of the Royal Society 21* (1975): 389–405; F. Morton, 'A short history of chemical engineering in the North West of England', in Furter, *A Century of Chemical Engineering, op. cit.*, 19–27.

[130] J. Kenner, 'Proposed course in chemical engineering', Manchester Municipal College of Technology (typescript, 9 Dec 1932). Kenner surveyed courses at Karlsruhe, Dresden, Zurich, Hanover and Breslau, as well as the London colleges, to determine whether they were designed to produce 'process chemists' or 'engineers'.

[131] William Cowen was appointed Lecturer in Chemical Engineering, because of his engineering background. He held a BSc from Owens College (which followed five years of engineering training), and had worked a further five years at ICI, Billingham. Cowen left in 1944 for an industrial post.

[132] Although a sub-committee recommended introduction of an honours course in chemical engineering (Minutes of Board of Faculty of Technology, Oct 1933, pp. 61–2).

[133] Kenner opposed the chair on the grounds that chemical engineering was a better education, and that any division of department should be 'vertical', i.e. in terms of course level, rather than 'horizontal' by subject division. Minutes of the Chemistry Sectional Committee (MCSC) (1929–39), University of Manchester Institute of Science and Technology Archives (UMISTA), F6.

percent of whom had trained initially as chemists, were attending the full time courses at the five institutions of university rank recognised by the IChemE.[134] These students were not fully trained as chemical engineers at the completion of their courses. They were, however, expected to possess sufficient practical skills in design and an awareness of the limitations imposed on designers by economic considerations to benefit from a period of industrial experience. Their design skills were informed by the theory of unit operations and, with the exception of the course at Birmingham, this was tested by means of a specific problem in design. The comprehensive nature of the design problems meant that considerable periods of time were spent teaching subjects drawn from the cognate branches of engineering and from chemistry.

Yet most employers during the difficult economic times between the wars were not particularly sympathetic to a new species of specialist. Nor did the manufacturers' organisations – the ABCM and BCPMA – have any clearly stated educational policies by the 1930s, although in 1934 the BCPMA invited professors of chemical engineering and applied chemistry to a conference to discuss mutual co-operation. But *The Chemical Age*, the weekly journal aimed at professional and managerial grades in the chemical and allied industries, proved more enthusiastic. It eagerly supported the IChemE's policy on examinations, partly by reprinting the early Home Papers and encouraging, in 1933, the Institution to recommend courses of study.[135] By 1937 it was enthusiastically commending the new undergraduate course at the University of London as a means of moving away from the unsatisfactory practice of training chemists with a 'smattering' of chemical engineering. Nevertheless, the realities of employment in the interwar years were disengaged from such concerns. The trickle of graduates, while beneficiaries of a crystallising Institutional and academic identity, entered a workplace largely unreceptive to representatives of a new profession.

---

[134] J. H. West, 'Engineers trained in chemistry' *Chem. Age 32* (1935), 180. There were nine honours degrees awarded in chemical engineering, and 61 full-time and part-time students attending postgraduate courses. U.G.C. *Returns From Universities and University Colleges 1938–39* (London, 1940), Tables 7, 8. To these figures must be added the numbers attending the undergraduate courses in Birmingham (Oil Engineering), London, Manchester (classified as Applied Chemistry) and Glasgow. The following are estimates of the output of graduates in chemical engineering between the world wars: Imperial College (1924–30), 36; (1930–8), 63 (based on 50% of number of college diplomas awarded in chemical engineering and in chemical technology); total 99; University College (1928–38), 159; King's College (1929–39), 24; Birmingham (1922–39), 125 (5–10 per annum); Manchester (1934–45), 47; Glasgow (1923–39), 3. Total, about 460. The membership of the IChemE in 1939 was 1018.

[135] Its publishers, Benn Bros., had been promoting the profession actively through its books and periodical since the early 1920s. *Chem. & Indus.* also published the Associate Examination papers, leading one reader to encourage the Institution to supply the answers so that aspiring chemical engineers and 'others engaged in the chemical industry' could learn the economic and design principles that were tested ['Enquirer', *Chem. & Indus.* 50 (1931), 799].

CHAPTER 5

DAMPENED ASPIRATIONS

Behind the discipline-building activities of educational planning and setting entrance standards lay the expectation that qualified chemical engineers would find a ready place in receptive industries. J. W. Hinchley had voiced this argument in 1928 when, speaking to American chemical engineers, he avowed disingenuously that 'there was no demand in Great Britain for chemical engineers – no demand whatever', while arguing that demand in America had been 'manufactured by the production of good students'.[1] Thus Hinchley's long-standing strategy (pursued first within the CEG and then via the IChemE) of developing teaching resources rather than promoting employment was made to seem justifiable and, indeed, natural. While members appear to have supported this line for its anticipated professional result, it is equally apparent that shorter-term goals were not being met. Relying on academic training as no earlier engineering group had done, chemical engineers attempted to construct their authority based on claims of natural competence for those tasks requiring scientific principles applied to the decidedly chaotic conditions of the British chemical industry. But the early expectations of the Institution were not met; the economic climate did not favour such perturbations to the labour process. Nor did a critical mass of employers find the arguments persuasive. While jobs could be found, they very frequently were of a nature or remuneration below graduates' expectations. Finding an occupational niche for the profession proved contentious and problematic.[2] The period was, indeed, marked by mixed fortunes: rising interest in the specialism did not translate directly into chemical engineering posts. It took the threat of another world war, and the renewed backing of the state, to achieve greater acceptance. This chapter therefore examines more closely the question of employment and occupational identity in firms large and small.

---

[1]   J. W. Hinchley, discussing A. H. White, 'Chemical engineering education in the United States', *Trans. AIChE 21* (1928), 55–85.

[2]   'The central phenomenon of professional life is the link between a profession and its work'. A. Abbott, *The System of the Professions: An Essay on the Division of Expert Labour* (Chicago: Chicago University Press, 1988), p. 20.

THE CHEMICAL ENGINEER AT WORK

The workplace during the early interwar period proved unreceptive for chemical engineers. In an atmosphere of declining state-industry involvement largely decoupled from academic programmes, the chemical engineer found a place in industry with difficulty. Attitudes among industrialists varied from cautious support to outright rejection of the territorial claims of the chemical engineer. The source of these opinions spread from individuals to the companies they influenced. New technological innovations and developments in industrial organisations, as outlined in Chapter 3, had been early sources of enthusiasm in the formative years of the IChemE. Yet many such technical innovations failed to create jobs for chemical engineers beyond short-term pre-commercial development. Technological change brought improvements in competitiveness of what were often marginal enterprises, but did not often entrain posts for more than a handful of skilled designers. Capital investment could be attracted by promising technical developments, but was difficult to sustain during this period of depressed markets. Even from the relatively successful chemical engineering programme at University College, London, for example, large firms seldom recruited chemical engineers in more than ones or twos during the 1930s.[3]

Application forms reveal that new members came from disparate industries. Through the 1920s, nearly a third described themselves as being employed in the 'chemical process' industry, and another quarter, like their first President, Sir Arthur Duckham, worked in the coal gas/coal tar/coke oven trade. The only other significant minority – comprising another fifth of the members – were those employed in mining, metals or foundries.[4] The end of munitions production had been tempered by the rise of some related industry. Besides those factories sold for little more than scrap, some explosives plants were readily converted to civilian products.[5] A few, for example, were converted to rayon production after the war, for which the production and raw materials were not dissimilar to gunpowder production. Similarly, National Dyes in Brimsdown reprocessed Ordnance factory waste to produce dyestuffs until the supply dried up in mid 1920s; they then turned to titanium applications and

---

[3]   The largest firms recruiting at UCL were: ICI (Ardeer [Explosives]; Huddersfield, Billingham, Manchester [Dyestuffs] and Widnes); Anglo-Iranian Oil; Kodak; George Kent; Glaxo; British Oxygen; Unilever; Babcock & Wilcox; Shell; United Dairies; Albright & Wilson; and Triplex Safety Glass.

[4]   By the 1930s, new members from the metals-related industries declined, while those in petroleum firms – almost non-existent during the first decade – rose to become second only to the chemical process industry. Coal products employed steadily fewer members until the period immediately following the second world war, when coal research enjoyed a resurgence. See Appendix.

[5]   Gretna was, however, the focus of considerable parliamentary debate after the war owing to labour protests and accusations of wastage concerning its closure and sale en bloc. Under the Coalition government the decision was taken in 1921 to dispose of the facility. Suitable offers not forthcoming, the ether, fuel, distillation and other plants were sold individually, and the land was transferred to local authority and auctioned piecemeal under subsequent Labour and Conservative governments. See *Parliamentary Debates 159-167* (1922–6).

mutated to become National Titanium Pigments. To many such transformed plants came experienced managers and designers from the ordnance factories.[6]

As often as not, firms made do with surplus plant and available manpower. The theme, as during all periods of financial straits, was improvisation.The chief engineer at one aniline factory, for instance, a practical man whose early training had been among steam-operated machinery,

> achieved wonders with second-hand plant. They had the knack of fitting in reaction vessels in odd corners and connecting them to other pieces of plant by delivery pipe which sometimes needed to be fabricated in bizarre, contorted shapes in order to fit. These units however usually 'worked' and served the Company well for the time they were needed.[7]

Other chemical firms sprang up to exploit market niches and developed *ad hoc* in response to outside forces. The history of the Saltend site, near Hull, is a good example. First developed in 1914 as a storage depot for fuel by two oil companies, the site later hosted a molasses-based distillery founded in 1922. This became British Industrial Solvents (BIS) and later a subsidiary firm of Distillers Company Ltd (DCL). IChemE founder member W. R. Ormandy, a consultant to DCL and a director of BIS, was a leading figure at the site from the late 1920s, when he closed the research facility in London to set up a development laboratory and pilot plant at Saltend.

Yet the employment of some graduates of chemical engineering programmes was not synonymous with the employment of recognised 'chemical engineers'. Despite a growing and diversifying chemical operation at Saltend, for example, the first two recognised chemical engineers on a staff of 29 were hired only in 1946.[8] During the war, the smaller BIS works in Surrey employed two Imperial College chemical engineering graduates – but as works manager and shift chemist.[9] The working culture at the collection of chemical firms on the banks

---

6   Queen's Ferry, for example, provided the designers and operators for the Bernard Laporte Co., in Luton, employed for new barium production processes. D. J. Oliver, 'John Sutherland, Master Chemical Engineer', unpublished typescript, 8 Jan 1997; 'The development of titanium pigments', (unpublished typescript, 17 Jan 1997); letters to Johnston, 2 Jan 1997 and 21 Jan 1997.

7   E. N. Abrahart, *The Clayton Aniline Company Limited 1876–1976* (Manchester: Clayton Aniline Co., 1976), pp. 44–5. This use of experienced but unqualified labour was a continuation of pre-war practice. For example, the first works foreman for Castner-Kellner, 'primarily a mason and bricklayer ... was responsible for the operation and erection of much of the early experimental apparatus, and later took charge of the erection and maintenance of the ever-increasing electrolytic plant' [*Fifty Years of Progress 1895–1945: The Story of the Castner-Kellner Alkali Company*, CRO DIC/BM 20/ 135, p. 39].

8   Ormandy (1870–1941) had erected synthetic phenol plants while Chief Research Chemist at Pilkington Bros during WWI, subsequently headed a cable and rubber company and taught science at a technical school there before returning to 'chemistry' and consultancy. At least two IChemE members there, S. T. Card and J. Howlett, were employed initially at Saltend as 'works chemists' between the wars. Only post-second world war employees (e.g. D. Ormston, W. M. Crooks, J. W. Pease and E. A. Chapman) listed their occupations as 'chemical engineers'. See AF 458, 1463, 2591, 3842, 4992, 5095, 6808.

9   J. Solbett interview with Johnston, 16 Dec 1997. Wilfred Cash and Solbett had both studied under S. G. M. Ure at Imperial College.

of the River Humber was one of segregated and rigidly defined job categories divided between fitters, semi-skilled and locally-trained 'process men' responsible for plant supervision, and development department chemists. Despite a handful of home-grown consultants such as Ormandy, operations relied heavily upon experience from abroad: the first distillery had been founded by an American, and BIS employed a Canadian manager. Chemical process expertise was, as often as not, imported via the German, French and Swedish workers who installed and sometimes operated imported units of plant, and seldom identified as 'chemical engineering'. A formally trained class of permanent design staff had no place in this environment. Employment at Saltend fluctuated cyclically in response to changing markets for alcohol and acids, ranging from a couple of dozen workers to hundreds. During nearly half of 1932, BIS ceased production completely as demand for its products dried up.[10]

With the owners and managers of chemical companies among the founders and first council members of the Institution, it might be expected that they would be significant employers of chemical engineers. As suggested by Ormandy at Saltend, this does not, in the early years at least, appear to have been the case. And the firm of George Scott & Son, which comprised a significant nucleus of chemical engineers and founder members of the IChemE north of the English border, in some respects represents a best-case example.[11] The firm, which had made its name in the Victorian gas industry, was run between the wars by James MacGregor as Managing Director and George W. Riley as Technical Director, both of whom were on the first council of the IChemE. The company, part of a network of Scottish firms controlled in large part by an extended family, had been active in designing, building and operating munitions plants during the first world war. After the war the firm shifted its focus to plant for oils, fats and waxes but sought opportunities wherever they appeared. MacGregor and Riley, both known as mechanical engineers, were keen to form and promote the new Institution, and during the 1920s several of its senior partners were practising chemical engineers.[12] Among some fifty

---

[10] A. Wilkinson, *Molasses to Acid: Saltend's First 75 Years Distilled* (Hull: BP Chemicals, 1997). The holdings of various companies operating at Saltend, including the German company HIAG, were eventually amalgamated by DCL, which in turn sold the operation to BP Chemicals in 1967.

[11] Founded in 1834, George Scott & Son was responsible for the design and manufacture of many chemical plants, and before the first world war had a laboratory to test heat transfer, fluid flow, mixing, agitation, size reduction and other processes. See W. E. Bryden, *Reminiscences of a 5/8 Ruddy Plumber, or Into the Caprolactum* (unpublished typescript, 1994), p. 9, and '100 years of chemical engineering: a retrospect of the early days of the firm of George Scott', *Chem. Age* 11 May 1935, 419–21.

[12] MacGregor (1878–1949) had taken evening classes in chemistry, physics, maths and mechanics at the Royal Technical College, Glasgow for a decade, gained experience at tar and ammonia distillation works for Scotts from 1901, and then become Managing Director of George Scott & Son (London) after a management buy-out in 1915. G. W. Riley (1883–1963) passed a 59 year career at Scotts. Among MacGregor's original partners was H. J. Pooley (1877–1941), who had studied at Liverpool University before joining Scotts in 1898, becoming works manager and claiming to be responsible for 'inception of most of Scott's present chemical plant designs and processes'. James Love Carlyle (1889–1952) had joined the firm in 1918 after studying engineering at Paisley technical college and working as a draughtsman for several

more junior employees was a single chemist in a small analytical laboratory although, during the 1930s, MacGregor took on immigrant engineers adept in plant design and a nephew trained in chemistry who had passed the AMIChemE examination.[13] Thus the chemical engineers at Scott & Sons were principally owners or partners who had gained expertise on the job. Few 'ready-made' chemical engineers found employment there or in other firms, at least with that title. An increasingly unstable and retracting trade situation proved a barrier to new categories of employee.

As suggested by the cases mentioned above, we cannot treat 'industrialists', even those in the chemical and allied industries, as an homogeneous group. Between the wars, many did not see any need to employ a specially trained 'chemical engineer' when the combination of an industrial chemist with a mechanical engineer was thought to be adequate. Courtaulds, for instance, had a long history of neglecting the employment of chemists, not to mention chemical engineers.[14] By contrast, other firms were keen to employ the small number of chemical engineers produced by the universities and technical colleges; Unilever was one such company, Glaxo another.

ICI as employment model

Explicit employment policies in such firms are difficult to discern, if indeed employers felt any need at all to react to the still indistinct and scarce 'chemical engineer'. This was equally true of the largest such firm in Britain: Imperial Chemical Industries. The most impermeable barrier to the employment of such specialists was the working culture of ICI. From its foundation in 1927, the firm retained in all of its manufacturing operations largely the existing division of labour between chemists and engineers. In the period 1928–38, for instance, only 13 of the 159 students of chemical engineering at University College, London were employed initially by ICI.[15] The ICI facility at Billingham, too, became notorious for an attitude of indifference to chemical engineers, although in fact a number was employed under other occupational titles. The firm and

---

firms from 1911. MacGregor, Riley, Pooley and Carlyle were founder members of the IChemE, with Pooley responsible for its Appointments Bureau. AF 53, 68 and 184; *Quart. Bull.* Apr 1949; *TCE* Feb 1964 CE23.

[13] Interview, W. E. Bryden with Johnston, 4 Dec 1997, and Bryden, *Reminiscences, op. cit.* See also SRO GD410 on the Balfour Group of companies. Bryden (b. 1913) had spent time at the Silverside works as a youth, and later studied chemistry at the Royal College of Science, but did not obtain a degree. Most of his career was spent at Balfour Scott, of which he eventually became Director. E. L. Rinman, a Swede specialising in paper pulp processes, and a Dr Bar, developing drying methods, immigrated to Britain from Germany in the early 1930s. Bar later formed Barr & Murphy [sic].

[14] Coleman, *op. cit.* pp. 54–5; 224; 226–8.

[15] William Cullen, 'A milestone', *Trans. IChemE. 16* (1938): 10–6; W. J. Reader, *Imperial Chemical Industries: A History* Vol. 1 (Oxford: Oxford University Press, 1970), pp. 249–57, 268–70, 354–63; K. Buchholz, 'Verfahrenstechnik (chemical engineering) – Its development, present state and structure', *Sci. Stud. 9* (1979): 33–62; 'Firms employing former students ... between July 1928 and July 1937' (1937) UCLA 32/3/3; Report on the Department of Chemical Engineering (1937–8), UCLA *32* (1938–9).

others like it, of course, had no need to intervene in the development of
university courses in chemical engineering since more than adequate supplies
of their preferred type of technical workers existed from other sources.

Why did ICI develop a reputation in occupational folklore for not recognising chemical engineers as a distinct specialism? The rhetoric of influential
individuals within the company played a part in this. The preference of its
President, Lord Melchett, for an industry-led state policy is described below –
a policy with which the first generation of organised chemical engineers, at
least, were somewhat at odds. The conglomerate was also closely modelled on
contemporary German practice, which was promoted by senior managers.[16] A
prominent exemplar was William Rintoul, its first Director of Research. Rintoul
(1870–1936) had been educated at Andersonian College, Glasgow, and joined
the Royal Gunpowder Factory at Waltham Abbey in 1894 as a chemist under
its superintendent, Sir Frederic Nathan, with whom he improved processes for
nitrating cotton and glycerine for explosives. When Nathan was hired by
Nobel's Explosives to manage their Ardeer factory in 1909, he arranged the
hiring of Rintoul as chief chemist.[17] Ardeer proved to be a training ground for
subsequent founder members of the IChemE. Under Nathan and Rintoul, from
about 1912, chemical engineers Hugh Griffiths and Dudley Newitt trained and
practised. Rintoul soon became Research Director and, when Nobel's was
merged with ICI in 1926, he continued as Director of Research for the expanded
company in London.[18]

For two men having such close careers and working experiences, Rintoul
and Nathan developed remarkably contrary views on chemical engineering.
Nathan was a founding member of the IChemE, its second president, and the
author of the first draft plan for the training of chemical engineers.[19] Rintoul,
by contrast, became the most vocal opponent of chemical engineering education
in the interwar years. Seemingly at every opportunity, he denigrated the categorisation of the specialism and the feasibility of inculcating the required training
in any academic programme. In 1924, he suggested that Nathan's IChemE
training plan would produce men 'little higher than what we, at Ardeer, call
Laboratory Assistants'. Chemical engineering was merely a variant of engineering chemistry, he claimed, and what was needed was bachelors-degree chemists
with some engineering training. In a 1931 meeting he reiterated that 'it was
probably impossible to produce, in any of our teaching institutions, a chemical
engineer'. Three years later, he took the initiative from the Institution by
promoting 'industrial chemists' as having natural responsibility for controlling
plant and manufacture or for the business side of industry.[20] Rintoul's beliefs

---

[16]   Before the first world war, advanced degrees – and working practices – in chemistry were
       commonly obtained in Germany.
[17]   F. D. Miles, *A History of Research in the Nobel Division of ICI* (London: ICI Nobel
       Division, 1955).
[18]   *Chem. & Indus.* 55 (1936), 709.
[19]   F. L. Nathan, 'Training of a chemical engineer', CM 15 Oct 1923.
[20]   IChemE, Extracts from replies received to the memorandum on 'The training of the chemical
       engineer', circa late 1924; 'The education and training of the chemical engineer' (discussion),
       Trans. IChemE 9 (1931), 14–20; W. Rintoul, 'Technical education as applied to the training of
       industrial chemists', *Chem. & Indus.* 53 (1934) 868–70.

appear to have passed directly to his successor at ICI. One of his subordinates, employed as a 'chemist' although trained in chemical engineering under H. W. Cremer, reported that Dr Applebey 'maintained that there was no need for such a thing as chemical engineering. All you had to do was put a chemist and an engineer together and there you were'.[21] Another, employed at ICI's Alkali Division in Northwich, Cheshire (originally part of Brunner, Mond Ltd), recalls that during the late 1930s chemical engineers were unheard of there: the company looked upon engineers (i.e. mechanically-trained engineers) 'as just necessary bodies, and their principle was that any new idea which came along, a chemist was always in charge of it, but they always put an engineer with him, right from the testtube side'.[22] This was undoubtedly fair for some projects, which demanded expertise in high-pressure engineering rather than its chemical or process aspects.

Nevertheless, this working culture was not entirely pervasive; a handful of ICI employees promoted a different attitude based on their own experience and training. In 1933, when George Sachs joined the Dyestuffs Division, Huddersfield, there was only one other chemical engineer, trained at MIT. But John McKillop, Chief Engineer of the Dyestuffs Division, was an early and enthusiastic member of the IChemE and employed an increasing number of trained chemical engineers.[23]

It has been widely claimed that ICI, the largest British chemical company, was atypical of business attitudes towards chemical engineers. Yet the case of Albright & Wilson, the Oldbury-based manufacturer of phosphorus compounds, shows a similar lack of take-up on the part of smaller firms. Contracts identifying new employees by job description and professional identity unambiguously map the firm's occupational categories. Founded in 1856, the company by the 1890s was hiring 'chemists' and 'engineers' as foremen to supervise its collection of 'process workers' and 'plumbers'. Plant design was in the hands of 'draughtsmen' supervised by the senior staff and directors. A single 'chemical engineer' was hired in 1902, but there is no evidence that he was seen as distinct from his colleagues.[24]

---

[21] Letter, P. H. Sykes to Johnston, 17 Dec 1996. Sykes completed a PhD in chemical engineering at King's College, London, in 1935.

[22] Interview, Sir Hugh Ford with Johnston, 2 Dec 1997. Ford studied mechanical engineering at Imperial College 1934–9, where he obtained a PhD before joining ICI to develop the first polyethylene production plant. Unappreciated among either the small development team or ICI management was the fact that W. R. D. Manning (1903–84), a Cambridge-education mechanical engineer who designed the pressure vessels, had joined the IChemE in 1928.

[23] Letters, G. Sachs to Johnston, 11 Dec 1996 and 23 Jan 1997. Sachs obtained a chemical engineering postgraduate diploma under M. B. Donald at UCL in 1933. McKillop (b.1882), who had taken a college engineering course, was employed as a draughtsman by a succession of steel companies before joining Nobels Explosives in 1911, where he designed, and later supervised, acids and explosives plants. See AF 60 (May 1922).

[24] BCA MS 1794, Box 50, 'Employment Agreements'; Box 59, 'Notes on Salaries 1915–1923'. Nine 'engineers' (mechanical, electrical and 'research'), nine 'chemists' and two 'draughtsmen' were engaged between 1892 and 1936. The 'chemical engineer' was Edwin Whitfield Wheelwright, who commanded a starting salary of £150 per annum. Such a salary was typical of those given new 'chemists' hired by the firm, but somewhat less than salaries offered to 'engineers'. By 1912 Wainwright earned £600 p.a., typical of Albright & Wilson's senior

While such classifications predate the IChemE itself, similar employment practices were in effect through the 1930s, when the firm acquired other struggling chemical manufacturers. It is also debatable whether the firm's specialisation in phosphorous products can be blamed for ignoring 'generalist' chemical engineers as employees; specialisation was, and remains, the dominant feature of small firms and the divisions of larger companies. The records of one of the acquired firms, Thomas Tyrer & Co in London, mirrored the employee categories of A & W. It employed two chemists and a laboratory assistant, a 'general mechanical plant details, lead burning, etc.' man, a 'works foreman', an 'engineer', a 'bricklayer', a 'carpenter and odd man', a 'stoker', a 'labourer', and 29 'process men and chemical labourers'. By the early 1930s, however, there were hints that a new division of labour was required. The Chief Chemist and Works Manager of Tyrer & Co (itself a title with a contested history in that firm) suggested to the directors

> a matter which has been frequently discussed between us during the last few months, that is, whether real economical advantages would be secured to the firm by the inclusion on the Staff of a member possessing the necessary qualifications, whose duties would consist in the main, of really active supervision of running plant and, as far as possible, all processes in operation.

He recommended a new kind of employee not dedicated to production or a single process, but who

> should theoretically be able to effect, (1) an improvement in the quality of the product; (2) a saving of process time; (3) an insurance against breakdown by a routine survey of operating plant.[25]

The proposed change in operations was sufficiently dramatic that the manager preferred 'that the decision on policy be left to the Board'. Nevertheless, the man they engaged, a 22 year old chemistry graduate, admitted that he had only elementary knowledge of chemical engineering (including 'some practical experience in the use of tools connected with the running of motor engines') but was keen to take 'a proper course of instruction'. Despite their seeking someone with 'knowledge of Chemical Engineering, Chemistry, and allied Subjects', the directors of the firm saw this as consonant with 'a workman ... a control man' willing to accept £3 10s per week.[26]

Thus ICI, far from asserting contrasting attitudes, represented the industrial consensus towards chemical engineering in the early interwar years – indeed, it inherited this perspective from the many constituent firms which it absorbed. Because of its size and organisational inertia, however, it undoubtedly retained a disdain for chemical engineers longer than did many smaller firms, which were more willing to adapt in the face of economic constraints.

----

     managers. These would be some 30 to 40 times higher in 1999 currency. See Appendix for typical salaries.

[25]   BCA MS 1794 Box 110, letter, F. Gosling to F. H. Chambers, 2 Mar 1934.

[26]   About £90 per week in 1999 price-adjusted figures. *Ibid.*, 'Staff and Workpeople'; F. Gosling correspondence re: A. W. Pross.

THE ECONOMIC ENVIRONMENT

What proved to be a difficult employment market for self-described chemical engineers could be blamed on the depression of the early 1930s as much as to problems of occupational colonisation. Indeed, economic causes began with the IChemE itself, which had been founded in a volatile and unpromising business climate. The young Institution's problems had started with the atrophy of corporatist aspirations in mainstream political and industrial circles that had been apparent from the early 1920s. At the war's end, when the Chemical Engineering Group was coalescing, the chemical industry had had reasonably optimistic expectations. The government had promoted British Dyes from 1915, leading to the consolidation of older firms. The explosives companies had similarly merged to form Explosives Trades Ltd in 1918 to become more competitive.

By the summer of 1920, however, the initially booming postwar economy was showing signs of faltering. The nadir came in mid 1921, the year that *The Economist* called 'one of the worst years of depression since the industrial revolution', when industrial production had fallen by nearly 20% and exports by at least 30%.[27] Albright & Wilson, for example, began 'wage reductions arranged by arbitration with [the] Ministry of Labour and the Joint Industrial Council, also by [the] Chemical Employers Federation agreement with Trades Unions'.[28] Most chemical manufacturers were affected. A not atypical situation was that of Alfred L. Booth, a member of the first IChemE council, who stated in his application form that

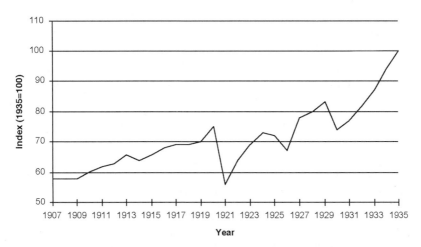

Figure 5-1 Interwar chemical production. *Source*: B. G. Reuben and M. L. Burstall. *The Chemical Economy* (London: Longman, 1973), p. 20, from 1935 Census of Production. See also economic data from Aldcroft, *The Inter-War Economy* (London: Batsford, 1970), 63.

---

[27] The 1921 depression was much deeper than the 1929–32 slump for industry, when production fell by only 11%. See D. H. Aldcroft, *The Inter-War Economy: Britain, 1919–1939* (London: Batsford, 1970), p. 30, and C. L. Mowat, *Britain Between the Wars* (London, 1955), pp. 125–6.

[28] BCA MS 1794 Box 59, 'Notes on Salaries 1915–1923'.

since 1920, owing to the abnormal trade conditions, the works has been practically stopped. I am the only member of the Fuel Staff to be retained.[29]

If such evocative admissions were rare in Institutional correspondence, over the next two years the poor prospects for the chemical industry were increasingly chronicled in the press.[30] By early 1923, there was a partial respite, thanks in part to the Safeguarding of Industries Act; foreign competition was reduced by high trade tariffs and the situation for the chemical industries became more stable.[31] Yet the chemical industry had another downswing in 1925–6.[32] The economic and political turmoil of the formative years of the IChemE are suggested by the fact that three general elections were held between 1922 and 1925.[33]

This thumbnail account must be qualified, however. As John Stevenson and Chris Cook have observed, the Thirties was a decade of incongruities, in which averages present a misleading picture. Exports halved in value between 1929 and 1931; general unemployment exceeded three million by 1933; some towns, dependent on old industries such as coal-mining or shipbuilding, saw three-quarters of their working population out of work. But for those still working (particularly those chemical engineers in Greater London), living standards and consumption rose gradually, leading to the growth of industries to supply them.[34] The chemical trade was, as a whole, less affected by depression than it had been during the brief, but deeper, 1921 recession. Production of rayon, the first successful synthetic fibre, for example, rose from 53 million lbs in 1927 to 115 million lbs by 1939.[35] By contrast, firms relying on older methods suffered.

---

[29]   AF 113, Aug 1922. The dismissal of more junior workers, a common strategy in times of economic hardship, is one explanation for the management-dominated profile of the IChemE membership during the interwar period.

[30]   For example, items in *The Times* on the chemical industry in 1921 exceed 40, a five-fold increase from the previous year. Most reported negative news, e.g.: 'Fall in chemical prices', 1 Jan 1921, p. 374; 'Unsettled conditions', 8 Jan 1921, p. 398; 'Keen foreign competition', 4 Jun 1921, p. 237; 'Idle markets', 2 Jul 1921, p. 317; 'No material change', 12 Sep 1921, 13.

[31]   The 'safeguarding of key industries' (i.e. economic protection of home industries through import duties), starting with discussion of the 1921 Act, was one of the most frequent topics of articles in *Chemistry & Industry* between the wars.

[32]   Aldcroft, *op. cit.*, 29, 47.

[33]   The David Lloyd George Coalition Liberal government, in which W. J. U. Woolcock had served, was replaced by the Conservatives under Andrew Bonar Law elected in November 1922, during which C. S. Garland was a government member. After Bonar Law's resignation owing to throat cancer, a minority Labour government took power under Ramsay MacDonald in the Dec 1923 elections. The government fell after scarcely 10 months and was replaced by Stanley Baldwin's Conservatives in October 1924, which remained in power for the next five years. Months before the economic slump in late 1929, MacDonald led another minority Labour government until the autumn of 1931, when a National government was formed. This administration, alternately under MacDonald, Baldwin and Neville Chamberlain as Prime Minister, falteringly governed the country until the early months of the second world war. See, for example, Mowat, *op. cit.*

[34]   J. Stevenson & C. Cook, *The Slump: Society and Politics During the Depression* (London: Quartet Books, 1977).

[35]   S. Pollard, *The Development of the British Economy, 1914–1990* (London: Edward Arnold, 1992), p. 45.

Scotts, for instance, closed its Silvertown works on the Thames in 1933 and, relinquishing nearly half its holdings to Henry Balfour & Co, moved production to Leven, Fife, where the older firm had a metals fabrication factory. There, they formed another offshoot company (Enamelled Metal Products) based on the technology of an American firm (Pfaudler Corporation), and sought to promote its capabilities in glass-coated vessels for the pharmaceutical and food industries.[36] Castner–Kellner, although opening a research laboratory in 1926, had been over-extended since its first world war production of poison gas, and closed its Wallsend, Gateshead and Glasgow operations between 1929 and 1932.[37]

Despite the repeated patterns of failure, merging and partial revival of such firms, the chemical industry showed general development, largely owing to the existence of the new combines such as ICI, Unilever and Courtaulds.[38] The expansion of chemical installations like those of ICI at Billingham also compensated for the decline of traditional industries in some regions.[39] Despite a 50% over-capacity by the early 1930s, when the Billingham workforce fell by some 7000, the site began the production of new fertilisers, petrol from coal and Perspex.[40] The market-sharing agreements between international producers also helped to cushion over-capacity. By 1933 production was back to 1929 levels, and increasing. Overall, the rate of growth was slightly behind the average for all industry in the interwar period. Internationally, the U.K. chemical industry was relatively stable, losing about 1.4% of world trade between 1929 and 1937.[41]

Institutional responses

The Great Slump quietly curtailed the initially high expectations of the Institution as well as its members. An atypical public admission of problems was the cancellation of a joint meeting with the AIChE – planned for London in 1932 and to have included 160 American delegates – 'owing to the National situation' (and perhaps as much to the deeper American depression).[42] One of the few other direct references to the economic crisis was a special IChemE committee set up in 1937 to promote industry in depressed areas. In its fifteen months of existence, the committee found it 'no easy matter' even to make

---

[36] Henry Balfour & Son had been formed in 1810 in Leven as iron founders, operating in the local coal industry.

[37] Castner–Kellner Ltd, *Fifty Years of Progress 1895–1945, op. cit.*

[38] Aldcroft, *op. cit.*, 199–200. On Courtaulds, see D. C. Coleman, *Courtaulds: An Economic and Social History* (Oxford: Clarendon Press, 1969).

[39] E.g. 'Chemical industry: Bromborough works closed', *Times* 30 Aug 1932, p. 12; the *Annual report of the chief inspector of factories and workshops for 1932* (HMSO, 1933) recorded that of 157 891 factories and 90 859 workshops, there was a net rise of 2537 factories and a net loss of 4855 workshops that year. See also Aldcroft, *op. cit.*, p. 94.

[40] Andrew Pettigrew, *The Awakening Giant: Continuity and Change in ICI* (Oxford: Basil Blackwell, 1985), pp. 123–4.

[41] Aldcroft, *op. cit.*, 249.

[42] Gayfere archive, box XIV/1 'IChemE Joint Activities'.

contact with the various government departments concerned and failed to accomplish anything, an indication of the lack of influence of the IChemE.[43] The trade depression and its repercussions for Institutional business went largely unmentioned in the activities of the IChemE. And there was a notable lack of discussion of what – to the members themselves – must have been glaringly obvious: the indifferent employment prospects for chemical engineers. While there are few Institutional indicators of unemployment, a pattern of inadequate posts, given members' education and training, can be inferred.

Brief references insinuated *inter alia* in Annual Reports and council minutes can be prised out and reconstructed only with difficulty.[44] Most of the infrequent mentions of jobs appeared in relation to the ill-fated Appointments Bureau. The Bureau activities reveal that, unlike educational planning, member employment suffered from short-term forecasting within the Institution. The Bureau itself established the extent of employer indifference to the specialism.

Intersecting Institution and employers: the interwar Appointments Bureau

Because professional identity was partly determined by occupational classifications, the IChemE took some interest in chemical engineering jobs. At the first meeting of Council in 1923, John Hinchley suggested that the Institution might be able to play a role in matching engineers to employers, after he had received a letter requesting help in filling a vacancy.[45] A few months later the council began to search for an Honorary Director of such a Bureau and, in early 1925, appointed H. J. Pooley as the better of two candidates.[46]

By that summer, the first employment offer, a plum £1000 per annum post from the Nitrate Producers' Association in Chile, had been circulated to all associate members; seventy-three members had registered with the Bureau by Christmas.[47] Prospects looked good, despite the temporary business difficulties exacerbated by the 1926 General Strike. Pooley suggested that 'once the coal industry has stabilised, the revival in trade which generally must follow will increase the scope and activities of the bureau. Even in the past stagnant year appointments to the total value of about £16 000 per annum have been made'.[48] A year later, with ninety names on the books – some 20% of the total membership – council heard that there were 'signs that trade was reviving',

[43] Gayfere archive, box III/5, 'Special Areas Sub-Committee'.

[44] For the source of quotations in this section, see IChemE *Annual Reports*, 1926–38, in *Trans. IChemE*, and CM 1926–36.

[45] CM 4 Jul 1923.

[46] Henry Jeffries Pooley (1877–1941) had spent his early career with George Scott & Son in London before beginning consulting in 1924. Besides being a founder of the IChemE, Pooley was a council member of the Association of Chemical Technologists and CEG and General Secretary of the SCI. See AF 68.

[47] The Chilean Nitrate Producers' Association post appears to have gone to M. B. Donald, later lecturer at UCL. No records of the Appointments Bureau are extant, however.

[48] IChemE Annual Report, *Trans. IChemE* 4 (1926), 8. This was about £325 000 in 1999 terms, or the salaries of one to two dozen persons.

and that the bureau would 'doubtless have more practical effect when industry recovers from the unfavourable conditions of the last few years'.[49]

By 1928, though, the problems with employing chemical engineers were beginning to appear endemic. The profile of available chemical engineers seemed to be a chronically poor match for employers' expectations. And Pooley, the mediator between the Institution and industry, was beginning to look for causes other than the general economic climate. The bureau, he reported, was filling fewer posts, and most of them were for foreign appointments. The following year, only 18 posts were filled, most of them junior. But the opposite requirement was blamed in 1930: the results in negotiating even junior appointments were 'not so satisfying' because 'demand noticeably emphasizes a *higher* engineering qualification than the average graduate possesses'.

For the remainder of the interwar period, the story of the Appointments Bureau was a chronicle of unremitting discouragement. In the depth of the Depression, the bureau's achievements contracted further, since 'the prevailing conditions in industry have influenced the volume of vacancies'. In contrast to earlier excuses, Pooley now noted pessimistically that the most junior posts offered were well *below* Graduate and even Student member qualifications (in any event, some junior members, at least, could not be coaxed to pay the annual fee for student membership even when entire classes were herded to apply by enthusiastic lecturers and IChemE officials such as S. G. M. Ure and M. B. Donald).[50] Hinchley claimed more optimistically than Pooley that 'obviously, during a period of trade depression the number of people requiring appointment must increase, yet the Appointments Bureau was functioning successfully, and where there were satisfactory candidates there was little difficulty in finding them suitable posts'. But 1932 was even worse: 'a year of record difficulty', when, 'as a result of the severe depression in industry, very few vacancies have been notified while, on the other hand, a larger number of members than usual is out of occupation from the same causes'.[51] The next year, the Bureau work was 'heavy but unsatisfactory'. Growth of the register of chemical engineers seeking employment was 'not surprising ... in view of the state of the industry'.

As in previous years, Pooley was praised in council as being tactful, sympathetic and energetic in a difficult task. He was, in fact, the fulcrum on which Institution hopes and industrial expectations balanced. The weight, however,

[49] At the end of 1927, there were 214 Members, 168 Associate Members, 61 Graduate Members and 14 Student Members.

[50] Sir Frederick Warner interview with Johnston, 19 Dec 1997. Warner, who like others studied for a chemical engineering diploma at UCL from 1931–4 largely owing to lack of employment opportunities, was unaware of the Appointments Bureau; he joined the local student chemical engineering society (the Ramsay Society) rather than the more costly IChemE. His first post was arranged by Donald. Similar instances of direct intervention by lecturers to find jobs for their graduating students were given by several interviewees graduating through the 1930s, none of whom mentioned the Appointments Bureau.

[51] Numbers were not reported, but it is likely that those registering exceeded 20% of the IChemE membership.

could not long be supported with the added pressure of economic depression. In 1934 there were more vacancies, but 'a very large number of unsatisfied names of members on the register'. In 1935, thirty-six vacancies, mainly junior, were filled, but 'openings suited to men of long and wide experience are still infrequent'; a year on, the number of posts filled was down by a third, even if some employers seemed 'more willing to offer salaries commensurate with the training of a properly equipped chemical engineer'. And in 1937, Pooley reported dismally that only 15 posts were filled because employers still too often required unique experience or exceptional training. The same was true a year later, because employers were still specifying qualifications 'too rigidly'.

No further mention of the bureau is extant.[52] The dissatisfaction of the large fraction of members seeking employment evidently became too much. When the Institution reported Pooley's death in 1942, his twelve years as Director of the Appointments Bureau were left unmentioned.

Extending influence: the 'Underwood–Garland scheme' for industrial development

The Appointments Bureau, mediating between the Institution's ideals for training and industry's lacklustre demands for cheap, specifically trained workers, was never portrayed as a vital service by the IChemE. There is no evidence, for example, that employers' vacancies were used to refine the Institution's training policy or job candidates' expectations. Nor was there much overt advertising to convince employers of the ostensible advantages of a generally trained and broadly experienced chemical engineer.

The Institution did, however, consider a more ambitious – and symbolically ostentatious – policy tying employment to industrial development. Although short-lived, its ultimate rejection by the IChemE council illustrates the aspirations of the early Institution and its slowly clarifying position by the mid 1920s. The episode also marked the first public exposure of A. J. V. Underwood, who was to become the most active protagonist of the profession over the next two decades.

In the Autumn of 1925, Underwood, a 28-year old Associate Member, proposed the formation of a 'Council for Industrial Development'. Trained in mathematics and civil engineering at the University of Leeds, Underwood had worked as a chemist for the Ministry of Munitions from 1916 to 1919, returned to Leeds for a Master's degree, and then practised as a 'research chemist and engineer' in Kenya, and as an independent consultant in Yugoslavia.[53]

---

[52]   The bureau appears to have been quietly wound up by September 1938, when the government requested the Institution to prepare a register of members in case of national emergency. Subsequent operation of such a bureau was precluded by the Institution's recognition as a registered charity for income tax purposes in 1940.

[53]   Arthur Joseph Victor Underwood (born Umanski) (1897–1972) became Honorary Secretary of the IChemE 1937–44, and practised as a consulting chemical engineer specialising in distillation, teaching part-time in the late 1920s and early 1930s at the London Hackney Institute and UCL. AF 274 Associate Membership application (surname Umanski), 1924 and Membership application (surname Underwood), 1929; R. Edgeworth Johnstone, 'Famous men remembered: A. J. V. Underwood', *TCE* Apr. 1985, 66–7; obituary, *TCE* May 1972, 206.

Underwood's proposal had started as a remedy 'for the present grave state of unemployment'. He presented his idea to Charles Garland, the Honorary Registrar of the Institution and Unionist MP in the previous parliament, as a solution hinted at a few days earlier by Prime Minister Stanley Baldwin. Baldwin had suggested that unemployment could be tackled by rejuvenating British industry, making it more efficient and scientifically based to improve competitiveness. Underwood quickly expanded his scheme to be based on an 'Industrial Development Council', envisaged as a non-official body ('to avoid the general prejudice against government control') that might receive government assistance.[54] He reasoned that the only basic remedy for unemployment lay in substantially reducing the costs of production by raising industrial efficiency. Reorganisation, by applying technical knowledge to problems and by replacing obsolete equipment would, claimed Underwood, stimulate the engineering, steel and coal industries.[55] This argument was a particular case of the widespread public rhetoric of the professional identity of chemical engineers being directed at industry and the state at this time.

Underwood suggested that the poor fortunes of chemical engineers were linked to the relatively high unemployment in smaller firms, which had been unable to afford research or replacement of plant during the trade depression of the previous five years. As a solution, he recommended a liberal interpretation of the new Trade Facilities Act to set up an advisory body, associated with the Board of Trade, that would provide loans to industry for new plant.

Chemical engineers were to be among the saviours of industry. For the chemical industry, chemical engineers alone constituted the requisite 'technical and financial experts' who could 'initiate general and particular methods of reducing production costs and achieving greater efficiency'.[56] Underwood proposed that the experts on the Industrial Development Council should be drawn from independent consultants or unemployed senior engineers. Given an official post, they would overcome firms' suspicions of bias, interference and loss of commercial secrets, much as sometime consultant and government alkali inspector G. E. Davis had done two generations earlier.

Thus, Underwood was promoting industrial collectivism mediated by technical experts – a revitalised technocracy with chemical engineers at the helm. The sources of his inspiration were rooted in his own working experience as a Ministry chemist and consultant chemical engineer. He cited the wartime system of inspection and control of factories, arguing that the 'remarkable results that were attained in many cases in reducing costs of production in spite of the general increase in the price of raw materials showed what could

---

[54] Letter, Underwood to C. S. Garland (n.d., circa Sep 1925). This and subsequent references in this section are to the Gayfere archive, box III/5: Miscellaneous Sub-Committees, file 'Industrial development sub-committee'.

[55] A. J. V. Underwood, 'A council for industrial development', unpublished typescript, Sep 1925.

[56] Yet firms were often more conservative than Underwood believed. For example, a letter of recommendation for Ernest L. Randall (1885–1971), member of the first IChemE council, stated 'My regret concerning him is that his ambition has led him to make too many changes'. AF 69.

be effected by the application of scientific method backed by adequate capital'. An example of the currency of this mood is given by Francis H. Carr who, on joining British Drug Houses, warned its chairman:

> Your works ... needs complete though transitional reorganisation ... Due to over-crowding of plant, the dirty and steamy atmosphere, the small-scale and laboratory type of control, I find it difficult from what I know of costing to see how high quality and profit-making can go together in the present set-up.[57]

Underwood went so far as to claim that a spirit of industrial co-operation would be reborn 'when it is generally realised that the present national situation is equally serious'.[58] Nevertheless, he was aware that his proposed advisory council in the postwar world would have to be clearly defined 'to prevent overlapping or friction' with other bodies. He saw particular affinities with the Department of Scientific and Industrial Research (DSIR) and Research Associations. Underwood's proposal raised the stakes for the new Institution from '*individual* health and education' to '*industrial* health and education', and positioned it as a key player beside industry and government.

The proposal found a strong supporter in Sir Frederic Nathan, second president of the Institution, not least because of his 'experience during the war of what can be done along the lines proposed'. He concluded that 'it could be a good thing for the Institution if it takes a lead in the matter'.[59] Nathan, with his extensive experience as an administrator and government committee member, soon extended the analysis, however. He noted the importance of labour organisations, raw materials and general charges on production costs. Sensitive to the boundaries they threatened to transgress, Nathan recommended that the Institution organise discussions with the ABCM, FBI and other bodies 'already dealing with specific aspects of the subject' on the importance of the scientific development of industry.[60]

A special committee was consequently set up by members of council having strong government and regulatory connections: C. S. Garland, W. J. U. Woolcock, Sir Alexander Gibb, W. Macnab and Nathan himself. It quickly became evident that the Institution would have to proceed cautiously in this heady but unexplored territory. Nathan, despite his earlier enthusiasm, back-tracked on whether the Institution could intervene 'on such ambitious lines'.

---

[57] Carr (1874–1969) after a grammar school education and two years as research assistant at the Imperial Institute, had become Chief Chemist for Burroughs Wellcome in 1898 and then a Director of Boots Pure Drug Co. during the first world war, before his appointment as a Director of BDH in 1920. Gaining his chemical engineering experience in the design of production processes for pharmaceuticals (including the first European insulin process), Carr was a founder of both the IChemE and ABCM, and held senior posts in the SCI, RIC, Chemical Society and Society for Analytical Chemistry. Carr papers, ICA B/CARR/3/44 and B/CARR/7.

[58] Underwood, 'A Council', *op. cit.*

[59] Letter, Nathan to Hinchley, 6 Sep 1925; F. L. Nathan, 'Synopsis of Mr A. J. V. Underwood's Memorandum', typescript (n.d.).

[60] F. L. Nathan, 'Note on proposals for industrial development and the application of science to industry', typescript, 27 Sep 1925.

The problem was primarily one of sphere of influence. There was already 'much being written and spoken about the development of industry on scientific lines'. Moreover, competitive interests in industry, and trade union rivalries, threatened to scuttle any co-operative scheme. The committee decided to sound out the DSIR and the Civil Research Committee of the Cabinet before going public.[61] Within weeks, though, the American Secretary of Commerce and engineer, Herbert Hoover, had coincidentally given an address on the same theme, and Underwood and Garland went ahead and contacted newspapers with their proposal, receiving mixed reviews.[62]

At this point, the initial impetus began to scale down. Nathan's discussion with H. T. Tizard, Secretary of the DSIR and Rector of Imperial College, suggested various ways forward, all of them more limited than what the IChemE had proposed. Firms applying for a grant might, for example, be required to furnish a report from the proposed body of experts judging its feasibility, but this would mean waiting for industry to take the lead, and would involve little co-operative participation. Alternatively, the Institution might persuade small chemical firms to place young university graduates in their workshops to suggest ways of increasing efficiency; this, however, would require precisely the industrial influence that the Institution was seeking to generate. A third way, suggested Tizard, might be for the IChemE to circularise the smaller firms to offer investigations of the efficiencies of their factories and management for a small fee. By furnishing certificates of the state of affairs of investigated firms, the Institution might help them secure loans from banks or government. Tizard considered this as the only practical consideration for the Institution. But again, this proposal placed the onus on the weak and little-known IChemE to promote industrial change, a requirement that Council members suspected was impracticable. And the only way of avoiding objections of firms might be to confine such inspections to power generating facilities, a considerable narrowing of the original aims.[63]

Thus, Underwood's grand scheme for reorganisation of the chemical industry had been mutated into a small-scale trial to be conducted voluntarily in small, and probably suspicious, chemical firms. But even with these considerable constrictions, the plan for action ultimately failed within the IChemE itself. The special committee noted that its 'panel of experts' would have to be selected with the greatest of care. As with Pooley and the Appointments Bureau, the scheme would depend entirely on them, and they would require tact and independence from commercial considerations.[64] But the Institution comprised disparate interest groups. Besides the politically-aware and 'reconstructionist'

---

[61] Minutes of special committee, 20 Oct 1925.
[62] 'Industrial waste elimination: "Transformation in U.S."', *Times* 5 Dec 1925; 'The Underwood–Garland scheme', *Chem. Trade J. & Chem. Engineer* 18 Dec 1925; 'A new proposal', *Observer* 3 Jan 1926.
[63] F. L. Nathan, 'Note of an interview with Mr. H. T. Tizard', 2 Nov 1925; minutes of special committee, typescript, 12 Nov 1925; 'Note of an interview between Sir F. L. Nathan, Prof. J. W. Hinchley and Mr. A. J. V. Underwood', typescript, 17 Nov 1925.
[64] F. L. Nathan, 'Industrial development', typescript, 4 Dec 1925.

minority with Ministry experience that made up the Special Sub-Committee, Council was dominated by consultants.[65] No sooner did the summary of the first special committee meeting reach Council than the plan was roundly criticised. The members decided that the proposal would 'interfere so largely with the interests of consulting Engineers as to be unworkable', and argued that such work would be better done by trade or Research Associations.[66] Over the next weeks, Nathan had no choice but to back away from the Underwood scheme; after consulting with the Federation of British Industry, he convinced the council members to pass action to them. The spark of political activism had been damped by the influential, if unrepresentative, consultant minority. In general terms, both the Appointments Bureau and the Underwood–Garland scheme are evidence of the failure of the IChemE to establish itself or its version of a professional identity for chemical engineers during the 1920s.

STATE AND INDUSTRIAL SUPPORT

The widespread sentiments of 'corporate bias' formed after the first world war, and suggested by the Underwood–Garland scheme, endured in modified form through the 1930s. L. P. Carpenter cites several influential individuals as favouring a corporatist state, including the second Lord Melchett, head of ICI; Lord Eustace Percy, Conservative Minister of Education 1924–9; and Roy Glanday, economic advisor to Federation of British Industry. Harold Macmillan and others worked on corporatist themes from 1927 to perhaps 1938, arguing that a series of healthy industries added up to a healthy economy, without the need for co-ordinating machinery. Industries would be made healthy by mergers, acquisitions, cartels and rationalisation.[67]

As epitomised by the Safeguarding of Industries Act, protectionism played an important role in boosting the chemical industry during the early postwar years. The lack of taxation on industrial alcohol, combined with a government subsidy from 1921, encouraged Distillers Company Ltd (DCL) to develop chemical products based on ethyl alcohol. Albright & Wilson's business was similarly aided by the government's imposition of a 33⅓% Key Industry Duty on the import of synthetic organic derivatives in 1932, allowing the firm to move production back to Britain after a 14 year sojourn in Canada.[68]

Closer to home, and shortly after the abortive planning for the Underwood–Garland scheme, Nathan devoted his Presidential Address to 'Industrial Efficiency', and how it could be promoted through co-operative arrangements

[65] The 12 members of council in 1925 included several who were, or had been, consultants: D. Brownlie, W. A. S. Calder, W. B. Davidson, Sir Alexander Gibb, F. A. Greene, H. J. Pooley and H. Talbot. 'Statement of the careers of candidates for election to the Council, third annual corporate meeting, 17th July, 1925', Gayfere archive box II/1, AGM 1923–39.
[66] CM 9 Dec 1925.
[67] L. P. Carpenter, 'Corporatism in Britain, 1930–1945', *J. Contemp. Hist. 11* (1976), 3–25; S. N. Broadberry, *The British Economy Between the Wars* (Oxford: Basil Blackwell, 1986), pp. 160–1.
[68] Fred Aftalion, *A History of the International Chemical Industry* (Philadelphia: University of Pennsylvania Press), 1991, pp. 137–9, 179.

of industry, government and chemical engineers.[69] Prevailing views in the profession and beyond were brought out in annual dinners of the IChemE, which gathered representatives of related bodies and could serve as a platform for planning initiatives.[70] J. A. Reavell, for example, called in his 1930 Presidential address for a partnership of industry and science through educational institutions, and the need for a trusting combination of industry, finance and science.[71] With wartime events behind them and economic failures at the forefront, however, the role of government was noticeably absent from his scheme. At the 1931 dinner, Sir Ernest Benn, J. A. Reavell and W. R. Ormandy reiterated the inadequacy of existing political parties, and the room for one centred, like the current identity of chemical engineers themselves, on questions of economy.[72] The guest speaker three years later was the Minister of Health, who came under fire from Lord Melchett for the lack of a government plan relating private enterprise to national policy. He raised the need for 'a different kind of organization, at present unknown ... applied not only to industry but to finance'.[73] Lord Melchett formed an Industrial Reorganisation League in the summer of 1934, and announced a bill that year in the House of Lords. Macmillan wanted an Enabling Act to permit industries to reorganise themselves. Initially, both industry and labour seemed to favour such organisation; the Federation of British Industry appointed a special committee to consider it, and seven of fourteen trade unions responding to an FBI survey said they needed more power to govern themselves. Nevertheless, opposition prevented the FBI from moving beyond the support of voluntary reorganisation of individual industries. Melchett appealed to the Board of Trade in 1935, but the government backed away from delegating statutory powers to industries to discipline their members, arguing that organised industry would jeopardise the interests of small firms and consumers.[74]

Thus, from the late 1920s the general trend was away from free competition and wholesale reshaping of the political and industrial landscape and towards a growing preference for a 'rationalisation' of the existing order, with the creation of much larger, oligarchical industries.[75] Science as the economic

[69] Sir F. L. Nathan, 'Presidential address', *Trans. IChemE* 4 (1926) 11–7.
[70] For example, the 1934 dinner brought together senior members of the Ministry of Health, National Physical Laboratory, Patent Office, the Wool Industries Research Association, Institute of Fuel, Institution of Petroleum Technologists, Institute of Marine Engineers, Institution of Engineers-in-Charge, SCI, Institute of Chemistry, Society of Public Analysts, British Association of Chemists, Coke Oven Managers' Association, Diesel Engine Users' Association, ABCM, BCPMA, Association of Constructional Engineers, Unilever Ltd, the Dyers' Company, and King's, Imperial and University Colleges, London. *Morning Post*, 17 Feb 1934.
[71] J. A. Reavell, 'The role of science in industry', Presidential Address 4 Apr 1930.
[72] 'Room for a new party', *Daily Telegraph* 7 Mar 1931; *The Times* 7 Mar 1931; *Engineering* 13 Mar 1931, 365.
[73] *The Times*, 17 Feb 1934.
[74] Carpenter, *op. cit.*
[75] See, e.g., John F. Wilson, *British Business History 1720–1994* (Manchester: Manchester University Press, 1995), pp. 141–50 and Leslie Hannah, *The Rise of the Corporate Economy*, London: Methuen, 1976, pp. 36–40, 74–6, 89.

saviour of the nation seemed a natural development in popular culture. J. B. Priestley wrote of one acquaintance with 'one of those enormous foreheads, roomy enough for an Einstein ... capable of rationalising huge muddled industries'.[76] The creation of ICI in 1926 (and discussed at greater length below) was indicative of the period. The new company merged British Dyestuffs, Nobel Industries, Brunner Mond and United Alkali to control a third of chemical production in Britain. Similarly, Unilever was founded in 1929 to merge British and Dutch interests.[77]

These attitudes were as much externally imposed as home grown. Cartels had existed in Germany before the first world war among producers of fertilisers, alkalis, dyestuffs and acids, and were extended internationally in the 1920s. The dyestuffs trade, for example, was apportioned between German, Swiss and French companies in 1928, and joined by ICI in 1931.[78] Albright & Wilson negotiated a monopoly in Britain on phosphorus production, and an international convention on nitrogen was agreed between ICI and three other companies in 1929. Such international cartels maintained cosy relationships and national monopolies for member firms. Nevertheless, rationalisation of the chemical industry was much discussed during the early Depression years as a wider government policy.[79] In 1932–3 the mood for reorganisation spread to the chemical societies and, the following year, to trade associations.[80] By the late 1930s some 1000 to 1200 trade associations existed in the manufacturing sector alone, and multinational firms important to the chemical industry, such as ICI, Unilever and Courtaulds, had networks of agreements across national frontiers.[81] The research associations and industrial cartels comprised a quasi-corporate organisation of the chemical industry, but were more concerned with promoting pragmatic efficiency than any economic philosophy.

Interest groups

Who, then, apart from a handful of individuals with influence, was concerned with the identity of the working chemical engineer and his subject between the wars, besides the Institution itself and educators with a vested interest in promoting 'their' discipline? An indication is given by participation in the 1936 Chemical Engineering Congress, consisting of a joint meeting of the IChemE and AIChE and an exhibition of chemical plant, held in association with the

---

[76] J. B. Priestley, *English Journey* (Toronto: Macmillan, 1934), p. 3.

[77] C. Wilson, *The History of Unilever* (London: Cassell, 1954).

[78] Firms such as Allied Chemical & Dye in America, Swiss IG, and Montecatini in Italy joined consortia in France and Belgium. See L. F. Haber, *The Chemical Industry 1900–1930* (Oxford: Clarendon Press, 1971), pp. 302–18.

[79] E.g. items in *The Times*: J. D. Pratt, 12 Oct 1929, p. 12; In parliament, 4 Feb 1930, p. 8; Sir H. Hartley, 'Chemical industry rationalisation', 9 Mar 1931, p. 14.

[80] See e.g. *The Times* 14 Mar 1934, p. 6.

[81] Pollard, *op. cit.* 75–92.

World Power Conference in London.[82] Eighteen government, industry and
institutional bodies participated, and seventeen British firms provided tours of
their facilities – with the notable exception of all branches of ICI and Unilever.[83]
An American editor observed

> Chemical shows are taken a bit more seriously in England than with us, I gather.
> The formalities of the opening by His Excellency, the President of the Privy Council,
> the Right Honourable J. Ramsay MacDonald, were indicative of the high regard the
> British government holds for such activities. And the eloquent old Scotchman gave
> everybody a real thrill with his appeal for a great new day of cooperation between
> science and industry on the one hand and government on the other. This fervid
> enthusiasm takes me back to our own chemical shows of 1918, '19 and '20, when our
> government was showing like interest in building an American chemical industry ...
> Certainly it makes the Chemical Show something more than a purely commercial
> marketing place for machinery.[84]

This apparent marshalling of support may have been a mirage, however. It is
doubtful that the wide participation represented anything more than a mere
convenience of association between quite disparate trade bodies. The categories
defined by the organisers – the IChemE, ABCM and IGasE – were largely
conventional, and had little explicit linkage with chemical engineering. The
original impetus appears largely to have been triggered by Sir Frederic Nathan's
association with fuel and power issues and his presidency of the IChemE. This
championing of chemical engineering was not repeated before the war and had
few discernible practical consequences.

Thus business interests developed little interest in chemical engineers as
employees, as efficiency experts or as facilitators of profit. For the most part,
the IChemE avoided direct appeals to firms and government, or confrontations
between their expectations and those of employers. It was a safer and more
reassuring policy to plan long-term training syllabi. In this way, results were
more open to interpretation; there was no direct conflict between what was
produced and what was required. This approach, however, insulated the
Council from the reality of the employment market, which found the available
crop of chemical engineers poorly suited to its expectations. By 1935, with the
total membership of the IChemE at about 850, the annual intake, apart from
the small number of university graduates, was only around ten new members
by examination and twenty who qualified by virtue of industrial experience.[85]
The issue of recognition in the *workplace* was largely sidelined for the more
easily addressed problem of recognition in *academe*. The re-emergence of gov-

---

[82] Gayfere archives, box XIV/1: 'IChemE Joint Activities'; *Transactions of the Chemical Engineering Congress of the World Power Congress* (London, 1937), Vol I, lvii-lxi.

[83] Among the largest firms were Distillers Co., Mond Nickel Co., the Gas Light & Coke Co. and Kodak.

[84] Letter, S. D. Kirkpatrick to M. A. Williamson, both of *Chemistry & Metallurgy*, 17 Aug 1936, Brennan archive.

[85] 78 Students, 174 Graduates, 285 Associates, 297 Members and 8 Honorary Members. Graduates of accredited teaching institutions were able to bypass some or all of the Associate Membership Examination.

ernment concerns about preparations for another war, however, was to reverse these priorities.

As chemical engineers had discovered two decades earlier, war changed everything. Indeed, the likelihood of war by the mid 1930s provided new opportunities for the Institution to press its professional claims and negotiate recognition. For the first time since 1918, the Institution was able to invoke the perception that the proficiency of chemical engineers was important – even crucial – for production in an environment of increasingly limited national resources. And unlike the earlier interwar period, when money was short, the state during the approach to war lubricated the economy from its monetary reserves. In this context of limited raw materials but abundant national determination, chemical engineers could nurture their self-portrayal as essential engineers of economy.

Nevertheless, the period in some respects represented a hiatus for the development of a professional identity. While chemical engineers, like those in most other skilled occupations, were almost all employed by the first year of the war, they remained largely undifferentiated from their co-workers having narrower backgrounds in chemistry or mechanical engineering. In the words of F. H. Rogers, President of the Institution in 1940, their professional qualities were 'latent in the community' but not 'officially recognised'.[86] The war put chemical engineering in a holding pattern while instigating new enduring liaisons between industry, the state and the Institution that were exploited only later.

Scale-up of production capacity

The next war began early for chemical engineers. Unlike the educational issues discussed in the previous chapter, which followed a steady course until the autumn of 1939, the industrial preparations for the second world war began a half-decade earlier. Organisation, or at least state scrutiny, of the country's technical capacity and its workforce was undertaken much earlier for the second world war than for the first. Most of the country's chemical engineers helped to design, supervise and populate rapidly expanded factories. The nature and organisation of these sites shaped the opportunities and wartime careers of the chemical engineers who became associated with them. To understand the destinies of the largely invisible chemical engineers during the war, then, it is necessary to continue the examination of their workplaces and employers.

After Hitler came to power and the Japanese occupied Manchuria in 1933, the British government began to take steps to reverse the gradual rundown of armaments provision. At the end of that year, the Supply Board appointed a small advisory group of industrialists to advise it on industrial matters.[87] The

---

[86]   *Trans. IChemE 18* (1940), 19–23.
[87]   Chaired by Lord Weir and comprising Sir James Lithgow and Sir Arthur Balfour, the committee submitted its report in Feb 1934.

great first world war explosives factories of Gretna and Queensferry were gone, having been dismantled and sold for scrap in the early 1920s.[88] The committee suggested that the three Royal Ordnance Factories (ROFs) of Waltham, Enfield and Woolwich would have to be supplemented by the larger engineering firms – those capable of undertaking both design and production – to satisfy plans for rearmament. The Royal Ordnance Factory organisation began to expand in 1934, at a time when public opinion was strongly against private arms manufacture and trading, and when armament firms were not, in any case, inclined to expand without government assistance. The Royal Air Force began to rearm late in 1936, soon followed by the other services. ROF Birtley, a 'last war' factory, was brought back into production, and the explosives factory at Irvine was reconstructed. Between 1936 and late 1938, ten new Royal Ordnance Factories were approved, and another fourteen in 1939. With this state-sponsored impetus, explosives production escalated in the years before the war.[89] Chemical warfare, although never used during the war, also absorbed considerable production capacity. The twelve agency factories approved between 1937 and 1942 were to take a major portion of the factory building programme, their cost exceeding that of all the agency factories for propellants and explosives.[90]

While the scaling up of chemical plants had similarities with that of the first world war, the planning and organisation were distinctly different. So too was the production environment for designers and plant operators a contrast from that of 1914. During the first world war, a strategy for industrial expansion had been rapidly developed: setting up agency factories, equipping engineering firms to employ their own factories, and expanding the Royal Ordnance Factories. Explosives factories had become, between 1915 and 1918, the most visible expression of the powerful Ministry of Munitions. Unlike projectile, heavy shell and filling factories, most of which were managed by industrial firms or boards of management, twenty explosives factories had been under the direct control of the Ministry, and nine under agency management. Thus all explosives factories had been government-owned, and a majority government-managed.

But in the second world war, there was to be no Ministry of Munitions, nor did production escalate as precipitously as during the last.[91] Instead, the state

---

[88]   Of 250 National Factories, only three – a filling factory at Hereford, part of an armament factory at Birtley and an explosives factory at Irvine – were retained through the interwar period as reserve factories. In 1936–7, these became the first of the new Royal Ordnance Factories. See W. Hornby, *Factories and Plant* (London, 1958), pp. 83, 110–20.

[89]   'The Royal Ordnance Factories (Engineering)' (typescript, 1951), *Historical notes on the R.O.F.s, particularly during the second world war*, PRO SUPP 5/1260; M. M. Postan, *British War Production* (London: HMSO, 1952), 10; 36; 392–3. Only two TNT factories were approved in Britain after 1939: an ROF at Sellafield, and an agency factory under ICI. Among the chemical engineers responsible for planning such factories were IChemE members P. J. Hinks, G. S. Witham and A. R. V. Steel.

[90]   Hornby, *op. cit.*, 160–1.

[91]   The production of explosives and propellants assumed less political importance than it had during the first world war, largely because of production forecasting and the employment of continuous processes with a three-shift system (although batch processes were developed to

had the relative breathing space through the second half of the 1930s to scale up munitions production by tapping existing commercial capacity. A system of agency factories joined the ROFs. The agency system, in which private firms managed state-owned factories on behalf of the government, avoided objections of firms benefiting from factory expansion at government expense.

Transatlantic connections

Such agency arrangements of the British government extended to North America. By the spring of 1940, for example, the construction was underway of four large Canadian plants for cordite, TNT and nitrocotton, as well as another seven filling factories. In the USA, some fourteen contracts were awarded by the British government in 1940, mainly to the Du Pont company and its offshoot, the Hercules Powder Company which followed American anti-monopoly legislation, for major explosives plants in Tennessee and New Jersey. The British government thus had a direct managerial and technical link with American production practices, and more particularly with American chemical engineering methods. This was due more to coercion than to a desire for transfer of expertise: Du Pont and Hercules had been reluctant to extend their existing plants until the agency arrangement was offered.[92] Besides their concerns for loss of profit, the image of explosives manufacturers as 'merchants of death', developed by the popular press during the 1930s, explained their reticence to expand. Partly for these reasons, and because the American military was not yet drawn actively into the war, the U.S. government itself was slow to increase its stocks of explosives before Pearl Harbour. Thus only a few small plants were manufacturing explosives and propellant powder in 1940, when British contracts expanded production. But over the next two years chemicals plants were erected across the country in a series of waves. Many of these were organised as so-called GOCO (government owned, company operated) plants, the American equivalent of the agency factory. Despite the relatively high visibility of chemical engineers in America, the chief operating contractors for the Ordnance Department and Chemical Warfare Service – the DuPont, Atlas, Hercules and Trojan companies – found their staffs stretched to breaking point to man the new plants. Other non-explosives manufacturers such as the Quaker Oats Company and Sherwin-Williams Paint Company were drafted in to operate plants, chosen for their reputation of good management, efficiency and financial stability rather than chemicals expertise.[93]

---

be more economic by the end of the war). UK peak output (reached in 1943) scarcely matched that of 1917, but was maintained longer. See Hornby, *op. cit.* 109–10.

[92] H. Duncan Hall, *North American Supply* (London: HMSO, 1955), pp. 104, 290–1; H. Duncan Hall and C. C. Wrigley, *Studies of Overseas Supply* (London: HMSO, 1956), pp. 82–4, 86–8.

[93] Harry C. Thomson and Lida Mayo, *U.S. Army in World War II: The Ordnance Department: Procurement and Supply* (Office of the Chief of Military History: Washington, D.C.), pp. 105–12. On GOCO organisation, see Leo P. Brophy, Wyndham D. Miles and Rexmond C. Cochrane, *U. S. Army in World War II: The Chemical Warfare Service* (Office of the Chief of Military History: Washington, D.C.), pp. 254–8.

Back in Britain, explosives and other chemical factories were to be administered by the Ministry of Supply, established a month before the outbreak of war.[94] From February 1942, the Ministry of Production took over several of its planning functions.[95] Neither the Ministry of Supply nor Production was the equivalent of the earlier Ministry of Munitions. Tasks were to be divided largely between existing firms, some of them American or employing American patents. Chemical Construction Ltd, for example, which licensed processes from American Cyanamid and other companies, built nitric acid and sulphuric acid plants up and down the country. Simon Carves, another large contracting firm, competed with them using a Monsanto process for sulphuric acid production. Even ICI, a relative giant, was quick to exploit existing US expertise. A tetraethyl lead plant constructed by ICI Alkali Division, for example, was 'as nearly as possible, a replica of one that had been built by Du Pont de Nemours in the USA, and they had provided the flowsheets and drawings', as had Dow Chemical for a bromine-from-seawater plant.[96] Other firms, such as Monsanto Chemicals Ltd in Ruabon, North Wales, had become Americanised in the interwar period. What became the Monsanto site had been used since 1867 by the British chemical manufacturer Robert Graesser, and merged with its competitor Monsanto Chemical Works in 1920. 'English staff were gradually reinforced by Monsanto men from the States, the business became more and more American in method and direction and a few years later ... entire control passed into Monsanto hands'. By 1930 its director was an American.[97] When American personnel and consultants at these and similar companies decamped in the first year of the war, British chemical engineers took over roles such as site engineers and shift chemists, responsible for supervising plant construction from existing plans, and then administering the process operation and men.[98]

Allocation of engineers

New factories also were designed and constructed by British firms such as that of past IChemE President Alexander Gibb, which was responsible for eight chemical and ordnance filling factories within the first 18 months of the war. Gibbs' consulting company employed chemical engineers such as Herbert Cremer, who had gained experience in the previous war, for senior administrative roles on their munitions plant contracts.[99] But resources were spread thinly.

---

[94] The Ministry of Supply limited its responsibilities mainly to Army supplies; as in the first world war, the Admiralty controlled naval production. Ordnance factories were under the control of a Director-General of Munitions Production.

[95] E.g. the Supply Depts, together with the Ministry of Labour and National Service and the Board of Trade, were assisted by union and employer representatives in the local organisation of Area Boards, which served as the interface between official policy and actual work. The organisation of these boards later came under the Ministry of Production.

[96] IMechE Hinton, A.2.

[97] Letter, S. D. Kirkpatrick to M. A. Williamson, 19 Aug 1936, Brennan archive.

[98] John Solbett and Sir Frederick Warner interviews with Johnston, 16 Dec 1997 and 19 Dec 1997, respectively.

[99] G. Harrison, *Alexander Gibb* (London: Geoffrey Bles, 1950), pp. 203–5.

The Balfour Scott company, for instance, one of a handful of chemical engineering specialist firms (including APV and Kestners) before the war, found itself
scrambling to fulfil contracts in non-chemical and non-process production:
wings for Hurricane aircraft, anti-aircraft guns, mountings and barges. While
'chemical engineering' work (as defined by the firm) expanded – through
Ministry of Supply contracts for munitions and specialty chemicals, and for
such projects as the design and planning of a never-constructed poison gas
factory – it did not do so in proportion to other war fabrication work. During
the peak wartime employment of some 1500 men (including crews seconded
from various ICI factories), a dozen or so chemical engineers were kept
busy supervising construction and commissioning the chemical plants they
designed.[100]

Besides the erection of new plants, and the scaling up of the production
capacity of existing facilities, the early part of the war absorbed plant personnel
with protective measures. Unlike the first world war, when sabotage by spies
was feared more than the few impotent Zeppelin raids, aerial bombing was
now the primary threat.[101] According to the 'Pink Area' scheme, petroleum
and chemical storage facilities on the south-eastern sides of the country were
shielded and prepared for immobilisation in case of invasion.[102] The Fawley
oil refinery near Southampton, for example, was protected by blast walls,
bomb-damaged in 1940 and 1941 and eventually shut down in 1943 when the
supply of raw materials through the vulnerable sea channels became too
sporadic.[103] Also responsible for a significant expansion in the workforce was
the 'shadowing', or dispersal to less vulnerable parts of the country, of chemical
plants. For instance, as a major constituent of explosives was ammonium
nitrate, several new factories for ammonia and ammonium nitrate were erected
to duplicate the production facilities at Billingham, situated on the vulnerable
north-east coast of the country. Similarly, toluene production, essential for
TNT, was distributed through about twenty small tar by-product plants.

Other production vied with explosives, with which chemical engineers had
become so closely identified during the first world war. Aviation fuel was by
far the largest volume chemical product, with Britain supplying fuel for most
of the Allied aircraft for Europe. As had been the case with munitions, demand
for fuel was grossly underestimated: during the peak of bombing raids, the
quantity consumed was some seven times higher than that envisaged before
the war.[104] ICI was a major producer of aviation fuel, using the technique of

---

[100] Scotts had fully merged with Balfour in 1940, and James MacGregor, its M.D., resigned in
1944. Interview, W. E. Bryden with Johnston, 4 Dec 1997, and W. E. Bryden, *Reminiscences
of a 5/8 Ruddy Plumber, or, Into the Caprolactam* (unpublished typescript, 1994). For an account
of plant commissioning for Balfour Scott in the immediate postwar period, see J. W. Nelson,
Essays of a Miner's Son (unpublished typescript, 1989).

[101] On spy fears at Ardeer, see letter, F. L. Nathan to D. Lloyd George, 27 Sep 1915, Nathan
papers FLN3.

[102] A. Agnew, 'The U.K. petroleum industry in war', *J. Inst. Petroleum 34* (1948), 411–2.

[103] Frank Mayo, *The Beginning of Fawley Oil Refinery* (unpublished typescript, 1996).

[104] Agnew, *op. cit.*, 402–16.

petrol production from the high pressure hydrogenation of coal and creosote oil which they had started using in 1935. A new plant, based on the hydrogenation of petroleum gas oils, was erected at Heysham under government control and operated by Shell, Trinidad Leaseholds and ICI.[105] Billingham and Heysham provided most of the aviation fuel until the end of the war. The Heysham process was particularly important, because it introduced the production of chemicals derived from petroleum and natural gas. Development of these new synthetic processes and fuel refinements required considerable research on plant design (such as furnaces and chemical reactors) and fundamental hydrocarbon reactions.[106]

ICI's war

As suggested by these examples, the principal government collaborator in chemical factory design and production was ICI, largely because of its incorporation of the strategically important Brunner Mond and Nobel Industries combine after the previous war. As the successor to some forty first world war explosives companies, and the major source of expertise in the high-pressure production of ammonium nitrate (with TNT, the most important explosive product of the period), the resources of ICI were inevitably of paramount importance to waging the war.[107] The Explosives Group naturally was busy with the re-armaments programme, and ammonium nitrate plants were supervised by Billingham personnel, but the heaviest load was on the General Chemicals Group which constructed and operated poison gas factories. All ICI divisions participated in the agency work of constructing and operating chemical plants, and a number of senior staff was seconded to the Ministry of Supply.[108]

ICI thus acted as a crucial link between the state and chemical engineers, mediating government directives and engineering actions. Its disposition is significant: this role of state-appointed intermediaries in 'sending demands up, passing directions down' has been identified as a key characteristic of corporatism.[109] In this respect ICI's operations reflected the politics of its formation in

---

[105] The Board came into existence at the height of the Munich crisis in 1938; at the outbreak of war, the oil companies surrendered their independence for the duration.

[106] The British coal-based chemical industry, producing such essential war products as synthetic fuels and rubbers, nevertheless lagged considerably behind developments in Germany, spurred by the goal of industrial self-sufficiency or 'autarchy'. R. Holroyd, 'The acceleration of progress in the chemical industry due to activities during the second world war', *Chem. & Indus.* 14 Jun 1969, 766–70; Guenter Reimann, *Patents for Hitler* (London: Victor Gollancz, 1945).

[107] Postan, *op. cit.*, 177–8, 399; Hornby, *op. cit.*, 107–11. Explosives and propellant production remained under the authority of Royal Ordnance Factories, ICI and various agency factories through the war. The ROFs manufactured somewhat more than half the propellant and less than half the planned explosives production; additional requirements were produced in the North American factories.

[108] IMechE Hinton A.2.

[109] P. J. Williamson, *Varieties of Corporatism* (Cambridge: Cambridge University Press, 1985), pp. 7–12.

1926, when the state had rejected collective and interventionist arrangements for the more easily managed relationships possible with large industrial combines (and which could compete more effectively internationally with new giants such as IG Farbenindustrie).[110] As mentioned above, the organisational structure was quite distinct from that of the first world war, with such chemicals and explosives factories now usually constructed and managed by firms in the chemical industry – a large portion of them by ICI. In the two years before the war, ICI undertook to design and build eighteen factories, and had built 25 by 1945.[111]

The apportioning of responsibility to ICI varied from site to site and with the progress of the war. For example, the mustard gas factory M.S.F. Randle was constructed by the Special Products Dept of ICI (General Chemicals) and operated by a separate 'Z' division of the Castner–Kellner Alkali Company, a part of ICI formed in 1936. Although the chemical plant was erected by direct labour under the supervision of ICI engineers, the technical information for the plant design was obtained from the Chemical Defence Research Establishment. ICI had to submit its designs and drawings to government departments until 1940, when the government liaison was transferred, but afterwards took full responsibility for the design of the plant.[112] Seconded ICI managers complained bitterly of the meddling of middle-level civil service clerks in the early part of the war, and of the 'endless chain of people who had to give approval', longer even than the steep hierarchies within ICI itself.[113]

While agency factories were owned by the state and financed from public funds, they differed from direct government control in an important respect: the personnel employed at such chemical factories were hired and managed by the agent. Because of its agency role, ICI wielded even more influence over chemical engineers than it had in times of peace. This operated in two ways. First, by occupational definition: the company continued to categorise its personnel as either 'chemists' or 'engineers', not 'chemical engineers'; second, by exclusion: ICI shaped the work force by relegating to peripheral industries those chemical engineers who could not be so divided. Employment as 'shift chemist' or 'general engineer' in a small chemical firm was thus a common wartime assignment for members of the IChemE. The many self-professed chemical engineers actually employed by ICI – some 9% of corporate members of the IChemE in 1936, and rising during the war – were selected for their collection of individual skills rather than any professional cachet. Thus the factory designs, operating methods, status and conditions of employment of many chemical engineers were determined largely by the principal agent, ICI.[114]

---

[110] Reader, *op. cit.*, 475–7; Reimann, *op. cit.*, 82, 128, 131.
[111] Reader, *op. cit.*, Vol 2 (1975), 256, 300, 473. See also Postan, *op. cit.*, 387–8, 433–4. All but one of the ammonium nitrate factories were agency factories managed by ICI.
[112] F. H. Bramwell, 'History of ICI's war effort: General Chemicals Division' (typescript, 2 Oct 1944), PRO SUPP 5/1003.
[113] IMechE Hinton A.2.
[114] Hornby, *op. cit.*, 156–7.

PROFESSIONAL IMPLICATIONS OF THE WAR

If the occupational status for IChemE members was often prosaic, the preparations and subsequent waging of the war provided occasions for slights and successes for the wider ambitions of the nascent profession. There was, however, an interregnum in executing more ambitious professional plans. In the first weeks of the war, Council suspended all standing committees except the Nomination Committee: significantly, only the accreditation functions of the Institution were deemed indispensable. Many of its members were immediately transferred to Government work, while others had Territorial and other military commitments; a few younger members were called up for active service. The embargo on public assemblies announced by the government caused an abrupt end to ordinary meetings (although these resumed in a more limited form in the 'quiet period at the opening of 1940').[115] Forward planning at the Institution also ceased: H. W. Cremer dropped his work on an 'ideal' undergraduate course syllabus as well as plans for National Certificate courses. And the expansion of teaching was truncated abruptly; in fact, teaching programmes were to constrict and retreat in the face of bombing and shortages.[116]

Besides triggering *ad hoc* responses to unpredictable events, the hostilities opened opportunities for the Institution, particularly in achieving professional recognition. State support was nevertheless dilatory, and the status of chemical engineers in industry went largely unacknowledged. Only with difficulty was the wartime context employed to further the ambitions of the IChemE's leadership. Officers of the Institution, however, were arguably more visible and influential during this period than at any other. Two episodes illustrate the difficulties encountered in their attempts to win official recognition for the profession: labour classification and state bursaries.

---

[115] *Trans. IChemE 19* (1941), 17, 19–24. Throughout the war, joint meetings were held with the Chemical Engineering Group of the SCI, and various regional meetings of the Graduate & Students group continued. Most authors had to withdraw offers of papers, or adapt to more relevant topics such as 'Industrial protection from air attack' and 'Ventilation and the blackout'.

[116] Deciding to keep the London headquarters, C. J. T. Mackie (Secretary from 1924 until 1950 (d. 1952) undertook Air Raid Precaution (ARP) duties. While the Institution's premises survived with only broken windows, aerial bombardments in 1940 killed John Hinchley's widow and damaged the Chemical Engineering Department at UCL, which had been the home of the first undergraduate degree course for scarcely three years. The Ramsay Laboratory of Chemical Engineering was destroyed by another raid in May 1941, closing it for the duration of the war. Bombing of UCL had been widely expected because of its proximity to London's St Pancras, Euston and King's Cross train stations. In anticipation of bombing, the six students of Donald's 1940–1 class (three of them from ICI, two Leverhulme trust students, and one Czech evacuee) undertook all their practical work before theoretical studies, in case their equipment should be destroyed mid-way through the course. The Institution's printers were also bombed and burned out that year, along with stocks of the *Transactions. Quart. Bull.* Jan. 1941; IChemE Annual Report for 1941; J. Solbett interview by Johnston, 16 Dec 1997; R. S. Tailby interview by Divall, 29 Jun 1992; H. W. Thorpe, 'Fifty years of chemical engineering at the Ramsay Laboratory. 1916–1939: The early years', *TCE* Jun 1974, 383–5 and A. J. Carter, *TCE* May 1978, 409.

Availability of manpower

Just as the government shaped pre-war manufacturing capacity, so was the workforce organised. As early as 1936, the Committee of Imperial Defence at Whitehall had begun taking an interest specifically in the availability of chemical engineers in Britain. In March, the IChemE President, Herbert Levinstein, and Underwood, the new Joint Honorary Secretary, met with the committee and promised them a memorandum. Underwood later vaunted the IChemE as having been the first engineering institution 'to take up with the government the role to be played by technical people in the event of an outbreak of war'.[117] Levinstein had had direct experience of chemical engineering during the Great War: as Managing Director of Levinstein Ltd, his expanded dye-making company had manufactured explosives and mustard gas, and in the months following the war, he had been part of the British mission examining German chemical factories.[118] War planning was clearly on his mind in 1936 as he drafted a Presidential address on the subject of 'The chemical engineer in face, and in times, of emergency'.[119] The five-page committee memorandum to the Committee of Imperial Defence correspondingly drew upon 'the lessons of the great war' which demonstrated 'an immediate need for soundly-trained chemical engineers'.[120]

Embarrassingly for the identity of their profession, however, Levinstein and Underwood cited ICI, a company that recognised no chemical engineers as such, as 'the first source to be tapped'. Nor was there any obvious surplus of chemical engineers upon which to draw. From the 460 corporate members in Britain at the end of 1936, the authors listed a preponderance in the fuel industries, general chemicals and plant manufacture, most of whom would have to remain in place in the event of war.[121] At that time of expanding war preparations without an immediate urgency, they stressed the centrality of education for generating more 'scientifically trained chemical engineers', and of the Institution itself in setting standards. Underwood went further, however:

---

[117] 'Seventeenth Annual Meeting', *Trans. IChemE* **17** (1939), 8.

[118] H. Fletcher Moulton, *The Life of Lord Moulton* (London: Nisbet, 1922), 224–61; H. Hartley, *Report of the British Mission Appointed to Visit Enemy Chemical Factories in the Occupied Zone Engaged in the Production of Munitions of War* (Min. of Munitions, 1919).

[119] Gayfere archive box XIV/1, file 'Committee of Imperial Defence, 1936–8'. The Presidential Address was eventually retitled less provocatively 'Chemical industry and the outlook for Europe' [*Trans. IChemE* (1937), 12–6], but remained concerned wholly with recalling the experiences of the first world war to promote central organisation of the chemical industry in preparation for the next.

[120] Letter, H. Levinstein to C. Longhurst, 19 Nov 1936, and 'Memorandum on chemical engineers available in times of National Emergency', p. 1. Both from Gayfere archive box XIV/1.

[121] The distribution of members was 39 at ICI, 27 in government service, 15 at Unilever and 24 in teaching. 103 members were employed in other gas and fuel industries, 80 in general chemical manufacture, 55 in plant design, 30 as various types of consultants, 17 in textiles, 15 each in the metal and food industries, 9 in mineral oils, and a further 42 'men of wide general experience' who were unclassified. Of members within the listed industries, 18% held administrative posts and 17% were engaged in research and design. Thus two-thirds were classed as relatively low-status plant operatives.

he made clear his own vision of a hierarchy of relevant professions. Appending a paragraph to Levinstein's draft, he pointed out that 'the large number of chemists engaged in teaching' should be considered 'a potential source of men suitable for employment in minor chemical engineering positions' only if they were given vacation training. Chemists, and particularly those without industrial experience, were to be considered only as a last resort.[122]

For Levinstein, then, the coming war offered an opportunity to press educational programmes and assure full and appropriate employment of his members. But for Underwood, by contrast, it signalled the chance to press jurisdictional authority with an attentive sponsor. With Underwood's additions, the conflict over the role of chemists and chemical engineers in industry was brought to government attention – if only fleetingly – for the first time.

Professional classification

By 1938 the threat of war was more apparent, but the role of chemical engineers still was not. Some members, such as 36 year old A. C. B. Mathews, were eager to be employed to best advantage, and wrote urgently to the IChemE:

> I am keen that my capabilities be used and ... am nearly signed on irrevocably with the R.N.V.S.R. [Royal Naval Voluntary Service Reserve]. Before it is too late I would like to know whether the Institution is running an active list of members and arranging their allocation.

Underwood replied reassuringly – some eight months later – that 'I feel it would be perfectly reasonable to do the same as in the last war, that is to say, if a special occasion arose in which your services were required in some technical occupation, arrangements could be made for your release'.[123] Underwood could do little more; the Institution's offer to the Committee of Imperial Defence 'to carry out any work in the national interest' had not been taken up by them.

The need for a list of chemical engineers, however, was recognised by the government late in the summer of 1938, when the Ministry of Labour sent a request to British engineering institutions to survey their members 'in case of national emergency'.[124] Spontaneous offers of service such as Mathews' flooded in to the Ministry after the Munich crisis in September. Underwood accordingly

---

[122] Underwood stands out as an alert guardian of professional boundaries. A year later he grumbled that, in a joint pamphlet on engineering institutions, 'chemical engineering has been rather reduced to a subsidiary part of gas works operation'. Letter, Underwood to H. Beaver, 16 Nov 1937, Brennan archive.

[123] Letter, A. C. B. Mathews to H. J. Pooley, 25 Jan 1938; Underwood to Mathews, Sep 26 1938, Gayfere archive box XIV/1. Amphlett Christopher Bucklett Mathews, having received his chemical engineering diploma under Hinchley and becoming a Graduate member from 1926, described himself as a chemist in charge of a trials laboratory and paint shop owned by ICI. When he became a full Member in 1943, he was still at ICI Paints.

[124] The Ministry had invited university teachers and researchers to register in May, to which some 3000 responded. See H. M. D. Parker, *Manpower: a Study of War-Time Policy and Administration* (London: HMSO, 1957), Chap. 19: 'Professional and Scientific Manpower'.

drafted notices that month to Institution members requesting that they register information on details such as experience, salary, availability and preferred location of work. On a wider front, the government drew up a Handbook of National Service detailing the skilled occupations most needed. The Handbook was distributed to British households, and backed up by a broadcast speech by Prime Minister Chamberlain and a rally at the Albert Hall in January 1939.[125]

This incipient channelling of patriotic enthusiasm soon foundered on details, however. According to the Ministry's classification scheme drafted the previous summer, chemical engineers were to be categorised as a sub-class of 'chemists', not as autonomous 'engineers':

> Chemists have been divided into four broad groups: (a) agricultural chemists, (b) research workers in pure chemistry, (c) chemists and chemical engineers in industry, (d) teaching.[126]

Underwood reported his discussion with the Ministry of Labour to the IChemE Council in January. Despite further discussions, the situation was no better by April. While the government continued to neglect chemical engineers, other professions did not hesitate to usurp the occupational niche they had staked out. The Institution of Mechanical Engineers instructed its members that those engaged in the design and construction of chemical plant should be classified within a category under *mechanical* engineering. Combined with the Ministry's ruling that chemical engineers engaged in plant operation be grouped together with *chemists*, Underwood pointed out, once again, that 'chemical engineering would seem to be divided between chemistry and mechanical engineering', and protested that IChemE members would be forced to register either as chemists or mechanical engineers.[127] Neither provided an acceptable demarcation of the chemical engineer's unique role. 'This classification seemed to suggest that a chemical engineer was a rather secondary variety of industrial chemist', recalled Underwood later.[128]

Council members interpreted this pigeon-holing as a dangerous precedent threatening to subjugate the profession permanently. The President, William Calder, Charles Garland and Underwood met with Industrial Chemistry Committee at the Ministry of Labour in early May 1939, but the Committee confirmed its previous decision about the place of chemical engineers in the Central Register, simply dividing one of the sub-headings into two.[129] The

---

[125] Parker, *op. cit.*, Chap. 3.

[126] Letter, Min. of Labour to IChemE, Aug 1938, Brennan archives, box 2 of 2.

[127] CM 19 Apr 1939.

[128] A. J. V. Underwood, 'Reflections and recollections', *Trans IChemE* 43 (1965), T302-T316; quotation p. T308.

[129] The committee comprised F. H. Carr (Chairman of BDH); W. Cullen (IChemE President, 1937–8); C. H. Desch, Metallurgy Dept, NPL; H. J. T. Ellingham, Imperial College; Dr J. J. Fox; Prof C. N. Hinshelwood, Oxford University; L. H. Lampitt; W. H. Mills; Sir Robert H. Pickard; R. E. Slade (ICI Research Controller); and several representatives of the Treasury and Ministries of Labour and Supply. There is no evidence extant that either Carr or Cullen championed the status of chemical engineers from within the committee, nor that they were lobbied by the IChemE council. Nor can we be at all certain of Carr's commitment to the IChemE (although he was an ardent promoter of chemical engineering as early as 1909): his

IChemE representatives were angered by the response. An entry in the minute book, subsequently crossed out, records their conclusions that:

> ... the Industrial Chemistry Committee hold the opinion that chemical engineers and chemists were interchangeable and that, as in the two separate groupings originally proposed, many of the sub-headings were identical or similar, it was expedient that only one professional group should be established. The Industrial Chemistry Committee were in no way interested in the past activities in this matter of the Institution.

Council was unanimous that the 'proposed method of classification was contrary to the national interest and unacceptable to the Institution'.[130] It recommended an approach to the Ministry of Labour to transfer chemical engineers to the Engineering Committee of the Central Register, which they now felt was a more appropriate body than the Industrial Chemistry Committee. A month later, in a special meeting, Council further advised Underwood to talk to the 'Mechanicals' to gain their support for a definition of chemical engineers as a distinct engineering profession.[131] There was thus an explicit identification of chemistry alliances as dangerous and engineering liaisons as serviceable, perhaps because of the preponderance of designers and consultants over academics in Council. This seed bore fruit: in a series of interviews and correspondence with other engineering organisations and the Ministry that Spring, Underwood succeeded in having the Registry classification amended to move the profession to the general engineering section of the Register. The re-categorisation was publically and rather theatrically trumpeted as a victory for the profession.[132] The grudging official acceptance was nonetheless in marked contrast to events in America, where the government had accepted chemical engineering as a distinct specialisation without debate.[133] Interestingly, too, the British War Office requested that its own criteria for classification be accepted by the IChemE: Council consequently agreed that the Army Special Certificate would allow admission to IChemE Graduate Membership.[134]

Applying such classifications proved difficult in practice. While the Ministry of Labour officers responsible for the Registry understood the nuances of skilled industrial trades, they had difficulty assessing the qualifications of professional applicants such as chemical engineers.[135] The following year the register, originally planned only for Corporate members, was expanded to include Graduates

---

papers dealing with Institutional matters were destroyed by a V-2 rocket. PRO LAB 8/156; Carr papers, ICA. B/CARR/6.

[130] CM 8 Apr 1939.

[131] CM 8 May 1939.

[132] E.g., *Chem. Trade J. & Chem. Engineer 104*, June 30, 1939, 601–2. See Brennan archive box 2 for congratulatory messages.

[133] National Roster of Scientific and Specialized Personnel, *Description of fields of specialization in chemistry and chemical engineering*, (Washington, D.C., 1944). See also F. J. Van Antwerpen and S. Fourdrinier *High Lights: the First Fifty Years of the American Institute of Chemical Engineers* (New York: AIChE, 1958), 117–20.

[134] 'Annual Report of Council', *Trans. IChemE 19* (1941).

[135] Parker, *op. cit.*, pp. 319–22.

because the government wanted to reserve specialised manpower from the age of 21.[136] M. B. Donald, Joint Honorary Secretary of the Institution, reported that 'quite a number of members' were on the Central Register but others not, because 'a surprisingly large proportion ... were already engaged on work of national importance'.[137] Donald's surprise suggests not only that the occupational fate of the IChemE membership was largely unknown to the Council, but that expectations were not high. Indeed, the Central Register was nothing like an employment exchange. With most of the registrants already in posts, and with some major employers like ICI not employing many, the Register proved an ineffective vehicle for employing chemical engineers.[138] Many fell between the cracks. S. R. Tailby, for instance, a 22 year old chemistry graduate and interrupted UCL PhD student in 1939, found himself turned down for general military service by an allocation board meeting at Imperial College as too highly trained, but soon became frustrated by the similar lack of offers from industry. Others, such as John Solbett, an Imperial College chemical engineering graduate originally from Bucharest, found the Ministry of Labour and National Service (MLNS) merely excluded him from sensitive posts, rather than helping him to find one.[139] F. H. Rogers, then IChemE President, noted that, despite the Institution's efforts, he 'did not think the Central Register had been used to any great extent so far as the Institution members were concerned and would have liked it to have achieved a great deal more'.[140]

But who were these members for whom the council sought occupational recognition? While the war produced a discontinuity in employment and status for chemical engineers, its effects on the rate of application and profile of new members of the Institution was more subtle.

The fraction of self-described chemical engineers was rising steadily. And while 'chemists' were still prominent, unspecified 'engineers' had almost disappeared compared to pre-war definitions. Evidently engineers had realigned themselves as 'chemical engineers', while chemists within the IChemE preferred

---

[136] Yet Herbert Levinstein had protested that inexperience precluded any 21 year old from being in any sense 'indispensable to industry', and that reservation would lead to a class of unjustly privileged young men. Indeed, he argued that no man below 25 could be called a 'chemical engineer' according to his Institution's membership policy. What was needed in the event of war, he argued, were experienced charge hands and foremen, a point disputed by both a *Times* editorialist and a government spokesman. *The Times* 23 May 1939, 12.

[137] M. B. Donald (1897–1978) taught at Imperial College for the remainder of the war following the death of S. G. M. Ure and the bombing of UCL in 1941. He was a founder member of the Institution, and helped edit the first issues of its *Transactions*. Joint Honorary Secretary of the Institution between 1937 and 1949 with A. J. V. Underwood, he engaged in explosives development during the war. See, for example, 'Memoirs of the early days', *TCE* Apr 1972, 136; 'Some memoirs of the early days', *J. Ramsay Society (UCL) 20–21* (1973), 75–90; J. Mullin, 'Famous men remembered', *TCE*, Jul/Aug 1985.

[138] 'Notes on the operation of the chemical sections of the chemical register of the Ministry of Labour and National Service' (1940), PRO LAB 8/156.

[139] S. R. Tailby and J. Solbett interviews by Johnston, 22 Oct 1997 and 16 Dec 1977, respectively.

[140] '18th Annual Meeting', *Trans. IChemE 18* (1940), 19–23.

[141] Sample: 70 of a population of approximately 325 new corporate members. For comparison with other periods, see Tables 4-2, 6-1, 7-1 and 8-1.

Table 5-1 New corporate membership by occupation, 1940–5[141]

Self description	
Chemical engineer	23%
Chemist	39%
Engineer	4%
Manager	23%
Educator	1%
Consultant	3%
Technologist/assistant	4%
Metallurgist	1%
Sales representative	1%

to differentiate themselves as a distinct category. This suggests a shifting of hierarchies, with 'chemists' and 'chemical engineers' vying for highest status.

Qualifications continued to rise gradually for new members. Over two-thirds (69.6%) now held at least one degree, and one in five had received a Master's or Doctorate. The war altered the geographical distribution of new members, too. Recruitment from overseas fell by a third, and significant numbers of chemical engineers moved to the British midlands and north-east to man wartime industries.

The demand for engineers and scientists generally continued to far outstrip supply. Enrolment of chemical engineers on the Central Register became compulsory in mid 1940. Members such as Norman Swindin, who had not felt 'able to agree to enrolment' earlier, were now asked by the MLNS to specify their competence in 24 different occupations ranging from teaching to the rubber industry.[142] In March 1942, however, the Institution's efforts to employ the Register as a vehicle for professional recognition were short-circuited: reservation by occupations was discontinued and replaced by reservation by individuals. The old Central Register was to be reserved for only 'highly qualified engineers and scientists', while a new Appointments Register dealt with those 'whose scientific attainments were not of a sufficiently high standard to justify their enrolment on the Central Register'.[143] As an occupational group, chemical engineers were implicitly, seen as neither highly qualified nor in short supply. It was deemed suitable by the MLNS, therefore, that the classification of 'chemical engineer' disappear, leaving only case-by-case evaluations.[144]

But reports of the misuse of chemical engineers persisted. Donald complained, later that year, that there was a continued lack of contacts between the universities and industry owing to government inaction: the forms filled in by university

---

[142] Letter, IChemE to N. Swindin, 19 Jun 1940, N. Swindin papers, box D 2/4. Swindin (1880–1975) was a founder member of the Institution and long-time supporter of the British Communist Party. See his autobiography *Engineering Without Wheels* (London, 1962); F. E. Warner, 'Portrait of an individualist', *New Scientist* 20 Sep 1962; D. Freshwater, 'Famous men remembered', *TCE*, Nov. 1984.

[143] Parker, *op. cit.*, 322–3.

[144] '20th Annual Meeting', *Trans. IChemE 20* (1942).

staffs and commercial firms for the Central Register had not been followed up. Pre-existing research teams were split up 'and each individual had to look for himself'.[145] Official recognition of the profession was more a case of benign neglect than active promotion.

State bursaries

Besides continued uncertainty in the government recognition of working chemical engineers, there were problems with the support of training. In January 1942, G. W. Himus, the successor to S. G. M. Ure at Imperial College, complained to the President of the Institution:

> A member of the College staff, who I believe is on the Joint Recruiting Board, alluding to the general policy of that body, stated that the general policy of the Board was to differentiate between students whose training fitted them for immediate and useful participation in the war effort, and those who, when trained, would not be immediately required ... Chemical Engineers were associated with Geologists, Mining Engineers and Petroleum Technologists, since it was suggested that our course in Chemical Engineering dealt primarily with the *design* of chemical plant, and the present need was for men to operate plant already erected.[146]

Physicists and radio engineers, by contrast, were both in constant demand and recognised by the state as key personnel.[147] A government view that graduates in chemical engineering were not continually required may explain the fact that State Bursaries were not available to students of Chemical Engineering. With a distinct professional identity indiscernible within chemical plants, chemical engineers became only sporadically perceptible to government eyes as plant designers.

Even if chemical engineers were not explicitly valued, workers as a whole within the chemical industry were. A corporatist arrangement of representatives of employers, trade unions and government Labour Supply Inspectors was instituted to form regional Labour Supply (Chemical Industry) Committees to regulate the supply of labour. It quickly became clear that the chemical industry was to be a special case, and its essential employees were kept unavailable for reclassification unless justified on health grounds.[148]

Following urging by Himus, Prof. Egerton argued to Lord Hankey that chemical engineers were essential to the execution of the war. The policy on bursaries was reversed.[149] Even so, Donald continued discussions with government departments through mid-1943 for the granting of State Bursaries, and a deputation of Institution officers finally wrung the award of six postgraduate

---

[145] Discussion, H. W. Cremer, 'The development of new chemical processes', *Trans. IChemE* 20 (1942), 38–43.

[146] Letter, G. W. Himus to C. S. Garland, 14 Jan 1942, Brennan archive, box 2. Himus, a fuel technologist more than chemical engineer, had been a Council member since 1932.

[147] Parker, *op. cit.*, 325–7.

[148] E. Bevan, *Parliamentary Debates 398*, 15 Mar 1944, 260.

[149] Letter, G. W. Himus to C. S. Garland, 14 Jan 1942, Brennan archive, box 2.

bursaries for Imperial College from Lord Hankey's Technical Personnel Committee.[150]

The allocation of a mere six State Bursaries, and these only for postgraduate courses, was protested to no avail. The IChemE, Institute of Petroleum, ABCM and BCPMA remonstrated to the Ministry of Labour and National Service in 1945 that these were 'out of all proportion to the relative importance of Chemical Engineering'.[151] Despite, or perhaps because of, such lack of official concern for educational provision, an increasing number of chemical engineers was being trained via other routes, notably by correspondence courses and evening classes. The most significant correspondence course – the Technological Institute of Great Britain (TIGB), founded in 1929 as a means of preparing for the AMIChemE examination – was producing a significant increase in candidates for the IChemE examinations during the war.[152]

INDUSTRIAL PERCEPTIONS OF IDENTITY

The battles over the National Register and State Bursaries were symbolic, and widely recognised as such. An editorial in *The Chemical Trade Journal* argued:

> ... it must be remembered that systematic chemical engineering training in this country is still in its formative stages and plastic enough to be influenced by the views not only of the Council of the Institution but of practising chemical engineers and industrialists as a whole.[153]

But the views of industrialists threatened to mould and set this malleable profession into the wrong shape. Large firms proved to be a chronic source of difficulty for professional recognition. As with its powerful role in employment and job classification, ICI was influential through its public portrayal of chemical engineers. A speech at a joint meeting of the IChemE and Chemical Engineering Group on 'The future of the chemical industry' by Lord McGowan, Chairman of ICI, was pounced on by chemical engineers in the audience as slighting their profession.[154] While admitting the 'vital importance' of chemical engineering, McGowan continued:

> There is, however, perhaps less unanimity on the subject of whether chemical plant problems either of construction or operation can best be dealt with by a specially trained chemical engineer or by a team of two, an engineer and a chemist working together ... I think it probable that in the case of the smaller manufacturing units

---

[150] CM 16 June 1943, 15 Dec 1943. The deputation consisted of F. A. Greene (President), S. Garland (past President), H. Griffiths (Vice President) and M. B. Donald (Joint Secretary). See also 'Policy regarding the allocation of State bursaries' (1941–4), PRO LAB 8/538.

[151] Letter, IChemE to R. W. Luce, 24 Mar 1945, Gayfere archive box XIV/1 file 'Supply of C.E.s 1939–1946'. The IChemE supplied representatives to a Joint Committee of the Ministry of Labour during the war years.

[152] A founder of the TIGB, John H. G. Plant (b. 1906), had studied chemistry, metallurgy and chemical engineering at the University of Birmingham and worked at British Celanese and the Gas, Light & Coke Co. until the Depression. AF 876.

[153] *Chem. Trade J.* **104** (1939), 4.

[154] Lord McGowan, 'The future of the chemical industry', *Trans. IChemE* **21** (1943), 25–9.

whose technical resources are limited and where one man only can be allocated to a problem, a chemical engineer will be of vastly more value than either a chemist or an engineer. On the other hand where circumstances warrant it, and this applies particularly to the larger organizations, the ideal arrangement may be the employment of an engineer along with a chemist.[155]

Underwood, ever the guardian of professional identity, protested, this time to the new President, F. A. Greene, with a detailed attack on the ICI philosophy:

Lord McGowan ... not only expressed the view that both alternatives were equally satisfactory but, in addition, seemed to indicate that the employment of a chemical engineer instead of a chemist and a mechanical engineer was a course likely to be adopted by smaller firms. The inference is that chemical engineers are only employed as a sort of second-best by firms which cannot afford to do the job properly by employing two persons.

It seems to me that the statements made by Lord McGowan must almost certainly have an adverse effect on the Institution's activities, particularly those which are concerned with promoting chemical engineering education. The facilities for proper chemical engineering training at universities in this country are notoriously meagre. It will be a most effective argument against the provision of further facilities to put forward Lord McGowan's views that there is really no necessity for training chemical engineers if the facilities for training chemists and mechanical engineers are adequate. If chemical engineers are only required by the smaller industrial firms, the corollary is that the Institution's activities in promoting chemical engineering are of a semi-philanthropic character for the purpose of helping the poorer members of the industrial community rather than a real contribution to technical development.

... The employment of a chemist and a mechanical engineer instead of a chemical engineer is entirely comparable with the employment of a physicist and a mechanical engineer instead of an electrical engineer. Such conceptions are far from novel. In fact, about a hundred years ago it was similarly maintained that there was no such person as a mechanical engineer and that all that was needed was a combination of a civil engineer and a mechanic. The Institution of Mechanical Engineers was formed to disprove this thesis and its growth is adequate evidence as to which view was correct.

I feel that this matter is one of considerable importance to this Institution which, after all, owed its formation very largely to this painful realisation during the last war that a combination of a chemist and a mechanical engineer was far from being a suitable substitute for a chemical engineer. If you agree, I would suggest that this letter might be brought before the Council at its meeting next week to provide an opportunity of discussion whether any action might appropriately be taken by the Institution.[156]

The letter has been quoted at length to illustrate the trends in thinking that were to dominate Institution efforts after the war, pressed by its most ardent spokesman; namely, a need to stabilise the official recognition of chemical engineers in active opposition to prevailing beliefs; a requirement for significantly expanded education; and a change in occupational status for working

---

[155] McGowan, *Proc. Chem. Eng. Group* 25 (1943), 27.
[156] Letter, A. J. V. Underwood to F. A. Greene, 13 Oct 1943, Brennan archive, box 2 of 2.

engineers. Such privately-voiced concerns were little match for the persuasive rhetoric of the influential Lord McGowan.

PLANNING FOR A POSTWAR WORLD

As in the first world war, prospects for reconstruction were actively discussed. This time, however, the impetus for reorganisation came less from the Institution than from government. An over-arching postwar plan for the chemical industry was not evident; industrial planning generally was much more of a piecemeal and contested affair than it had been at the end of the last war. Nor was the significance of chemical engineering in the winning of the war so widely acknowledged as it had been a quarter century earlier. Chemical and the still secret nuclear engineering development activities at ICI, while promising to expand the *occupation*, did little to promote a *profession* of chemical engineering.[157] And the precarious government recognition of chemical engineers threatened to be toppled once wartime demand slackened. Officers of the Institution were therefore active in promoting their current status and eventual postwar role.

Two years after Pearl Harbour but with still two years of hostilities ahead, Underwood wrote to the Ministry of Labour, stressing that chemical engineers would be essential to the 'active survival' of a postwar British chemical industry:

> There is striking and depressing disparity between the conditions in this country and in the U.S.A. where facilities for training chemical engineers, the number of qualified chemical engineers trained annually in approved courses and the relative numbers of chemical engineers employed in industry are on vastly greater scale than in this country. I have no hesitation in saying that proper facilities for training chemical engineers in this country are totally inadequate.[158]

His new theme of the need for government-supported educational facilities was to be repeated and elaborated over the next decade.

Presidential addresses, usually reported in the general press, also had a rhetorical role in vaunting the accomplishments, future requirements and professional aspirations of British chemical engineers and the Institution – along with the more general themes of British adaptability and national fortunes.[159] F. H. Rogers warned in 1941 that war emphasised the centrality of technical

---

[157] Nuclear engineering is discussed in the next chapter.

[158] Letter, Underwood to F. M. H. Markham, MLNS Central Register, 6 Dec 1943, Brennan archive, box 2.

[159] On such wartime themes, and relative dearth of media portrayal of engineers during the second world war, see Angus Calder, *The Myth of the Blitz* (London: Pimlico, 1991). By early 1945, for example, with the end of the war clearly in sight, IChemE President F. H. Greene could begin publicising the war-time record of chemical engineering by citing his own involvement with a large underground factory, a story receiving enthusiastic coverage in London newspapers. Perhaps because the processes and location (later identified as Corsham, Wiltshire) remained secret, the stories promoted themes of British resolve more than the indispensability of chemical engineering. F. A. Greene, 'A by-way of chemical engineering', *Trans. IChemE 23* (1945), xii–xv; 'Secret plan beat the Blitz', *Evening News*; 'Hush-hush factory was built in a mine', *Star*; '80-acre factory under the ground', *Evening Standard*, all 13 Apr 1945.

advancement for survival, and that 'in the postwar period, our weak spots will be probed ... and commerce is war in trade'. He predicted a dire postwar situation in which run-down and corroded plants, superseded by new technical processes, could not be replaced owing to excessive taxation – a Britain, by inference, in which chemical engineers would be in great demand but economically out of reach. With nice irony, he advocated national self-sufficiency, 'free of unwise governmental and political interference', but combined with protectionism in the form of import barriers to provide the financial environment in which chemical engineers could solve national problems.[160]

Oliver Lyttelton, Minister of Production, had similar predictions but a different message when he addressed the IChemE Annual General Meeting in 1943. Success of an impoverished postwar Britain in export markets would rely upon using the 'resources and possibilities' of the Dominions and Colonies effectively, expanding the British chemical industry on its historical basis of coal, and by 'vigorous planning' of all aspects of the chemical industry on a 'national scale'. This would require industry to be 'prepared to co-operate without reserve, and to foster a closer relationship with the Universities and with the Government Research organisations', an objective on which, claimed Lyttleton, the Institution had been founded after the previous war.[161]

At the same meeting, Charles Garland, post-first world war politician and now President, also addressed reconstruction problems. He cited the approaching crisis of access to raw materials, complicated by new national boundaries and demands for colonial independence, and looked to chemical engineers to provide a new efficiency and economy of manufacture. Garland, a founder of the CEG and IChemE after the first world war during a period of optimism about corporatist solutions, now criticised national planning. In a chemical industry expanding organically, he asserted, 'planning can only mean damming the springs of inspiration and the rivers of progress – stultification and decay'. Moreover, claimed Garland, chemical engineering naturally demanded a system of payment by results rather than subservience to restrictive bureaucratic control. But the dangers of postwar prediction are illustrated by Garland's professed doubts that 'military explosives not dependent on nitric acid could be produced', 28 months before the public announcement of the atomic bomb.[162]

Frank Greene, another founder member who reached the Presidency, used his Presidential address the following year for more direct proposals. Urging closer links with universities, as already existed with the University of London, he called for the support of industry and the state in endowing university chairs of chemical engineering, and extension of the award of the 'war-time measure' of State Bursaries.[163]

[160] F. H. Rogers, 'Some post-war problems', *Trans. IChemE 19* (1941), xiii–xvi.
[161] O. Lyttleton, speech to IChemE AGM, 2 Apr 1943, Gayfere archive box II/2 'Annual General Meetings 1940–1958'.
[162] C. S. Garland, 'The chemical engineer in reconstruction', *Trans. IChemE 21* (1943), xiii-xvi.
[163] F. A. Greene, 'Our title: a reminder', *Trans. IChemE 22* (1944), xii-xv.

On the side of the manufacturers, the BCPMA was busy discussing the disposal of government surplus chemical plant, just as the government itself had been doing some 27 years earlier. The Association also proposed, as had the ABCM in the previous war, investigating chemical facilities in Germany after its occupation, and urged a postwar embargo on German manufacturers and the confiscation of German plant 'to make good bottlenecks in ... home productive capacities' and to 'secure freedom from aggression and increase our exports to the extent required to pay for the imports that are vital to the maintenance of our standard of living'.[164]

The predictions of the Institution officers, industry and government contacts revealed a near consensus on Britain's dire economic position at the end of the war. They differed dramatically, however, in technical predictions and proposals, and in evaluating the outcomes of the policies they promoted.

Despite its arguably successful wartime activities, the Institution was more often responding to outside events than acting decisively on its own initiative. Government and industrial recognition for the profession had been tentative, grudging and transitory. The small but important concessions – occupational classification and the provision of a handful of government bursaries – had been won with the support of allies at the IMechE, ABCM and BCPMA. And as in the final year of the first world war, when something approaching a consensus grew concerning the future of technical education and industry, 1945 arrived with chemical engineers eager to take advantage of larger-scale events. But still a small organisation maintained by honorary and elected officers leading a largely passive and unacknowledged membership, the IChemE had neither the resources nor the internal pressure to provoke change unaided.

---

[164] BCPMA, 'Disposal of government surplus chemical plant', C.159, 17 Oct 1944; 'Enemy chemical plant industries', C.164, 7 Dec 1944; 'Reparations', 14 Jan 1946, all in N. Swindin papers, box 44. On the occupation and dismemberment of the pervasive I. G. Farbenindustrie conglomerate, for instance, see Ernst Bäumler, *A Century of Chemistry* (Düsseldorf: Econ Verlag, 1968), pp. 1–8, and Tom Bower, *The Paperclip Conspiracy: The Battle for the Spoils and Secrets of Nazi Germany* (London: Paladin, 1987), esp. pp. 192, 220, 232.

CHAPTER 6

RAPID EXPANSION

During the decade and a half after the second world war, British chemical engineering was transformed. The chemical and process industries evinced an unprecedented expansion, and chemical engineers found new and more senior places within them. The discipline became firmly rooted in universities and colleges, increasing the number of degree-trained chemical engineers ten-fold. With this increased visibility, the profession and its Institution were both elevated in status and shifted decidedly towards engineering alliances. The scope and length of this chapter therefore reflect the importance of this period.

The most direct source of these changes was the altered postwar and post-colonial environment for industry, education and organisations. The war had forged new relationships between the state, industry and technical organis-ations; had encouraged the mobility of technical workers and hence a transfer of expertise across such domains; and, had introduced technological change by funding new development while bottling up the consequent commercial pressure by wartime production controls. Moreover, the six-year pause in higher educa-tion generated its own pressure, which the state was anxious to channel into technical education. Chemical engineers and their Institution consequently scrambled to participate in a rapid postwar expansion of the chemical industry and to colonise new occupational territory in other process industries. Some of that territory had lain vacant and unclaimed long before the war; other domains had been revitalised by imports; yet others had been newly opened as a consequence of wartime research. Although penetration of the profession into these areas was uneven and largely uncontrolled, for the first time it gave the specialism an industrial acceptance approaching its academic claims.

EXTENDING DOMAINS

Chemical engineers after the war found posts readily in expanding industries, particularly petrochemicals and synthetics. Plant contracting consequently swelled, too, as a specialist occupation.

The explosion of the petrochemical industry

The most significant new sector – in terms of new working methods, posts for chemical engineers, and economic expansion – was the petrochemical industry. During the wartime emergency, the refining of petroleum on a very large scale had developed rapidly. The growth of the heavy organic chemical industry based on petrochemicals was partly the result of the state's fiscal policy, and the government took an increasing interest in the workforce responsible for such newly vital industries. On both economic and technical grounds, the refining of petroleum and the manufacture of petrochemicals were best undertaken as continuous processes, and the Institution and others claimed that the training of a chemical engineer was admirably suited both to the design and the operation of the new plant.[1] In the period 1948–52, about half of all graduates in chemical engineering in Britain consequently entered the petrochemical firms, and by 1957, this sector recruited a higher proportion of chemical engineers to other qualified scientists and engineers than any other industry.[2]

If the employment boom was a postwar phenomenon, it had prewar origins. Undeniably, however, the production of 'petrochemicals', or chemical products based on petroleum-based raw materials or feed stocks, developed more slowly in Britain than in America, where oil had been pumped since the late nineteenth century.[3] Britain had no known natural reserves of petroleum, and developed its sources of supply largely from its possessions in Burma and, later, in what was then Persia.[4] The conversion of the British naval fleet from coal to oil before the first world war impelled further development. The first petroleum refinery in the country, distilling petroleum from Borneo, was set up in 1915 on the Welsh coast, having been transferred from Rotterdam and recommissioned in the short space of nine weeks.

Post-first world war expansion of refineries was cautious and diffident. One of the few British refineries was set up at Fawley, near Southampton,[5] and another was constructed at Llandarcy, Swansea, from 1917 and commissioned in 1921. From the following year it supplied most of the products sold by

---

[1] *Heavy Chemicals: A Report of a Productivity Team Representing the British Heavy Chemical Industry Which Visited the United States of America in 1952* (London: British Productivity Council, 1953); J. A. Oriel, 'The Chemical Engineer in Plant Operation and Management', *TCE* Apr 1955, xxxviii-xlii; R. H. Simpson, 'The chemical engineer in a medium-sized company', *TCE* Oct 1960, A45–A49; J. M. Coulson, 'Engineers and human service', *TCE*, Jul/ Aug 1973, 360–6, 389; W. S. Norman, 'Some aspects of progress in chemical engineering', *TCE* Jun 1974, 391–4.

[2] A. N. Holmes, 'Process economics in the chemical industry', *TCE*, Aug 1950, A47-A52; John Oriel, 'Petroleum and the chemical engineer', *Trans. IChemE* 35 (1957), 174–80; 'The supply and distribution of chemical engineers in Great Britain', *TCE*, Oct 1958, A41-A45; J. T. Davies, 'Chemical engineering education for the future', *TCE*, Sep 1965, CE232.

[3] On the rise of the American specialism, see E. W. Constant II, 'Science in Society: Petroleum engineers and the oil fraternity in Texas, 1925–65', *Soc. Stud. Sci. 19* (1989), 439–72.

[4] T. A. B. Corley, *A History of Burmah Oil Company 1886–1924* (London: Heinemann, 1983); D. Yergin, *The Prize* (London: Simon & Schuster, 1991).

[5] Frank Mayo, *The Beginning of Fawley Oil Refinery* (unpublished typescript, 1996).

British Petroleum, which until then had been obtained from the USA. Although then developing a cracking process at their research laboratory at Sunbury upon Thames, the company purchased an 'off the shelf' process from the American contractor M. W. Kellogg, and another from the American Universal Oil Products, in 1927. At Sunbury, staff had risen from two in 'a dingy cellar laboratory' in 1917 to 46 by end of 1924, although results were handicapped by 'the lack of accurate instrumentation, the absence of standardised testing procedures and the inadequacy of chemical engineering to permit the controlled degree of experimentation which was needed'. Such limitations were also felt by workers studying cracking at other firms, such as Dr H. M. Stanley at the laboratories of Distillers Company in Epsom during the early 1930s.[6]

The Twenties nevertheless saw petroleum refining transformed from an art to a science. According to two American commentators, 'chemical engineers discovered the petroleum industry and the industry discovered chemical engineers'.[7] While this was primarily an American phenomenon, new members of the IChemE employed in the industry shot from none in 1925–9 to some 25% during the following five years. Nevertheless, the Institution shared such specialists with the more occupationally-focused Institution of Petroleum Technologists, formed in 1914.[8] The huge demand for aviation fuel during the second world war decisively expanded the capacity of the oil industry in Britain. After the war, collaborative ventures of oil companies sponsored new refineries on British soil, notably the BP Isle of Grain refinery on the Thames estuary, the Shell Stanlow refinery, and the extension of the Esso refinery at Fawley. The bulk of the capital investment was made between 1949 and 1953.[9] British companies were spurred to further develop indigenous petroleum refining after the war by the 'dollar gap': prewar oil and petrol had been purchased largely from American suppliers. Mobil, for example, which faced a rapidly rising demand for its lubricants, was compelled to construct a refinery at Coryton, along with a distribution network for the petroleum that was its by-product.[10] Largely as a result of American investment and plant contracting (described below), the British petroleum industry boomed after the war. The end of petrol rationing was another important factor. Where three-quarters of the refined petroleum had been imported in 1939, Britain by 1954 had eight major and

[6]   R. W. Ferrier, *The History of the British Petroleum Company* (Cambridge: Cambridge University Press, 1982), Vol. 1, pp. 453–5, 459, 483.

[7]   W. A. Myers and J. C. Martin, 'Chemical engineering and petroleum', in: W. T. Dixon and A. W. Fischer (eds.), *Chemical Engineering in Industry* (New York: AIChE, 1958), pp. 80–91.

[8]   *Petroleum: Twenty-five Years Retrospect: 1910–1935* (London: Institution of Petroleum Technologists, 1935). Membership in that institution reached some 1300 members by the mid 1930s, when the IChemE membership was about 800 in all grades.

[9]   By 1956 there were 15 refineries in the UK. During that decade, production increased from 1488 thousand tons per year to 28681 tons. D. Burn, 'The oil industry', in: Duncan Burn (ed.), *The Structure of British Industry* (Cambridge: Cambridge University Press, 1958), Vol. I, pp. 156–217.

[10] Ralph Harris and Arthur Seldon, *Advertising in Action* (London: Andre Deutsch, 1962), p. 244.

four smaller oil refineries, producing 38% more than needed for national consumption, and making it the fourth largest refiner in world.[11]

Following closely behind the expansion of oil refining was petrochemical production. Where before the war the European annual production of petrochemicals was only 10 000 tons (scarcely 0.5% of America's output), by 1950 Britain alone was producing 10.4 million tons.[12] Until the war, the British chemical industry had relied on fermentation alcohol, coal and tar derivatives for organic chemical production. But wartime demand and new products altered priorities. ICI, previously a general chemicals manufacturer, had become a major producer of wartime aviation fuel at Billingham.[13] ICI's oil hydrogenation facility, and another set up by the ICI/Shell/Trinidad Leaseholds consortium at Heysham in 1941, had provided the basis of a petrochemical industry. Shell also began producing a detergent based on 'wax cracking' the following year.[14] After the war, ICI sought associations with petroleum refiners to provide its feed stocks for nylon and polyethylene. Similarly, Distillers Company Ltd (DCL) found sugarcane molasses to be increasingly uneconomical as a raw material for its solvents; it associated with Anglo-Iranian Oil Co to form British Petroleum Chemicals in 1947.

Petrocarbon, a company formed after its directors had built petroleum refineries in Italy, Belgium, Spain, Portugal and Sweden before and after the war, decided in 1945 to build the first large European petrochemical works, or 'chemical oil refinery', in Manchester. But chemical engineers, as such, were not the first choice as designers of such facilities (for instance a physicist who had researched gas separation processes, Martin Ruhemann, was engaged to develop a plant to produce ethylene, propylene and other hydrocarbon fractions). The production of chemicals that could be built up from hydrocarbons was complex and relatively new. The oil was separated into various components by conventional distillation and 'cracking', or thermal dissociation by a series of furnaces, to yield a series of lighter compounds, which were separated, reacted with other compounds and combined to synthesise new chemical products. Consequently, unlike earlier British refineries, which typically employed one or two site chemists to lay out and survey the works, Ruhemann was part of a team of chemists, process engineers and oil consultants from several European countries.[15] The works, by then termed a 'petroleum-chemi-

---

[11]   J. H. Dunning, *American Investment in British Manufacturing Industry* (London: Allen & Unwin, 1958), pp. 61–2. The raw material for a mere 11% of all organic chemical production in 1949, petroleum was used for 87% in 1968. B. G. Reuben and M. L. Burstall, *The Chemical Economy* (London: Longman, 1973), p. 33.

[12]   This increased to 30.3 million tons by 1960 and 116 million by 1970. Willem Molle and Egbert Wever, *Oil Refineries and Petrochemical Industries in Western Europe: Buoyant Past, Uncertain Future* (Aldershot: Gower, 1983), pp. 16, 56.

[13]   Sir Andrew Agnew, 'The U.K. petroleum industry in war', *J. Inst. Petroleum 34* (1948), 402–16.

[14]   Peter H. Spitz, *Petrochemicals: the Rise of an Industry* (New York: John Wiley, 1988), pp. 364–7.

[15]   The manager, Henry E. Charlton (b. 1903) had designed oil recovery and refining plants with his employer and later partner, George Moore, from 1928. He had begun his career as a works engineer, having trained part-time in electrical and mechanical engineering. AF 2693.

cals plant' was commissioned in 1950 to become the first significant petrochemical producer in Europe, although never a great commercial success.[16]

The development of new petrochemical products transformed the chemical market, as much by replacement as by expansion. The business in solvents from molasses, for example, was undermined by the development of alcohol from ethylene. The alcohol market itself eroded in the early 1950s. Similarly, the dehydration of alcohol for use as motor fuel – an important process in the early years of the war – came to an end. The coke oven industry withered with the employment of oil to replace coke in blast furnaces. The restriction of the coal and steel industry, through the 1950s, contributed to a further loss of markets.

Petroleum refining and petrochemicals began to transform occupational opportunities within a few years. Major firms investing in this new form of organic chemicals manufacture included ICI, British Hydrocarbon Chemicals (a joint venture of BP and DCL), Courtaulds, Shell, Monsanto and DCL. Through the 1950s the petrochemicals industry assumed an increasing importance for chemical engineers, promoted by such men as J. A. Oriel, President of the IChemE in 1957 and a Vice President of the Institute of Petroleum, who had spent his career at Shell and was to engineer its funding of a chair of Chemical Engineering at Cambridge.[17] A 1958 'Chemical and Petroleum Engineering Exhibition' – the first dedicated exhibition since the prewar World Power Congress, and sponsored jointly by BCPMA and Council of British Manufacturers of Petroleum Equipment – assembled, said an editorialist for *Chemistry & Industry*, 'the largest gathering of chemical engineers this country has known'.[18]

IChemE members in the petrochemical and refining industries rose from 10% in 1950 to some 17% by 1963. According to J. B. Brennan, then General Secretary of the Institution, chemical engineers typically made up one worker in ten in the petroleum industry, working beside four chemists and three mechanical engineers.[19] Yet professional identity gained no firm foothold in these developments; chemical engineers did not stand out among their colleagues. British Petroleum, for example, which employed 62 IChemE members

---

[16] M. Ruhemann, 'Early memories of Petrocarbon' (unpublished typescript, c1983); 'First in Manchester' (unpublished typescript, 3 Feb 1986); H. E. Charlton, 'Chemical engineering aspects as applied to the building of a petroleum-chemicals plant', paper to North-Western branch of the IChemE, 28 Apr 1951. On Ruhemann (b. 1903), see 'A physicist discovers chemical engineering', *TCE* Dec 1988, 94–5.

[17] A graduate in chemistry and mathematics from the Universities of Cardiff and Cambridge, John Augustus Oriel (1896–1967) worked at the Central Laboratories, Shell Haven Refinery and Suez Refinery as chemist, works chemist and manager, respectively. As manager of the Shell Refining and Marketing Co. from 1938 he extended the Stanlow and Shell Haven refineries and established the Thornton Research Centre. H. Hartley, 'The John Oriel memorial lecture', *TCE* Sep 1969, 138–45. F. Mayo, 'The petroleum industry', *Symposium on Chemical Engineering Education* (London: IChemE, 1957).

[18] *Chem. & Indus.* 10 May 1958, 537.

[19] J. B. Brennan, 'Chemical engineering manpower in the chemical industry'. *TCE* Apr 1961, A44-A45; IChemE, *Careers in Chemical Engineering* (1961).

by 1963, continued to attribute its plant designs to 'chemists' rather than 'chemical engineers' in its literature.[20]

Synthetics

Like petrochemicals, the production of synthetic materials had prewar origins but came to employ a significant fraction of chemical engineers after the war. Rayon, based on cellulose, had been available from the 1890s and manufactured according to methods, and manpower, largely familiar to the traditional textile industry. Nylon, however, developed during the 1930s by Du Pont in America, required production processes drawn from the chemicals industry. The preparation of the first large-scale production plant for nylon required some 230 Du Pont 'chemists and engineers' at various stages. The process was licensed to ICI in 1939, and it established a new joint company, British Nylon Spinners, with Courtaulds that year.[21]   While nylon production had substantially American roots, its contemporary, Terylene, was a development of ICI. After development and appraisal work in the late 1940s, ICI set up a polymer-producing and spinning operation at Wilton, Yorkshire. The product had limited success. A contemporary economist observed that

> [o]ne cannot help suspecting that the main factor holding back the growth of both nylon and terylene output has perhaps, apart from the postwar British slowness in constructional engineering, been the fact that ICI as a whole is growing so rapidly.[22]

Yet ICI was the place to be for innovatory processes. Its numerous divisions had commercialised a spectrum of new synthetic materials, from low density polyethylene in 1939 to Terylene fibres (1949), Teflon (1950) and polyacrylonitrile fibres in 1960.[23] Chemical engineering problems included heat transfer (because the reactions could proceed rapidly), filtration and mass transfer. The need for continued and intensive physical research was obvious: the characteristics of melt flow were distinct and complex for each new material, and the design of process equipment often a black art. Nevertheless, many such synthetic materials could be processed on equipment adapted from that developed earlier for the rubber industry for mixing, rolling and extruding.

As with petrochemicals and the contracting industry (discussed below), foreign – particularly American – influences were important. Besides the licensing of products, American branch operations during this period served as a conduit for the transfer of technology and design expertise. One of the first British

[20]   British Petroleum Ltd., *Our Industry: An Introduction to the Petroleum Industry for the Use of Members of the Staff* (London: BP, 1958); H. Longhurst, *Adventure in Oil: The Story of British Petroleum* (London: Sidgwick & Jackson, 1959).

[21]   Douglas C. Hague, *The Economics of Man-Made Fibres* (London: Gerald Duckworth & Co, 1957), pp. 93–9. The nylon polymer produced by ICI was manufactured originally from benzene, a coal-tar derivative, although American plants from 1945 began producing it from the naptha fraction derived in distilling crude oil.

[22]   Hague, *op. cit.*, 125.

[23]   On polyethylene at ICI, see Spitz, *op. cit.*, 257–65 and interview, Sir Hugh Ford with Johnston, 2 Dec 1997.

polystyrene plants, for example, was a direct American import. The mechanism of such technology transfer followed an established pattern. Monsanto Chemicals Ltd in London acquired details of existing polystyrene plants at Springfield, Massachusetts and Texas City, Texas by sending a chemical engineer from the Chemical Engineering Group at their London office, who collected information for an adapted British version from the civil and mechanical sections of the Monsanto engineering department. Unlike its wholly British counterpart, ICI, Monsanto England had a chemical engineering group from the late 1940s. By the early Fifties it employed about a half dozen chemical engineers, all with university training in the subject. As with the polystyrene plant, most facilities that they built were adapted from the technology provided by the Monsanto American operations. From relatively sparse designs – sometimes 'a couple of sheets of paper and an instrument control diagram', reported one engineer – some modest scaling up or scaling down might be undertaken, followed by the specification of locally available equipment and construction details. Significantly, the working methods of the British chemical engineers were substantially the same as those of their American counterparts owing to frequent interactions. Engineers travelled between both countries to communicate design details and to resolve commissioning problems.[24]

Silicone production followed a similar path. The Ministry of Supply in 1943 invited the American Dow-Corning Corporation, which had been formed that year from Dow Chemical Company and Corning Glass to promote the development of silicones, to meet with the British firm Albright & Wilson Ltd. A&W became their agents, and in 1950 the two companies formed a new joint company (Midland Silicones Ltd), which built a silicone production plant in Barry, South Wales, on the site of a former magnesium-from-sea-water plant. Again, expertise was acquired in the form of 'know-how' picked up by British chemical engineers and chemists visiting the American facility.[25] With the postwar lack of resources, ingenuity in using available materials was important. The chemical engineer used existing foundations and a boiler plant for the new process, and added second-hand tanks and a steam engine operated in reverse as a steam compressor.[26] Yet pragmatic engineering was combined with innovation, particularly in the employment of automatic controls.[27]

This overview of extending postwar industries does not accord a prominent role to chemical engineers. This is not an oversight. While increasingly employed in such industries (and even forming the nucleus of a few industrial groups, such as at Monsanto) few were singled out as representatives of a new occupational class of specialist.

---

[24] C. S. H. Munro to Johnston, 16 Jan 1997. Munro (b. 1913) studied chemical engineering at Imperial College in the early 1930s under S. G. M. Ure.

[25] The chemistry of silicones had been elucidated at the University of Nottingham nearly a decade earlier.

[26] H. D. Anderson to Johnston, 17 and 25 Jan 1997. Anderson, joining as a student in 1938, was President in 1976.

[27] D. W. F. Hardie and J. Davidson Pratt, *The Modern British Chemical Industry* (Oxford: Pergamon Press, 1966), pp. 214–5.

The rise of plant contracting

Despite indicating a continued existence in the professional shadows, the cases above hint that chemical engineers were beginning to provide a technical competence recognised in certain spheres, particularly in the design of plant for new processes. Indeed, the employment of chemical engineers in general plant contracting was an occupational niche that developed significantly after the second world war. Here, they could argue more persuasively that they had unique skills beyond mere adaptation and extension. Contracting was, to a large extent, an outgrowth of the independent consulting work dabbled in by so many early chemical engineers such as George Davis, John Hinchley, William Macnab and Hugh Griffiths.[28] There were qualitative changes, however. First, before the second world war, such consulting had been small-scale and short in duration, usually supplementing more regular income; Hinchley, for example, complemented his 'bare living' as a teacher with consulting work, but was bitter about the low rates paid.[29] Davis, before him, had relied on publishing and commercial chemical analysis as much as consulting for income; the consulting side of the business inherited by his son Keville gradually decayed. Macnab undertook his sporadic explosives consulting in office space provided by J. Arthur Reavell, who himself had his Kestner evaporator company as major source of income.[30] Secondly, 'contracting' before the war was principally by firms which also fabricated equipment – Kestner, for example, and vessel specialists APV. Expertise in plant design was frequently found in such men associated with, or owning, plant manufacturing firms, and consultants often acted in conjunction with such suppliers. Griffiths, for instance, supplemented his teaching at Battersea Polytechnic by acting as a representative of German plant equipment manufacturers. Percy Parrish, another founder member of the IChemE active between the wars, similarly consulted while acting as an agent for equipment suppliers while employed by South Metropolitan Gas in London.[31]

Importantly for professional status, the postwar years saw a growing distinction between consultants, plant contractors and equipment manufacturers. On the one hand, the full-time consultant as specialist chemical engineer appeared. Firms such as Manderstam & Partners, and Cremer & Warner, joined the few established consulting firms such as Alexander Gibb & Partners that had occasionally taken on chemical engineering problems along with larger civil or

[28] Davis Bros., among the earliest practising consultants in 1890, were unusual in publicising their fees: one guinea for consultation; reports or investigations from five guineas; specifications for 5% of the plant cost, or at least 25 guineas; and supervision for 5 guineas per week. Davis archive, Science Museum library, DAV 3/4-1. The guinea, as indicator of Davis's professional pretensions, distinguished him from mere manufacturers and managers paid in pounds.

[29] E. M. Hinchley, *John William Hinchley: Chemical Engineer* (London: Lamley & Co., 1935), p. 43.

[30] J. A. Reavell, *The Kestner Golden Jubilee Book 1908–1958*, (London: Kestner Evaporator & Engineering Co., 1958), p. 8.

[31] Parrish had been trained at various technical colleges and spent his early career at Yorkshire chemical companies. AF 247; S. A. Gregory, 'Hugh Griffiths and consultancy', *TCE* Oct 1986, 3.

mechanical engineering projects before the war. On the other hand, plant contractors increasingly developed generalist expertise and dissociated themselves from direct agency status for specific equipment manufacturers. A handful of 'pure' contractors existed before the war – Woodhall Duckham (specialising in coke ovens), West's Gas Improvement Company (building retorts for gas production plants), and Simon-Carves, for example. Such firms, while usually having an assembly shop and a pool of labour and site engineers, did not supply material or equipment.[32] Consulting engineers had been active since the nineteenth century, although arguably less in the chemical industry than in mechanical and civil engineering. Consultants and contractors now positioned themselves more clearly between their customers – the chemical producers – and equipment suppliers.[33] Roles also became more sharply defined, with consultants typically undertaking a study of an aspect of design, while contractors more frequently engaged in over-all design work, the supply of the process technology, procurement, plant commissioning and sometimes operation. This distillation of tasks was quite different from that in the general construction industry, where design and construction duties traditionally were separate. For the new chemical engineering contractors, conceptual design, project engineering and administrative co-ordination often merged, with the hardware purchased from specialist equipment manufacturers.[34] Both contractors and consultants found competition from in-house engineering groups, which, after the war, still continued to take responsibility for their companies' plant designs and extensions but with an expanded importance and at an increased pace.[35] Many chemical-producing firms discovered that designing and producing their own complex, large-scale chemical and petroleum plants burdened them with a surplus of engineers. The case of Petrocarbon, mentioned above, is a good example of a successful solution: the company was conceived as a separate business entity distinct from its partner, Petrochemicals, formed to operate the plants it designed.

The change in the nature of chemical engineering contracting was partly due to the working practices developing among the American consultants and plant contractors coming to Britain to support the growing petrochemicals industry. The development of new process technology, such as cracking, between the

---

[32] For a personal account of the activities of such a site engineer, see J. W. Nelson, *Essays of a Miner's Son* (unpublished typescript, 1989).

[33] I. M. Sheldon, T. A. J. Cockerill and D. F. Ball, 'The UK process plant contracting industry', *Industrial Chemistry Bulletin*. Aug 1982, 116–24. For a wider economic study, see also I. M. Sheldon, *The UK Process Plant Contracting Industry: Structure, Conduct and Performance* (PhD thesis, Univ. Salford, 1981). The distinction between plant manufacturers and plant contractors is mirrored in their separate trade associations: the British Chemical Plant Manufacturers' Association (BCPMA) and Council of British Manufacturers of Petroleum Equipment (CBMPE) had origins and aims distinct from the British Chemical Engineering Contractors' Association (BCECA) founded in 1965.

[34] J. M. Solbett, 'Process engineers and others', *TCE* June 1979, 430–6.

[35] By 1971, nevertheless, Imperial College reported that some 12% of its chemical engineering graduates found employment with contracting companies. A. R. Ubbelohde, 'Chemical engineering and science-based education', *Chem. & Indus.* 31 Jul 1971, 878–80.

wars, along with the rising complexity of processes, provided paths to new competencies for a number of American engineering firms. The war, during which the USA provided petroleum to its allies, also helped cap its dominance of the market. Finally, the postwar economic void in Britain gave ample room to foreign investors. Several contractors – including Badger, Bechtel, Lummus, M. W. Kellogg, Ralph M. Parsons, Procon, and Stone & Webster – established UK operations in the late 1940s and 1950s as part of the postwar US capital investment in Europe. Foster-Wheeler, Babcock, John Thomson and other firms shared the boiler business. Such contractors were responsible for the significant refinery projects after 1945, and all but one were American. By the middle of the decade, American contractors were designing most of the equipment used in the British petroleum industry.[36]

Not all branches of the chemical and allied industries, and by extension chemical engineering, were dominated by the American example, however. While Union Carbide, Monsanto Chemical, American Cyanamid, Dow and Du Pont operated branch operations in UK, only Monsanto could be classed as a general producer of chemicals like ICI. There was virtually no US capital in coal-tar products such as coal distillates, dyes and dyestuffs, or in heavy chemicals such as acids, alkalis, explosives and salt production. Overall, American operations in Britain accounted for some 10% of the gross value of chemical and allied products by the mid Fifties.[37]

And while British process plant contractors adopted American practices, they started from a different background. Firms such as Woodhall-Duckham, Power Gas Corporation, Simon-Carves and Humphreys & Glasgow had specialised between the wars in the supply of equipment for gas processing plants. New British firms more closely followed the American mould. Costain John Brown Ltd, for instance, set up in 1948 to undertake the construction of oil refinery projects, became a major process plant contractor after absorbing firms having engineering expertise.[38]

Nevertheless, the plant contracting industry was profoundly shaped by American investment and design practice. British manufacturers gained considerably from American expertise in, for example, alloy-lined pressure vessels, specially welded fittings, pumps, valves and control gear, techniques of piping and erection, and special steels. Half of the British petroleum refinery equipment firms claimed to have benefited.[39] Transfer of design knowledge in petrochemical plant was slower to come. A 1955 report by the Organization for European Economic Co-operation observed that 'in some smaller refineries Europeans are now sometimes in charge of the engineering and process design', but that it would 'take European consultants a long time to establish the large-scale

---

[36] Some 350 UK firms supplied refinery equipment by 1955, but contractors were American-dominated. Dunning, *op. cit.*, 68–9, 208–11.

[37] Dunning, *op. cit.*, 59–60.

[38] Sheldon, Cockerill and Ball, *op. cit.*, Table 4. See also D. F. Ball, *Process Plant Contracting Worldwide* (London: Financial times Business Information, 1985).

[39] OEEC, *Oil Equipment in Europe* (London: Technical Assistance Mission, No 121, 1955), pp. 13–4 and 212.

research laboratories and pilot plants ... and to train staff of competent engineers and chemists specializing in petroleum work necessary for them to develop research on new processes comparable to that done by U.S. firms'.[40]

Besides new roles for chemical engineering contractors, the process plant industry itself changed quantitatively after the war. The scale of individual projects began to rise precipitously, eventually by more than an order of magnitude. This was a result both of larger individual units (and a transition to continuous, rather than batch, processes) and of a tendency to contract complete industrial projects rather than separate plants.[41]

EXPANSION TO NEW FIELDS

Petroleum, petrochemicals, and synthetics – cases of tardy expansion of a latent industrial potential – came to employ chemical engineers largely as a result of changed economic conditions and American expansionism after the second world war. But entirely new domains appeared and were supported by other sources. The most important of these endeavours, the British nuclear programme, began in rudimentary form during the second world war, providing the profession with firm footholds in postwar industries and a serious claim to a broader scope. The genesis of this new industry, and the development of its chemical engineering connections, is therefore worth exploring in some detail.

Creating a national nuclear programme

British chemical engineers became associated with nuclear energy at the earliest exploratory stages of the wartime atomic bomb project, and continued to insinuate themselves in the emerging multidisciplinary domain. The importance of chemical separation, thermodynamics and mathematical analysis to the project made the expertise of university-trained chemical engineers particularly relevant to the solution of its development problems.[42]

A special committee of the Ministry of Aircraft Production had been set up in April 1940 to investigate the feasibility of an explosive based on uranium. ICI had an early involvement via its assistance to individual scientists associated with the project, and by September of that year ICI representatives were on the Technical Sub-Committee.[43] Significantly, the Committee's report prepared that year identified one of the outstanding tasks as chemical engineering:

> As regards the manufacture of the ^{235}U we have gone nearly as far as we can on the laboratory scale. The principle of the method is certain, and the application does not appear unduly difficult as a piece of chemical engineering. The need to work on a

[40] *Ibid.*
[41] Solbett, *op. cit.*
[42] For a technical evaluation, see James W. Kuhn, *Scientific and Managerial Manpower in Nuclear Industry* (New York: Columbia University Press, 1966), p. v.
[43] R. W. Clark, *The Birth of the Bomb* (London: Phoenix, 1961), pp. 127–39, 155–8.

larger scale is now very apparent and we are beginning to have difficulty in finding the necessary scientific personnel.[44]

Between 1941 and 1943, it became increasingly clear that wartime Britain had neither the labour nor economic resources needed to develop the bomb or even uranium refinement, and the project concentrated instead on complementary work with Canadian and American groups. The 'Tube Alloys' Directorate, a division of the DSIR attached to the Ministry of Supply which succeeded the original committee, relied on the only two companies it had identified as having the necessary staff, knowledge and experience for the work: ICI and Metropolitan–Vickers.[45] ICI, via its Fertiliser and Synthetic Products Division at Billingham, the General Chemicals Division at Widnes, the Alkali Division at Winnington and the Metals Division at Birmingham, studied the production of heavy water, produced the chemicals for the diffusion plant and uranium metal for the first test reactors, supervised the production of special membranes for the model diffusion units, and operated the diffusion units.[46]

By mid-1944, a year before the Americans detonated the first atomic bomb, the Directorate allowed ICI to discuss with subcontractors their plans for postwar heavy-water and uranium hexafluoride production plants. By the summer of 1945 the ICI board agreed to undertake contracts for the construction of nuclear reactors, but they did not receive authority to proceed with buying materials or construction before war ended. The company also investigated options such as acquiring technical knowledge about heavy water plant design from Du Pont in America.[47]

Such research and planning work provided a wealth of engineering knowledge. The purification and separation of the necessary materials, and operations of the plants, required attention to corrosive problems, extreme cleanliness, high-quality vacuums, precise mechanical tolerances and auxiliary equipment to provide stable control. ICI thus became familiar with the principles and potential problems of chemical engineering for nuclear power through the last four years of the war and planned an active participation after it.

After the war, responsibility for continued development of nuclear energy passed from the DSIR to a department of Atomic Energy of the Ministry of Supply, until the United Kingdom Atomic Energy Authority (UKAEA) suc-

---

[44]   *Report by M.A.U.D. Committee on the Use of Uranium for a Bomb* (1941), in: M. Gowing, *Britain and Atomic Energy 1939–1945* (London: Macmillan, 1964), p. 397.

[45]   Hilary Rose and Steven Rose, *Science and Society: The Chemists' War* (Harmondsworth: Penguin, 1970), pp. 63–5.

[46]   W. J. Reader, *Imperial Chemical Industries: A History* (Oxford: Oxford University Press, 1975), Vol. 2, pp. 287–96.

[47]   For the excellent official histories of the AEA, see the works of Margaret Gowing: *Britain and Atomic Energy 1939–1945 op. cit.; Independence and Deterrence: Britain and Atomic Energy 1945–1952 Volume I: Policy Making; Volume II: Policy Execution* (London: Macmillan, 1974). Regarding Du Pont, see p. 434 of *1939–1945*. For subsequent organisation of the nuclear programme, see Central Office for Information, *Nuclear Energy in Britain* (London, 1969). Regarding ICI's postwar plans for research and power generation. PRO AB 1/331 *ICI Progress Reports (Gen Chemistry Div) 1941–43.*

ceeded it in 1954. The Ministry quickly set up, in 1946, an Atomic Energy Research Establishment (AERE) at Harwell, Berkshire and a separate production organisation which, over the following six years, established a number of factories for producing and processing nuclear fuels.

But where ICI took on atomic research as both a wartime duty and for the commercial promise of peace-time power generation, smaller firms were distinctly unenthusiastic. Government categorisation of such research mirrored the awkward pigeonholing of chemical engineers themselves during this period. An 'atomic pile', thought civil servants, 'did not fit easily into any of the existing industrial classifications, but the whole complex of plants belonging much more obviously to the chemical plant industry than to any other industry. However, the Ministry of Supply was advised that none of the British firms in that industry was capable of undertaking the work'.[48] Indeed, in September 1946, representatives of the Directorate of Atomic Energy and of the Contracts Branch of the Ministry of Supply discussed chemical plant requirements with the Council of the BCPMA. Despite the government aims of 'starting virtually from scratch ... to lead the world in atomic energy development in the shortest possible time', which would necessitate 'the expenditure of scores of millions during the coming five to ten years and involve an immense amount of original designing and equipment', the Directorate had been 'extremely disappointed at the chemical plant industry's lack of response to its enquiries'.[49] Part of the firms' reticence may have been attributable to a general impression that continued Ministry work would be relatively unprofitable, secretive and hamstrung by red tape. Moreover, the urgency of development – for what eventually culminated in the nuclear weapons test at Montebello in late 1952 – provided 'far less certainty than any industrial firm would accept'.[50] The government itself was concerned that the planned construction of one or two nuclear reactors would make heavy demands on the capacity of the chemical engineering and heavy electrical industries which were of great importance to the revival of the export trade.[51] Interestingly, the government dialogue was with plant manufacturers rather than with professional designers or their Institutions: the Directorate was prepared 'to sub-delegate whole sections of the design and supply of plant to the industry'. The IChemE did not figure in the discussions at all.

Yet a vacillating flirtation with chemical engineering continued as the programme developed. The nucleus of a chemical engineering division at AEA formed at Chalk River, Ontario, in 1946, where a 'small group of chemical

---

[48]   Gowing, *op. cit.*, Vol. 2, 155.

[49]   BCPMA memo on atomic energy, 4 Oct 1946, Swindin papers, box 44. The same situation endured through the next decade; when the UKAEA tried to interest chemical firms in the nuclear industry, they were warned by the BCPMA that consortia 'would be impracticable and that even so only two or three firms would likely be interested.' Only when consortia of firms were proposed did a few decide to take the commercial risk. PRO AB 38/83 (1958).

[50]   L. Owen, 'Nuclear engineering in the United Kingdom – the first ten years', *J. BNES 2* (1963), 23–32 and 296–8.

[51]   Gowing, *op. cit.*, Vol 1, 168.

engineers [or 'chemists', depending on the source] had built a primitive pilot plant in a tar paper covered tower on the banks of the Ottawa River', the British contingent developed a flowsheet for industrial separation uranium and fission products.[52] Sir John Cockcroft, the Director of Harwell, noted that 'the Government, through the Ministry of Supply and the Department of Atomic Energy, had become a substantial user of chemical engineering' with 'a very flourishing Chemical Engineering Division' at Harwell which included nine Associate Members and eight Student Members of the Institution by 1949.[53] The original Chalk River Division moved to Britain in 1950, where it eventually grew into two branches, concerned with chemical engineering operations (e.g. heat transfer and hydrodynamics of gas-coolants) and process technology (e.g. graphite technology, chemical recovery and the fixing of fission products in glasses).[54]

In parallel with atomic research, however, went production – a domain that made much less use of chemical engineering. This was largely due to the continuity of working culture imposed by Christopher Hinton, responsible for designing and managing the factories of the postwar nuclear programme.[55] Having studied mechanical engineering at Cambridge and become an exceptionally young Chief Engineer of the Alkali Group at ICI in 1930, Hinton (1901–83) had briefly supervised cordite production, and then directed the munitions filling factories for the Ministry of Supply for the duration of the war. Indeed, several personnel influential in the wartime Royal Ordnance Factories adopted equally important roles in the nascent nuclear programme. For example, Henry G. Davey, who had taught chemical engineering at the Trefforest School of Mines before the war, became Superintendent of ROF Drigg and then Superintendent of Works at the first nuclear pile at Windscale; Robert Alexander, from ROF Drigg and ROF Sellafield, became Assistant Works Manager at UKAEA Capenhurst; Stanley F. Hines, employed as a chemist at ROF Pembrey, transferred to Capenhurst where he became Works General Manager.[56] The core group that was to direct the industrial programme

---

[52]   H. K. Rae, 'Three decades of Canadian nuclear chemical engineering' and N. L. Franklin et al., 'The contribution of chemical engineering to the U.K. nuclear industry', both in William F. Furter (ed.), History of Chemical Engineering (Washington: American Chemical Society, 1980), pp. 313–34, and 335–66, respectively; Sir John Cockcroft, Chem. & Indus. 17 Nov 1956, 1342; Gowing, op. cit., Vol. 2, 405.

[53]   For Cockcroft, see 'Interpretation of scientific knowledge', Chem. Age, 30 Apr 1949, 613–4. Prominent among the employees was the head of the Harwell Engineering Division, Harold Tongue (1894–1960), who had trained at the Manchester Municipal College of Technology, gained wide industrial experience the Chemical Research Laboratory, authored two books on chemical engineering before the war, and been at the Montreal laboratory of the Tube Alloys project during the war. AF 915.

[54]   A. S. White, 'Chemical engineering research at A.E.R.E.', TCE Dec 1962, A66–A73; C. M. Nichols and A. S. White, 'Chemical engineering and atomic energy', Chem. & Indus. 17 Jan 1963, 51–5. Quart. Bull. IChemE, No. 93 (Jul 1948).

[55]   For a first-person account, see Owen, op. cit., and IMechE Hinton A.4–A.6. 'Lord Hinton of Bankside', Biog. Mem. Fellows of the Royal Society 36 (1990) and P. Pringle and J. Spigelman, The Nuclear Barons (London: Sphere, 1982).

[56]   For Davey (b. 1908), see AF 2187 and 7508; Alexander (b. 1915), see AF 2232 and 7988.

over the next decade was made up of long-time associates of Hinton. They had worked under him first in the Alkali Group at ICI Northwich, then supervised cordite factory construction for the Ministry of Supply in the early years of the war, and later still managed the filling factories.[57] Not surprisingly then, atomic production was centred on such familiar sites: the headquarters of Hinton's industrial group had also been the temporary headquarters of Ordnance Supply, and was a former Ministry of Supply shell-filling factory at Risley, near Warrington; the first uranium fuel plant (erected at Salwick, near Preston, renamed Springfields, and gradually taken over from the ICI caretaker staff) and the first nuclear reactor and chemical separation plant (at Sellafield, Cumbria – renamed Windscale) had been wartime ordnance factories. Hinton recalled:

> The initial atomic energy establishments had been represented to me as munition factories for the manufacture of fissile material for bombs ... and, having learned by experience that the strategical location of factories for the next war is normally based on experience in the last, I guessed that the most acceptable factory sites would be in the North West.[58]

The continuity of purpose, working culture and personnel profile were thus firmly reinforced.[59] This grafting of working culture from ICI stock, resulting in the partial exclusion of chemical engineers, was remarkably different from the American case, where Du Pont and its overtly chemical engineering teams had been contracted to take responsibility for production reactors and chemical separation early in the Manhattan project.[60]

The distinct national differences between the two nuclear programmes go beyond managerial culture and extend to political organisation. On the other hand, the continuity in the British case is evident. There are interesting parallels between Hinton and the first world war work of K. B. Quinan, described in Chapter 3.[61] Like Quinan, Hinton had managed part of a major commercial firm before his war and worked for the government on the highest-priority and most time-critical military project of his time, on problems largely linked

---

[57] Eight engineer/designers from ICI Northwich and the Filling Factory organisation joined Hinton, one of whom, Charles J. Turner, had designed the first polythene plant with W. R. D. Manning at ICI before the war, and was responsible for the first chemical separation plants of the atomic programme.

[58] IMechE Hinton A.2 p.127.

[59] The same was true in the USSR, where in 1945 Stalin directed the People's Commissar of Munitions to provide atomic weapons in the shortest possible time. Richard Rhodes, *Dark Sun: The Making of the Hydrogen Bomb* (New York: Simon & Schuster, 1995), p. 179.

[60] Rodney P. Carlisle and Joan M. Zenzen, *Supplying the Nuclear Arsenal: American Production Reactors 1942–1992* (Baltimore: Johns Hopkins Press, 1996). See also W. K. Davis and W. A. Rodger, 'The chemical engineer and nuclear energy', in: W. T. Dixon and A. W. Fisher, Jr. (eds.), *Chemical Engineering in Industry* (New York: AIChE, 1958). Even theoretical physicist Eugene Wigner had trained in chemical engineering in Berlin in the 1920s [Obituary, *The Times*, 9 Jan 1995, 19].

[61] There was continuity, too: C. S. Robinson (1887–1969), who had worked with Quinan at the Ministry of Munitions and in South Africa, was Hinton's superior in the Filling Factory Directorate. AF 225.

to chemical engineering, but remaining little-known outside his immediate circle. Civil servants, probably like their predecessors at the first world war Department of Explosives Supply, saw Hinton's Industrial Division of the Department of Atomic Energy – founded two years before Quinan's death – as a 'strange experiment in which civil service machinery was used to carry out a large industrial enterprise'.[62]

There were other similarities with the first world war situation. In a nascent nuclear programme in which 'chemical engineering difficulties' were identified as being a central concern, accredited chemical engineers were decidedly side-lined. And both Quinan and Hinton were reluctant chemical engineers, urged by others to support the profession. Despite Cockcroft's publicly marking out nuclear engineering as an important domain for postwar chemical engineers, not all nuclear sites – particularly the design and production facilities – made use of them. This is partly attributable to occupational classifications. Exactly as in Quinan's first world war factories, posts were defined as various grades of either 'engineer' or 'chemist' linked to the Civil Service pay scale.[63] Reactors and chemical separation processes were designed in Hinton's Chemical Plants Design Office at Risley. Harwell (initially, at least) employed 'chemists', Risley 'engineers', and the awkward 'chemical engineer' was dispossessed.

Like Quinan some thirty years earlier, Hinton surrounded himself with practi-cally-trained men who lacked formal chemical engineering credentials. Of the nucleus of men whom Hinton brought from the munitions filling factory organisa-tion in 1946, three had worked with him since his prewar ICI days; two had university degrees. Only one of the first dozen had had direct experience with atomic energy.[64] Harold Disney, the chief designer of the first large gaseous diffusion plants, for instance, had worked at a railway wagon building works and taken evening classes to HNC level before the war; James Kendall, responsible for the first Harwell and Windscale reactor designs, had no orthodox training in engineering, having managed a building firm before joining the wartime filling factory group.[65] Backing up this highly experienced but motley team were the industrial chemists at ICI General Chemicals and Billingham who had studied uranium purification and gaseous diffusion, respectively, during the war.[66]

In at least one respect, Hinton was at a severe disadvantage with respect to

---

[62]   Gowing, *op. cit.*, Vol. 2, 67.

[63]   Staffing shortages were exacerbated by the restrictions of salary scales and job categories maintained by the Ministry of Supply. These were eased with the creation of a separate Authority, when research chemical engineers were graded on the more inclusive category of 'Scientific Officers'. By 1958, annual salaries for Harwell employees had risen to £815 – £2080 as Engineers or £645 – £2080 as Scientific Officers. These should be multiplied by 12 for their 1999 equivalents. Owen, *op. cit.*, 26; interview, P. N. Rowe with Johnston, 10 Mar 1998; PRO AB 17/231, *Careers in Nuclear Engineering*.

[64]   Dennis Ginns, an engineer at Chalk River. See Owen, *op. cit.*, 23.

[65]   Gowing, *op. cit.*, Vol. 2, 22, 31; Lord Hinton of Bankside, *Engineers and Engineering* (Oxford: Oxford University Press, 1970), pp. 3, 5, 39.

[66]   Most design groups at Risley had one or two Designers, five to ten Draughtsmen and an occasional Chemical Advisor. 'Plans for staff organisation, 1946–1950', PRO AB 19/56; Owen, *op. cit.*, 24, 28–9; IMechE Hinton A.3.

his predecessor. Where Quinan had kept large collections of files on chemical engineering plants and processes around the world, Hinton arrived at Risley armed with none.[67] He later wrote:

> In most design offices, when the plant flowsheet has been drawn, it is possible to take, from the files, drawings which have been used for other plants and to modify these or use them as a basis for the design of some of the vessels that are needed ... The trouble did not lie in the fact that we were incapable of doing this work when starting from a clean sheet of paper, it lay in the fact that this took time and that experienced engineers (of which we were desperately short) had to do work which could otherwise be done by draughtsmen. I asked my old Division of ICI if they would let me have designs of typical vessels; they refused, claiming that their processes were secret. Considering that I had asked for drawings of vessels that might be used on any chemical plant and which had no secret features, I thought this was singularly unhelpful.[68]

Like Quinan a relentless manager and private man, Hinton had no links to chemical engineering organisations until he had achieved public recognition: he joined the IChemE in 1954, after having been knighted for his atomic work.[69] Neither his subordinates nor other key persons in the programme – e.g. Robert Spence, who had headed the Chalk River 'chemistry' group and later seeded the Harwell chemical engineering team, Charles Turner, responsible for the chemical plant designs, nor David Deverell, chemical engineering supervisor at Aldermaston – became members of the Institution. At least through the early 1950s, Hinton's organisation made little use of the new Chemical Engineering Division administered by Arthur S. White at Harwell.[70] The large and expensive laboratories there had been intended to develop new chemical separation processes, but turned instead to research on solvent extraction and novel designs for reactors. Other long-term problems of nuclear engineering, particularly the corrosion of pipes and vessels aggravated by high temperatures, erosion and radiation effects, were tackled less systematically.

---

[67] PRO SUPP 10.

[68] IMechE Hinton A.3 p. 152.

[69] For Hinton on chemical engineering, see 'The part played by the chemical engineer in bridging the gap between research and plant construction', *Trans. IChemE 32* (1954), 205–9; 'The manufacture of uranium metal', IChemE Manchester branch meeting, Jan 1953; 'Some aspects of chemical processes ancillary to atomic energy', IChemE Manchester branch meeting Jan 1955; 'The chemical separation processes at Windscale works', Castner Memorial Lecture, Feb 1956. IMechE Hinton A.3. Hinton (later Sir Christopher and eventually Lord Hinton) had been denied the authority from his superiors at least once to present a 'full technical report to a learned society' because of concerns for secrecy. Gowing, *op. cit.*, Vol. 2, 129.

[70] The head of the Chemical Engineering Division for twenty years, White (1903–87) had neither a research degree nor wide research experience, having obtained his chemical engineering diploma from UCL in 1934 after losing his job during the depression, and then working at ICI Dyestuffs in Manchester. White nevertheless encouraged his staff to publish and to join the IChemE, and was active in the Institution. 'Harwell', said one associate, 'bristling with whizz kids and academic protégés, needed a sound and experienced engineer to keep their feet on the ground'. P. Rowe, 'Obituary: Arthur Southan White', *Diary & News*, April 1987, 3; interviews, Johnston with P. N. Rowe (10 Mar 1998) and Sir Frederick Warner (19 Dec 1997); AF 3300.

There is some evidence that Hinton, at least initially, was not eager to employ chemical engineers. When in 1950 a letter from an associate in the Department of Atomic Energy suggested replacing a departing physicist with 'a chemical engineer with some knowledge of atomic energy', Hinton instead proposed an ex-ICI Dyestuffs mechanical engineer.[71] Similarly, Hinton hesitated to appoint Frank Kearton, then 'a young chemical engineer at ICI Billingham' to lead diffusion plant development. With basic training 'as a chemist and not as an engineer', Hinton believed that Kearton lacked the required management skills, and could have lowered morale and delayed progress despite his valuable and recognised technical competence.[72]

The absence of chemical engineers was not entirely due to the perspective Hinton carried away from ICI. There was a true shortage of qualified chemical engineers, and a reticence by many to enter the security-bound, military-oriented and under-paid civil service immediately after a long war. As a result, a high proportion of the professional appointments of Hinton's group were of men already in the Civil Service, whom they knew personally, or 'who had been put in touch with us by old friends'.[73] Had it not been for the 'extreme shortage of chemical engineers', and the 'great disparity in the size and strength of the chemical engineering industry in the two countries', argued the official historian Margaret Gowing, Britain would not have been limited to spending money on its nuclear programme at scarcely 1/35 the rate of the USA.[74]

While the British industry 'had spent its first 8 years almost entirely in meeting military demands', and 'worked behind a curtain of secrecy', the expansion of the programme to include civil power gave more freedom to its workers.[75] After the introduction of the power-generating nuclear reactor at Calder Hall in 1956, chemical engineering employment accelerated. A government report that year found some 103 chemical engineers at the UKAEA, filling a variety of roles there and in the British nuclear industry.[76] They were also becoming a significant presence within the IChemE. Appropriately for the country that seemed to be leading in commercial nuclear power production, a 1958 survey found over 5% of IChemE members to be working in the atomic energy industry, ahead of those in the plastics, dyes and food industries.[77] By the early Sixties, the Institution counted 125 of its members employed at the nuclear plants and research centres of the UKAEA, making the Authority

---

[71] 1 Mar 1950 letter Eric Welsh to Hinton, PRO AB 19/66.
[72] IMechE Hinton A.4. Kearton (1911–92), having joined ICI in 1933, worked on the atomic energy project 1940–5. He went to Courtaulds in 1946, where he rose quickly to senior management.
[73] *Ibid.*
[74] Gowing, *op. cit.*, Vol. 2, 343.
[75] C. Hinton, inaugural address, *J. BNEC 1* (1955), 1–2..
[76] *Scientific and Engineering Manpower in Great Britain* (London: HMSO, 1956); K. B. Ross, 'The chemical engineer in industry', *TCE* Oct 1960, A38-A39.
[77] 'The supply and distribution of chemical engineers in Great Britain', *TCE* Oct. 1958, A41-A45, and J. B. Brennan, 'Chemical engineering manpower in the chemical industry', *TCE* Apr. 1961, A44-A35.

second only to ICI in its employment of chemical engineers.[78] The pace of expansion could not be kept up indefinitely, however. Hinton himself, disillusioned by the inflated technical and economic claims for British reactor designs being made by the government, and by moves to rapidly commercialise the industry, left the UKAEA in 1957 for the Central Electricity Generating Board. His support for American reactor designs contributed to infighting and delayed decisions, shaping the expansion of the industry until the British version was adopted officially in 1965.[79] A. S. White's Chemical Engineering Division at Harwell expanded to some fifty staff – some 1% of total UKAEA employment – by the late 1950s. Thereafter, following White's retirement in 1968 and later reorganisation of Harwell from subject-based to project-based administration, the Chemical Engineering Division disappeared. Its building was demolished in the late 1990s, by which time several Presidents of the Institution and professors of chemical engineering had worked at the UKAEA at some point in their careers.[80]

Professional fallout

Thus the occupation of chemical engineers expanded into the nuclear industry. But what of their professional aspirations? Some within the IChemE attempted early on to extend its jurisdiction to this new domain of nuclear engineering. As early as November 1945, a mere three months after the public revelation of atomic energy by the bombing of Hiroshima and Nagasaki, Council member E. A. Alliott had suggested 'establishing a representative committee on development of atomic (nuclear) energy'.[81] Sir Harold Hartley, speaking at a BCPMA dinner, pressed further that the older specialisms were threatened by the new: 'with the lure of nuclear physics and electronics, classical physicists were a dying race', he claimed, and 'chemical engineers had to do the research'.[82]

---

[78]   IChemE, unlabelled volume tabulating members by company affiliation, 1963. The distribution was 35 at the Atomic Energy Research Establishment, Harwell (headquarters of the AEA Research Group); 24 at the headquarters of the AEA Reactor Group and Production Group, Risley; 19 at the Atomic Weapons Research Establishment, Aldermaston; 20 at Preston (uranium metal manufacture); 7 at Seascale; 5 at the Atomic Energy Establishment, Winfrith, Dorset (power generation research); and 15 elsewhere. This was a high point for chemical engineering contributions; the sum had been almost identical (127) in 1960. Nevertheless, a later booklet could still claim that Harwell 'comprises one of the largest teams of chemical engineers in the country.' *Research Careers in Chemistry and Chemical Engineering – Harwell and Amersham* (UKAEA, 1968), pp. 10–12, PRO AB 17/483.

[79]   Pringle and Spigelman, *op. cit.*, pp. 274–5.

[80]   These included John Collier (Pres. 1995); Peter V. Danckwerts (Pres. 1965 and prof); N. L. Franklin (Pres. 1979 and prof); G. R. Hall (prof); Geoff Hewitt (Pres. 1989 and prof); Robert Edgeworth Johnstone (prof); John Menzies Kay (prof); Peter Rowe (Pres. 1987 and prof); W. L. Wilkinson (Pres. 1980). For Franklin (1924–86) at the UKAEA, see the extensive PRO holdings (PRO AB 38).

[81]   Letter, Alliott to Council, 6 Nov 1945, Gayfere archive box III/5; CM 21 Nov 1945. A Nuclear Processing Group was founded in 1978 when Council approved Subject Groups; the AIChE, by contrast, formed a Nuclear Engineering Group as early as 1954.

[82]   H. Hartley, *Chem. & Indus.* 25 Apr 1953, 404–5; H. Hartley, 'The place of chemical engineering in modern industry', *School Science Review* No. 126, Mar 1954, 199–202.

Indeed, Hartley saw nuclear engineering and chemical engineering as keys to
the country's economy. He interpreted both as a new opportunity to assert
national prestige against foreign competition, writing to a manufacturer

> what worries me is the cool assumption that the USA is going to export packaged
> power reactors about the world just as they have done in the case of oil refineries.
> Chemical engineering at long last is going ahead in Britain and I hope we shall soon
> be in a position to break the American monopoly in the latter case and that we
> shan't allow them to get a monopoly so far as power reactors are concerned.[83]

By the mid-Fifties, the BCPMA itself was eager to embrace the new technology, becoming a founder member of the Nuclear Energy Trade Associations
Conference. The colonisation of new territory by British chemical engineers
also had a precedent: chemical engineers in America had been prominent in
the growing nuclear industry, and, in 1955, nuclear workers there affiliated
with those in the oil and chemical industries.[84] That year, at the invitation
from the Civils, Mechanicals, Electricals and the Institute of Physics (which at
that time represented primarily applied physicists in industry rather than the
academic-based physicists catered for by the Physical Society), the IChemE
collaborated to form a joint body for the advancement of nuclear technology:
the British Nuclear Energy Conference (BNEC).[85] Significantly, it was 'not be
regarded as a completely new technology, but rather as an extension and
development of existing technologies'.[86] This seeming promotion of the IChemE
to a privileged status with larger bodies was short-lived; by 1961 its representation had been diluted by the addition of the Physical Society, Iron and Steel
Institute, Institute of Metals and Institute of Fuel.

Unsurprisingly, a key player in the BNEC and its chemical engineering
connections was Christopher Hinton, its first chairman.[87] He personified this
union of disciplines, being a Fellow of the Royal Society and member of the
ICE, IMechE and IChemE – at least belatedly. In 1954, when both were

---

[83] 20 Jan 1955 letter Hartley to S. R. Armsdon of CJB, Hartley papers box 275.
[84] The dominance of American chemical engineers in some aspects of the US nuclear programme
      was not enduring, however. Du Pont, with its culture of 'almost haphazard, cut-and-try
      methods' was replaced by General Electric and its electrical engineering 'systems approach'.
      Instead of viewing the nuclear reactor, as Du Pont did, as a versatile and malleable collection
      of plant for producing a product, GE saw it as a product in its own right, akin to the industrial
      transformers and generators that had made its fortune. There was thus a 'change of engineering
      style ... of the new profession of nuclear engineering from its World War II roots in chemical
      engineering to the electrical engineering-dominated style of the 1960s and later'. See Melvin
      Rothbaum, *The Government of the Oil, Chemical and Atomic Workers Union* (New York: John
      Wiley, 1962); H. M. Vollmer and D. Mills, 'Nuclear technology and the professionalisation of
      labor', *Am. J. Sociology* 67 (1962), 690–6; Carlisle & Zenzen, *op. cit.*, 45–58, 88–90, 109–20;
      quotations p. 109.
[85] CM 19 Jan 1955.
[86] Hinton address, *op. cit.*
[87] Initially, Cockroft, Ross and White of Harwell were also members of the governing board.
      PRO AB 19/83. Hinton had earlier been the figurehead President of the Nuclear Engineering
      Society, Windscale', an employees' group sanctioned by the Department of Atomic Energy of
      the Ministry of Supply.

involved in organising the BNEC, Harold Hartley had written to Hinton suggesting that 'your name on our list would be a real help to the Institution'.[88] Hinton promptly joined, and began promoting the Institution and chemical engineering education. He appears to have been motivated principally by his chronic problem of manning the overstretched UKAEA Industrial Division. In August of that year he complained to Hartley that American universities and colleges of technology were running training courses for 'nuclear engineers', but designed for men already employed in industry.[89] Hinton thought that a postgraduate school of nuclear engineering would be difficult to establish because of the insufficient number of staff in what he identified as the appropriate specialism: chemical engineering. He tried, instead, to interest existing chemical engineering departments 'to broaden ... the application not to chemical plants, but rather to process engineering'.[90] Hartley had been seeking to extend the profession in this way through his presidency of the IChemE, and urged Hinton on.

University departments, however, were not immediately eager to co-operate. Hinton's discussions with the Manchester College of Technology fizzled out when the work of preparing a syllabus of nuclear engineering overwhelmed the enthusiasts there. When Hinton approached Prof. T. R. C. Fox at Cambridge, who had served on the Atomic Energy Council, he found his proposal for a UKAEA-funded research reactor at Cambridge coolly received. Supporters John Menzies Kay and John Cockcroft reported that nuclear engineering courses within the chemical engineering department were impracticable given the already heavily-loaded syllabus, noting that Fox was 'hopelessly over-cautious about the whole matter', although his counterpart in the Engineering Department was 'really quite keen'.[91] The second half of the decade appeared more propitious, however. King's College, Newcastle, discussed a nuclear engineering course, and Kay was appointed Professor of Nuclear Technology at Imperial College.[92]

[88] Hartley, 23 years Hinton's senior, became from this time something of a confidante, arranging a visit from the Duke of Edinburgh (for whom he was a close adviser), pressing the IChemE and BNEC links, and later advising on a career shift. PRO 19/40, 8 Dec 1954 letter; IMechE Hinton A.170, J.24; Hartley papers Box 144.
[89] American universities had, in fact, initially been hampered in teaching nuclear engineering. Academics working on the Manhattan project had been drawn mainly from departments of Physics and Chemistry. Seven schools offered nuclear engineering courses in 1950, but 132 by 1962. For most of 1950s, Atomic Energy Commission programmes were the major source of supply of nuclear engineers. Kuhn, op. cit., 51–3.
[90] 4 Aug 1954 letter Hinton to Hartley, PRO AB 19/84. See also AF 7551 and IMechE Hinton J.24.
[91] PRO AB 19/84.
[92] Minutes, meeting of the first Advisory Panel in Chemical Engineering, King's College, Newcastle, 20 Feb 1957. Coulson (1910–90), head of chemical engineering, noted that many students now entered nuclear engineering. Other panellists agreed that, as the marine industry would likely be the next to take up nuclear power, a nuclear engineering course in the chemical engineering department would be particularly relevant for Newcastle. On the Imperial College department under Kay's successor, see G. R. Hall, 'Nuclear technology in a university environment', Imperial College of Science and Technology inaugural lecture, 9 Jun 1964. Kay (1920–95) had taught chemical engineering at Cambridge from 1948 and been Chief Technical Engineer at Risley from 1952.

This seemingly vacant professional territory was soon invaded, however. Given the active participation of Hinton and the major engineering institutions in education and the BNEC, the proposed formation of a separate and unaffiliated Institution of Nuclear Engineers (INucE) in 1958 provoked a letter from the Institution of Civil Engineers noting their 'disquiet'. Hinton, for example, had opposed the tendency for Imperial College to promote nuclear energy as a new discipline, and sought to block more public professional claims.[93] The ICE suggested that each of the members of the BNEC might publish a statement in their journals to discourage members from participating in the affairs of the INucE. After checking with their solicitors, the IChemE council agreed. 'The Council sees no reason for the formation of new societies in nuclear energy', ran the statement, because 'the new industry did not bring into being a new kind of engineer, but was a new challenge in the application of existing branches of engineering'.[94] The INucE, undeterred, formed in 1959, with two corporate classes and eight non-voting grades to encompass qualified, unqualified, individual and corporate members. As the IChemE itself had argued some four decades earlier, the Secretary of the INucE was eager to stress

> that we are a professional body, and that our work is confined to the spread of knowledge in the field in which we serve no matter how diversified the field becomes. We are in no sense a splinter institution. We are a new creation in a new and expanding field and we are tied to none.[95]

The older Institutions consolidated their opposition to the INucE in 1960 by founding the British Nuclear Energy Society (BNES), which would be open to qualified engineers from the constituent societies and to others 'actively engaged' in the technical aspects of nuclear energy and ancillary subjects. In a direct challenge to the broad church of the INucE, it claimed:

> One of the most important roles of the Society is to provide a channel for the exchange of experience and the cross-fertilization of ideas between all engaged in the field of nuclear energy ... our members should include not only physicists and engineers, but biologists, medical men, agriculturalists, lawyers and accountants.[96]

The new BNES found a home at the headquarters of the Institution of Civil Engineers but failed to limit the influence of the INucE. The breadth of interests and initially lax standards of entry promoted the expansion of the INucE through the 1960s, attracting a disparate collection of members of generally lower formal attainment than their counterparts in other Institutions.[97]

---

[93] PRO AB 19/40, letters, O. A. Saunders (Mechanical Engineering Dept, Imperial College) and Hinton, Jan 1956.
[94] CM 16 Jul 1958; 'British Nuclear Energy Conference', *TCE* Oct 1958, A47. The same point was made by G. R. Hall, *op. cit.*
[95] 'From the Secretary', *Nuclear Energy*, Feb 1966, 31.
[96] Sir Leonard Owen, *J. BNES 1* (1962). The new Society had received some 1700 applications by late 1961.
[97] By 1968, 34% of Members and 15% of Associate Members of the INucE held bachelor's degrees or above, and 43% and 18%, respectively, were corporate members of other engineering institutions. Its council, observing that 'in the early days of any foundation it must assemble its main roots from the best talent available at the time', decided that 'future applicants must

Embracing biochemical engineering

Chemical engineering was expanding in other directions as well. By the mid-twentieth century there was a growing list of processes for producing chemicals using biological operations. 'Scientific' brewing methods, for example, based on the specialism of 'zymotechnology', had flourished from the late nineteenth century. In Britain, several strands of research combined in the period before the first world war to promote what has become known latterly as biochemical engineering or biotechnology. A rubber shortage between 1907 and 1910 encouraged searches for synthetic alternatives, notably at the large organic chemistry department under William Perkin, Jr. at Manchester. Chaim Weizmann, a young member of the department, pursued fermentation of agricultural produce as a promising route. Collaborating with industrialists and seeking to build 'a new English industry', by the beginning of the first world war Weizmann was Reader in the new subject of 'bio-chemistry' at Manchester.[98] His discovery of a bacterium that could yield acetone and butanol from starch was quickly commercialised by Nobel's explosives because of the value of the chemicals for explosives production. An acetone plant at the Royal Naval cordite factory at Holton Heath was in production by 1916, and the importance of the Weizmann process was trumpeted by Churchill, first Lord of the Admiralty. Thus fermentation vied with distillation as a key unit operation in wartime production. After the first world war, when petroleum reserves appeared limited, the production of synthetic fuels again helped to promote interest in biological methods.[99] Among those enthused by the potential was Sir Harold Hartley.[100] Nevertheless, the large manufacturers such as ICI paid little attention to such processes in the interwar period, when the employment of coal-tar as a raw material appeared most promising.

The second world war revitalised the development of biochemical processes. The production of synthetic rubber and drugs assumed significant dimensions by the end of the war. The case of penicillin, in particular, again illustrates how chemical engineering methods were adopted and adapted gradually, and how technology transfer operated between countries and disciplines. The antibiotic effect of the penicillium mould was described by Alexander Fleming in 1928, and benzylpenicillin was isolated and made in small batches in England from 1940 at the School of Pathology at Oxford University by a group under Howard W. Florey. ICI started a pilot plant operation in late 1942, after the formation of a consortium of British companies known as the Therapeutic Research Corporation, or TRC, the previous year.[101] Penicillin production

---

expect to be set a stiffer test of their experience and knowledge. This is scientific progress.' ['From the Secretary', *Nuclear Energy* Nov/Dec 1967, 153 and May/June 1968, 85]

[98] Weizmann (1874–1952) became the first President of Israel in 1948.

[99] Robert Bud, *The Uses of Life: A History of Biotechnology* (Cambridge: Cambridge University Press, 1993), pp. 17–21, 37–45.

[100] See, for example, H. Hartley, 'Agriculture as a source of raw materials for industry?', *Journal of the Textile Institute 28* (1937), 172.

[101] L. Bickel, *Rise Up to Life: A Biography of Howard Walter Florey* (London, 1972).

became the last major use of agency factories by the Ministry of Supply, which approved three major factories in 1943 and 1944 for its production.[102] In 1942, Florey had arranged for American manufacturers to produce the antibiotic. Unlike the British scientists, who had employed vast series of shallow bowls to grow the mould, the American chemical engineers developed a continuous process based on deep tanks of corn steep liquor aerated by oxygen and stirred by paddles.[103] Because of such innovations, the USA was able to supply ten times as much penicillin to the UK in 1944 as could be produced by indigenous firms. By the following year, however, British production exceeded American imports.[104] Nevertheless, American pharmaceutical firms such as Pfizer, Merck, Squibb and Eli Lilly became major antibiotics manufacturers in the postwar years based on such new technology.

One of the first penicillin production plants in Britain was set up by Distillers Co. Ltd in the last year of the war. The company, which specialised in alcohol production and fermentation processes, collaborated with an American company, Commercial Solvents, which was beginning to make penicillin on a large scale. Its transfer of technology is another example of the contingent and precarious evolution of such projects. Three Distillers employees – a biochemist, a research chemist and a chemical engineer – were sent to collect information about the American plant, constructed by E. B. Badger & Co. in Indiana. The visiting British chemical engineer stayed seven months, during which the team was unable to buy equipment owing to the project's relatively low priority for both the American and British governments. He expedited payment and acquisition by referring to himself not as a 'chemical engineer' but as 'Director of Procurement', and by returning to Britain with British Air Force trainees via Canada. The American equipment brought back included filters for fermentation, new centrifuge designs, a high vacuum system and instrumentation not available in Britain. The resulting British plant was a close copy of the American original. The American chief engineer supervised construction, and a second American engineer replaced him until the plant's start-up. The Distillers plant was not extended; Squibb set up their own plant in Britain using a different process. Distillers, however, repeated this importation process by purchasing the rights to a streptomycin production process from another American company.[105]

---

[102] William Hornby, *Factories and Plant* (London: HMSO, 1952), p. 166.

[103] Developed in 1941 by the American Department of Agriculture in Peoria, Illinois, this process was based on a similar method for fermentation yielding gluconic acid. A. L. Elder, 'The role of the government in the penicillin program', pp 1–12.; R. D. Coghill, 'The development of penicillin strains', 13–22; D. Perlman, 'The evolution of penicillin manufacturing processes', 23–70; and, E. J. Lyons, 'Deep tank fermentation', 31–6, all in: Albert L. Elder (ed.), *The History of Penicillin Production*, Chemical Engineering Progress *Symposium Series 66* (1970); K. B. Raper, 'The penicillin saga remembered', *American Society of Microbiology News 44* (1978), 645–53; Bud, *op. cit.*, 103–5; Hans T. Clark *et al.* (eds.), *The Chemistry of Penicillin* (Princeton: Princeton University Press, 1949).

[104] M. M. Postan, *British War Production* (London: HMSO, 1952), p. 358.

[105] G. H. Hill to Johnston, 15 Jan 1997. Hill obtained a chemical engineering postgraduate diploma from Imperial College in 1935.

This building upon prior experience was typical of early British antibiotics production. Beecham, an established firm, was led into penicillin production in the early 1950s through its search for an improved manufacturing process for tartaric acid. One of the original members of the Oxford group that had first refined penicillin suggested a plant that could produce both tartaric acid and penicillin.[106] Among the other synthetic pharmaceuticals that originated in the UK and employed British chemical engineers was Mepacrine, an anti-malaria drug, and diphtheria vaccine, both from 1940, and Griseofulvin, manu-factured by Glaxo, UK from 1952.

After the second world war, American chemical engineers were active in appro-priating such processes as a 'natural' part of their discipline. In 1947 the US journal *Chemical Engineering* suggested that the 'biochemical engineer' was a variant of the chemical engineer, and gave its award to Merck for its development of the industrial fermentation process to produce streptomycin.[107] Harold Hartley, too, repeated his pre-war message as President of the IChemE.[108]

The growth of a disciplinary niche for biochemical engineering was more grad-ual. A handful of biochemical engineering departments emerged in Britain. At University College, London, M. B. Donald focused a group within the chemical engineering department, and he and biologist E. M. Crook, also at UCL, consid-ered founding a journal of biochemical engineering in 1957. At the University of Birmingham, on the other hand, a biochemical engineering department was cre-ated from the former British School of Malting and Brewing. Ernst Chain, one of the Nobel prize winners for the discovery of penicillin, headed a purpose-made institute at Imperial College.[109] Somewhat later, a biochemical engineering course was instituted at University College, Swansea.[110]

There was, nevertheless, a hesitant union between the biological and chemical engineering domains. As shown by penicillin production apparatus, the pharma-ceutical and brewing industries adopted chemical engineering techniques dubi-ously and without attributing central importance to a chemical engineering perspective to problem solving. The APV company, for example, which had supplied aseptic aluminium vessels for Weizmann's acetone-butanol process during the first world war, set up a chemical engineering department in 1940. Penicillin manufacture pressed the department to reorganise towards the end of the war, when F. E. Warner, eventually to become an elder statesman of the profession, assumed responsibility for the provision of heat exchangers, pipelines and fittings, pumps and control gear for the 'appalling cumbersome' bottle-culture penicillin plants also adopted by Boots (other firms used quite different equipment: Borroughs-Wellcome grew penicillium on surface culture, and ICI developed a process using flat trays). Indeed, Warner cited this as an example of the different

---

[106] Bud, *op. cit.*, 105–6.

[107] *Ibid.*, 101.

[108] H. Hartley, 'Chemical engineering at the cross-roads', *Trans. IChemE 30* (1952), 13–9.

[109] E. M. Crook, 'How biotechnology developed at University College, London', *Biochemical Society Symposium 48* (1982), 1–7. R. Bud, *op. cit.*, 155.

[110] Started in 1970 as a joint honours degree in chemical engineering and microbiology, it became a Biochemical Engineering option of the chemical engineering degree in 1974.

outlook of engineering fields in a 1951 BBC European Service broadcast on 'A new field in engineering': mechanical engineers solved the problem by using great quantities of bottles automatically transported, he argued, while chemical engineers built tanks to hold as much liquid as 80000 bottles and grew mould by stirring it with air.[111] APV was building a penicillin extraction plant for Wellcome, at Langley Park, as American information dried up in the face of postwar competition.[112] Materials were also in short supply. APV's request for stainless steel to build a new design of plant based on submerged fermentation was refused as being too long-term.[113] The British market evolved rapidly, with some firms negotiating licences from American companies and others merging. Glaxo, Boots, Wellcome, BDH and Allen & Hanbury increasingly controlled the market. At APV, penicillin contracts dried up. Although Warner was made manager in 1947 of a new department to handle food, fruit juice, wine and pharmaceuticals, after his departure APV maintained separate chemical engineering and food processing groups linked only by their shared expertise in distilling plant, divisions which endured through the 1960s.[114]

PROSPECTS FOR CHEMICAL ENGINEERS

While chemical engineers figure in each of these accounts of emerging and expanding industries, they do not dominate them. The chemical engineers in these new industries remained largely invisible until their growing numbers – as 'designers', 'engineers', 'chemists', 'scientific officers' or 'managers' – led to consolidation into technical groups within firms restructuring after a rapid initial growth.[115]

ENGINEERING A DISCIPLINE: EDUCATION AT MID-CENTURY

The members of the IChemE were not wholly overlooked by unappreciative employers, as might be inferred from the accounts of occupational changes described above. If chemical engineers were swept up as initially minor players in expanding industries, they later had considerable success in shaping their

---

[111] *Quart. Bull. IChemE* Oct 1951.

[112] Sir Frederick Warner interview by Johnston, 19 Dec 1997. Warner (b. 1910) had obtained a chemical engineering diploma from UCL in 1933 and worked during the war as a site engineer for Chemical Construction. AF 1265, 1320.

[113] The unavailability of plant owing to the waiting list for stainless steel also prevented significant chemical engineering development at ICI Plastics immediately after the war [S. R. Tailby interview with Johnston, 22 Oct 1997].

[114] G. A. Dummett, *From Little Acorns* (London: Hutchison Benham, 1981), pp. 81, 114, 132; G. A. Dummett to Johnston, 2 Apr 1997; G. A. Dummett, 'Chemical engineering in the biochemical industries', *TCE* Jul-Aug 1969, 306–10. See also Bud, *op. cit.*, 118–20.

[115] IChemE, 'Prospects for chemical engineers in the chemical industry', July 1946 (unpublished); D. J. Oliver, 'Chemical industry pay and conditions 1930 to 1939', (unpublished typescript Jan 1997), and A. Cluer, 'Careers in chemical engineering', speech at King's School, Chester, 30 Mar 1960; revised as IChemE pamphlet, 1961.

[116] Sample: 163 of a population of approximately 820 new corporate members. For comparison with other periods, see Tables 4-2, 5-1, 7-1 and 8-1.

Table 6-1 New corporate membership by occupation, 1945–56[116]

Self description	
Chemical engineer	35%
Chemist	15%
Engineer	14%
Manager	25%
Educator	4%
Consultant	1%
Technologist/assistant	6%
Metallurgist	1%
Sales representative	0%

evolving disciplinary presence. The postwar era saw strenuous attempts by the Institution to establish educational programmes and to influence 'manpower' planning. The period proved crucial for the development of chemical engineering as an academic discipline. In parallel with the rapid growth of chemical production over the following decade, educators and IChemE officials struggled to increase the number of chemical engineers and their uses.

Influential employers, in need of competent specialists, were broadly approving of these moves. The financial support of a number of prominent firms helped secure the establishment of several new academic departments in the 1950s, as well as the rejuvenation of some of the older ones.[117] The sums involved (Table 6-2) were large compared with those donated to the universities

Table 6-2 Financial donations to university departments of chemical engineering[118]

Year	Source	Amount	Beneficiary	Purpose
1945	Shell Group	£435 000	Univ. Cambridge	Chair, scholarships
1945	Courtaulds	£110 000	Imperial College, London	Chair
	Lord Brotherton	£55 000	Univ. Leeds	Building extension, lectureship, scholarships.
1956	several	£306 000	Univ. College, London	General Engineering Building
1956	Shell Group	£1100	King's College, London	Studentships
1957	ICI	£25 000	Univ. Sheffield	Building extension

[117] MEC (1944–5); P. W. Reynolds, 'Chemical engineering with ICI at Billingham' *TCE*, Mar 1965, CE55–CE56. By 1953, a survey of managers at ICI suggested that they favoured training scheme for chemical engineers involving a first degree in mechanical engineering, followed by a postgraduate course of one or two years' duration, preferably after 'a few years' of industrial experience. 'Training of chemical engineers', *Quart. Bull. IChemE*, Oct 1953, xxxi.

[118] 'Department of Chemical Engineering, The University of Cambridge', *TCE* Feb 1960, A35–A36; A. C. G. Egerton, 'The Department of Chemical Engineering and Applied Chemistry, 1939–1942', *J. Imperial College Chemical Engineering Society 1* (1945): 7–14; C. F. Kearton, 'The chemical engineer in the man-made fibre industry', *Trans IChemE 32* (1954), 216–21; A. N. Shimmin, *The University of Leeds: The First Half Century* (Cambridge: University of Leeds, 1954), pp. 105, 173; Annual Report, University College, London (1955–6), 6; 'Chemical engineering in Sheffield', *TCE* Aug 1957, A35.

for teaching in other branches of engineering, and they were thus significant signs of industry's new attitude towards chemical engineering.

The new departments established in the 1950s at King's College, Newcastle (then part of the University of Durham), at University College, Swansea, jointly at the University of Edinburgh and the Heriot-Watt College of Technology, Edinburgh, and at the University of Nottingham also benefited from financial donations from industry.[119] However, the bulk of the recurrent funding of departments came from the government, via the University Grants Committee (UGC). This funding was in part the result of a political consensus that the state should shoulder a much higher proportion than before the war of the cost of all higher technological education. But the expansion of chemical engineering was far greater than that in any other engineering discipline. This was partly the result of the unparalleled success of the IChemE in influencing governmental opinion.

The experience of the war was crucial. Following the fight for the distinct classification of chemical engineering and the granting of state bursaries, chemical engineers were nominally put on the same basis as chemists, physicists and radio engineers. In 1943, the Minister of Production, Oliver Lyttelton (later Lord Chandos), expressed approval for the attention to academic and practical training given by the IChemE.[120]

However, the committee of inquiry established in 1944 by the Ministry of Education into the provision of higher technological training after the war (the Percy Committee) had not specifically considered that for chemical engineers.[121] Consequently, the IChemE, alone among the professional societies of engineers, sought the support of the Ministry of Labour and National Service, and that of the Board of Trade.[122] In March 1945, it presented, with the Institute of Petroleum, the ABCM and the BCPMA, the result of a survey of 90 chemical manufacturers. This suggested that over 'the next few years' a minimum of 700–800 chemical engineers educated to the level of a degree would be needed.[123] The matter was considered by a sub-committee of Lord Hankey's inter-departmental Technical Personnel Committee (TPC), the membership of which was largely sympathetic to chemical engineering. The representatives of the universities were in particular all keen to expand teaching of the discipline; they were Sir Alfred Egerton, A. R. Todd (later Lord Todd of Trumpington),

[119] 'Trends in Technological Education', *Quart. Bull. IChemE* Jan 1955, xxxvi.

[120] 'Annual corporate meeting in London', *Trans. IChemE 21* (1943): xvii–xx. See also 'Annual corporate meeting luncheon', *Trans. IChemE 22* (1944): xvi–xvii. The report of the committee chaired by Lord Eustace Percy in 1945 identified as major problems the diversity of educational provision and the isolation of universities and technical colleges from employers. It urged the development of some colleges to university standard, a national body to co-ordinate technological studies, and a campaign to increase the prestige of the technical professions.

[121] *Higher Technological Education*, Report of a Special Committee on Higher Technological Education, (London: HMSO, 1945).

[122] M. Davis, 'Technology, institutions and status: technological education, debate and policy, 1944–56' in: Penny Summerfield and Eric J. Evans (eds.) *Technical Education and the State Since 1850* (Manchester: Manchester University Press, 1990), pp. 129–32.

[123] 'Technical Personnel Sub-Committee: Chemical Engineers Sub-Committee', PRO LAB 8/1657.

then recently installed as professor of organic chemistry at the University of Cambridge, and J. H. Wolfenden, a chemist from the University of Oxford who, during the war, had undertaken a study of the provision in the United States of teaching facilities for chemical engineering. On the basis of a very limited inquiry, the committee revised the estimate of the future demand for chemical engineers to a figure in excess of 1200 over the next four to five years.[124]

The sub-committee went further than was customary in giving advice on the curriculum. It rejected the continental approach to training based upon the study of the technologies appropriate to specific industries, and confirmed the IChemE's contention that chemical engineering should be regarded as a discipline with application to a wide variety of industrial processes.[125] It commended both the existing course at Imperial College and that proposed at Cambridge (which laid a greater stress on an initial training in mathematics, physics and chemistry). By implication, the committee also endorsed the IChemE's 'Scheme For A Degree Course in Chemical Engineering' (1944); this broadly followed the syllabi of the courses in the University of London.[126] These findings were sufficient to secure the support of the UGC for the expansion of facilities in the universities.[127]

Neither the TPC nor the UGC considered the giving of more detailed advice on the curriculum as falling within its powers. Thus the responsibility for the development of courses in chemical engineering remained with the same authorities as before the war: teachers, overseen by the educational committee of the IChemE and by sympathetic employers.

NEW ACADEMIC PROGRAMMES

The fifteen years following the war witnessed the most rapid expansion in chemical engineering teaching that the subject has known. The number of institutions offering chemical engineering courses doubled, with new offerings ranging from part-time diploma courses to degree programmes. This expansion of chemical engineering was part of a tidal wave of technological education in which engineering was to be taught as distinct scientific disciplines.

'Emergency' courses

The 1945 united front of the IChemE with the Institute of Petroleum, ABCM and BCPMA petitioning the Ministry of Labour and National Service for expanded training in chemical engineering influenced several actions. The petition noted 'with very great disquiet' the inadequate supply of chemical engineers

---

[124] *Ibid.*

[125] *Ibid.*

[126] *Scheme For A Degree Course in Chemical Engineering* (London, 1944) The syllabus was drawn up by William Cowen (Manchester Municipal College of Technology); H. W. Cremer; M. B. Donald; F. H. Garner (University of Birmingham); W. C. Peck (Battersea Polytechnic); and A. J. V. Underwood.

[127] 'Chemical Engineers: Proposed Conference of Representatives of Universities ...' (1946) PRO, LAB 8/1342.

to industry, which inevitably would cause a loss of postwar export markets for the chemicals and plant industries. The organisations pointed to the support of industry, which already had provided grants to Cambridge University and Imperial College. The IChemE and its co-petitioners recommended that 100 students per year be deferred from military service, and that as many state bursaries as possible be provided.[128]

The Technical Personnel Committee (Chemical Engineering Sub-Committee) correspondingly devised a scheme for special emergency courses in May 1946.[129] The special courses were consequently announced that year by the Ministry of Education to meet the 'great and increasing demand for the chemical engineer'. Planned to continue for a minimum of four years, these one-year, full-time courses were open to students having a prior degree in science or a Higher National Certificate (HNC) in Engineering or Chemistry. The Ministry of Labour and National Service tempted candidates with the promise of exemption from examination for Associate Membership of the IChemE (except for the Home Paper) and the likelihood of immediate employment upon completion.[130]

In 1947 such special short term courses were initiated at Battersea Polytechnic, South West Essex Technical College, Loughborough College and Bradford Technical College. The instructors, for the most part, had backgrounds in industry.[131] Similarly, the Ministry of Supply's contribution to the training of chemical engineers was a summer Practice School held at an underused Royal Ordnance Factory in 1954, and repeated annually.[132]

---

[128] Letter, IChemE, I. Petroleum, ABCM and BCPMA to R. W. Luce, Min of Labour and National Service, 24 Mar 1945; 'Technical Personnel Sub-Committee: Chemical Engineers Sub-Committee' (1945), PRO LAB 8/1657.

[129] *Technical Personnel Committee – Chemical Engineering Sub-Committee*, May 1946, Gayfere archives, box XIV/1. Members of the sub-committee included representatives of the Ministry of Labour, Ministry of Education, War Office, Board of Trade, Ministry of Supply, DSIR, Ministry of Fuel and Power, National Service Chemistry Advisory Committee to Technical and Scientific Register, and academics: Profs. A. Egerton and D. M. Newitt of Imperial College, A. R. Todd of Cambridge and J. H. Wolfenden of Oxford. There was no direct representation by the IChemE.

[130] MLNS, *Special Courses at Technical Colleges: Chemical Engineers* (June 1946).

[131] S. Russell Tailby (b. 1917), for example, had worked at ICI in explosives, then plastics, during the war, before coming to Battersea. He was fifth in line of succession from J. W. Hinchley, Hugh Griffiths, Frank Rumford (1900–83) and William C. Peck (1897–1983). T. K. Ross (1916–78), with an external MSc from London, had industrial experience ranging from trainee chemical engineer to works manager and R&D department before the South West Essex appointment. See F. R. Whitt, 'Early teachers and teaching of chemical engineering', *TCE* Oct 1971, 370–4; AF 6895.

[132] *Chem. & Industry* 13 Feb 1954, 167; 186. The tutor of the first four-week course (open to final-year Honours students) was E. S. Sellers (b. 1912) of Cambridge University; R. Edgeworth-Johnstone, a former VP of the IChemE and assistant Director of Ordnance Factories, was the Ministry supervisor. From 1958, the course was extended to six weeks. *Chemical Engineering Practice School Report, 1958*, Royal Ordnance Factories (Explosives) Bridgwater Report No G.268 (June 1959), PRO SUPP 5/1230.

Higher National qualifications

The establishment of a National Certificate in chemical engineering after the second world war was a crucial issue for the IChemE, and one that replayed most of the difficulties it had encountered with recognition and influence. The need for National Certificate courses was discussed as early as 1939, but left in abeyance during the war.[133]

The argument pressed by proponents within the IChemE council was that chemical engineering could not truly be recognised as a 'primary' discipline alongside mechanical, electrical and civil engineering until it, too, could offer students a higher national qualification. In the last year of the war, the subject of higher national qualifications had been again taken up by the Education

Table 6-3 Introduction of National Certificates in engineering[135]

Subject and Institution	Year
Mechanical engineering	1921
Electrical engineering	1923
Production engineering	1941
Civil engineering	1943
Chemical engineering	1952

Committee as the variety of national certificates rose (see Table 6-3), but little support was found. John Oriel, in particular, complained that they would pose a threat to the standards of the Institution.[135]

By the following year, however, the committee's views were shifting. The emergency courses initiated in 1946 were seen merely as a stop-gap in a serious situation of under-supply of trained chemical engineers. A longer-term solution to meeting the demand was sought in the form of training programmes for new National Certificates. An approach was made to the Ministry of Education, but the Ministry offered only the combination of a Mechanical Engineering and Chemistry certification. From the outset, there was no plan to award Ordinary National Certificates (ONCs) in chemical engineering. The Advisory Council on Scientific Policy, formed in 1947, was made responsible for examining the supply and demand of scientific and technical labour. That year, the Ministry of Education arranged for 'supplemental endorsement' of National Certificates in Mechanical Engineering. This allowed the certificates of candidates who had completed a supplementary course in chemical engineering to be countersigned by the IChemE. This was followed up in 1950 for supplemen-

---

[133] Letter, W. Cumming to Council, 15 Feb 1939; CM 15 May 1939. A committee on National Certificates was set up under Cremer, which reported that decisions on HNCs were usually a matter for educational authorities rather than professional institutions, and probably demanded more clerical work than could be justified.

[134] MEC, 1945.

[135] Adapted H. B. Watson, 'Organizational Bases of Professional Status: A Comparative Study of the Engineering Profession' (Ph.D. dissertation, U. London, 1976), p. 145.

tary endorsement of Ordinary and Higher National Certificates in Chemistry and Applied Chemistry. For the ONC endorsement, the student was required to spend a further 180 hours studying Engineering Science (applied mechanics, applied heat and engineering drawing); for the HNC endorsement, the extra hours had to be devoted to a chemical engineering course.[136]

By 1950, the National Advisory Council on Education for Industry and Commerce (NACEIC) had conferred with the major engineering institutions and drafted a report on 'The Future Development of Higher Technological Education'.[137] Despite this state support, the IChemE did not recognise the HNC as equivalent to a degree for membership purposes.[138] By 1951, though, the Education Committee agreed that holders of the HNC with endorsements should gain partial exemption from the associate membership examinations.

In response to the NACEIC report, the Institution officers informed the Ministry that they felt that its definition of 'technology' was unclear. They asserted that technical colleges should teach 'secondary' technologies 'related to immediate or special work in industry' and leave to universities the 'primary' technologies 'more closely related to fundamental science' and 'applied to a wide range of industries'. While the primary technologies were suited to a degree course, the teaching of secondary technologies had been complicated by the confusion between the needs of technicians and technologists. Council initially recommended an extension of the prewar arrangement in which such secondary technology courses followed on from degrees in pure science and were recognised by postgraduate diplomas or master's degrees. Within a year,

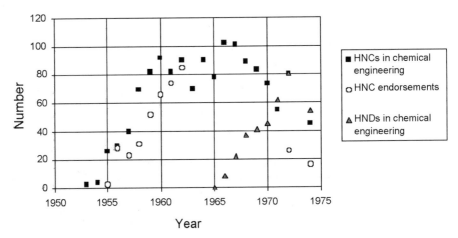

Figure 6-1 Higher National Certificate (HNCs) and Diplomas (HNDs). *Source*: IChemE Annual Reports.

---

[136] Ministry of Education, *Chemical Engineering*, Admin. Memo. 353 (17 Mar 1950). 180 hours was considered to require one extra year of evening study.

[137] Accompanied, that year, by '*Present and future supply and demand for persons with professional qualifications in chemical engineering*' (London: HMSO, 1950).

[138] Nor did the Institution recognise the new Dip.Tech. as being equivalent to a degree. For the situation in other Institutions, see Watson, *op. cit.*, 145–7.

however, Council members were coming round to supporting the granting of the 'special awards' that they had earlier rejected.[139]

M. B. Donald vaunted the HNC programme as a means of broadening membership without lowering standards (indeed, some HNC students later did postgraduate studies). Discussion grew wider after an IChemE committee chaired by John Oriel published its *Scheme for a Part Time Course in Chemical Engineering* (1952).[140] The committee called for a smaller curriculum than previous diploma courses, and tailored to local needs. Nevertheless, some complained that, at three years, the proposed HNC in chemical engineering was longer than most others.[141] Despite the absence of serious objections, representatives of industry, government departments and higher education were divided on their opinions of course content, the form and feasibility of providing teaching facilities, and the calibre of the resulting product. Further publications in the *Transactions* rehearsed the controversy.[142]

The Ministry of Education finally launched the scheme in 1952. The new qualification was linked to IChemE membership: holders of the HNC in chemical engineering qualified for Graduate Membership, and those having an HNC endorsement in chemical engineering were exempted from all but the Home Paper (design section) of the Institution examination. By December 1953, HNC courses in chemical engineering were established or under consideration at eleven technical colleges. But enrolments were so poor that the Ministry of Education was unwilling to fund chemical engineering laboratories, and without laboratories certificates could not be granted. To avoid this vicious circle, Council members proposed various solutions. Frederic Garner of the University of Birmingham suggested more publicity in industry; Oriel encouraged the assistance of the ABCM or BCPMA to stimulate interest; but others noted that convincing school-leavers of the advantage of a part-time course in chemical engineering was difficult at best, because no distinct trade or post could be identified for certificate holders.[143]

The ABCM and BCPMA, in their turn, urged a survey of training facilities to quantify the situation. The results quickly confirmed that technical college courses were not popular. While university degree and postgraduate diploma courses were subscribed at nearly full capacity, their technical college equivalents filled only about one-quarter of the available places.[144] Explanations were varied. The Institution contended that the problem was caused by a lack of

[139] '*Comments by the Council on Draft Report T757/137*', 14 Apr 1950; letter, J. B. Brennan to Permanent Secretary, Min. of Education, 12 Jan 1951.

[140] See also IChemE, 'The education of the chemical engineer by part-time study', *Trans. IChemE* 29 (1951), 346–55.

[141] The HNC courses in Civil Engineering and Gas Engineering also ran over three years.

[142] E.g., Earl of Halsbury, 'Higher technological education: a review of the controversy', *Trans. IChemE 30* (1952), 277–85.

[143] CM 16 Dec 1953.

[144] IChemE, '*Survey of Chemical Engineering Training Facilities in Great Britain*' (1953), unpublished Council discussion paper, Gayfere archive. Of a reported capacity of 319 places for first degree and postgraduate diploma courses at universities, 295 were taken; for technical college courses merely 34 of 133 places were filled.

science teachers in grammar schools to provide preparation and career advice. One college complained of difficulties attracting a lecturer, and that the need to follow a six year part-time course 'does not attract Students who have yet to undertake National Service', while another ascribed the low numbers squarely to the need to pay a £125 non-resident's fee for students entering the county. The status of the HNC programme was even more dire: of nine offered courses, five had no students enrolled in their part-time day courses, four had none for their evening courses, and Battersea Polytechnic had none at all.[145] What had been promoted as a viable scheme struck Ministry of Education representatives as being unwanted.[146] Those courses that were successful, such as that at Birkenhead Technical College in the Wirral, were able to attract a ready market of students from the nearby Shell, Unilever and other chemical industries.

In the Autumn of 1954 Harold Hartley, then President, wrote to all corporate members to promote HNC courses. Responses again were mixed. Of two dozen Members expressing an interest, several complained that no suitable courses existed in their region, or that the shortage of labour meant that men were taken on for training in plant work irrespective of their technical qualifications. Part-time courses demanded local facilities available to working students. While a few managers promised participation, most either requested further information or forecast one or two course attendees annually from their company.[147] Hartley, ever the publicist, wrote of the promise and bottlenecks of the profession in the *Financial Times*.[148] The following summer the next President, John Oriel, again sent an open letter to address the complaints and to emphasise that the part-time courses were widely available and suitable for day-release training programmes for small industry.[149]

By 1957, with over a hundred HNCs in chemical engineering or supplementary endorsements granted, the programme was still open to criticisms. One industrialist observed that the Ministry, for once, 'was ahead of demand' in providing the HNC certificate, for which the number of students was 'pathetically low'.[150] The Principal at West Ham Municipal College warned that an overhaul was needed, because entering students were not well enough 'conditioned'. He urged an ONC in chemical engineering as a preparatory qualification.[151]

Other initiatives were proposed. The Institution in 1956 had recommended a skeleton curriculum for works training by individual companies. This practical

---

[145] *Ibid.*, 8, 11.

[146] S. R. Tailby interview with Johnston, 22 Oct 1997.

[147] J. B. Brennan, Gayfere archive box 2, 'H.N.C. in Chemical Engineering', circa Dec 1954.

[148] H. Hartley, 'Chemical engineering' and 'Chemical engineers – are we training enough?', *Financial Times*, 19 May 1955 and 22 Aug 1956, respectively.

[149] J. Oriel, Gayfere archive box 6/Accred, 'New courses in chemical engineering', Council discussion paper, July, 1955.

[150] F. Caunce, 'Discussion of recruitment to the profession', *Symposium on Chemical Engineering Education* (London, 1957).

[151] E. A. Rudge, 'The Higher National Certificate in Chemical Engineering at West Ham College of Technology', *TCE*, June 1957, lii, xxxvii.

training was essentially a chemical engineering apprenticeship. The IChemE stressed, through, that such a course had to be 'complementary to the technical college course'. The plan called for a five year course including the major instructional components of plant fabrication, process plant operations, plant design, plant maintenance, laboratory, instruments, safety, organisation and management.

Students completing such a course were required to take the IChemE Home Paper only.[152] The Kestner Evaporator and Engineering Co soon offered a course drawing heavily on this model, although oriented towards its own requirements of mechanical shop skills.[153]

New and extended university programmes

On a much more ambitious scale were academic programmes instituted at universities and funded, in some important cases, by industry. By 1952 there were degree courses at eleven centres, along with diploma courses at another dozen.[154] From war's end to 1960 the annual production of chemical engineers from all types of course rose by five-fold to some 600 per year.[155] The rise of university courses was undoubtedly aided by the effort of the IChemE to recruit public school boys for the profession. The rise in the number of chairs illustrates a more dramatic colonisation of academic institutions: from a single professor of chemical engineering at the beginning of the war to fifteen by 1960.[156]

Among the more prominent new courses were those at Cambridge, Birmingham and Imperial College.

Cambridge

Brought into existence by the endowment of a chair of chemical engineering and provision of scholarships by the Shell Group, the department's running costs were also met entirely by Shell in the early years. This was largely due to the efforts of John Oriel, who had discussed the establishment of a chair before the war with Professor Norrish, and who revived plans in 1944.

There was considerable discussion as to whether the department should form part of the Faculty of Engineering or that of Chemistry, but eventually it was

---

[152] 'Works training', *Chem. & Indus.* 14 Jan 1956, 35.

[153] 'Kestner training scheme', *Chem. & Indus.* 5 May 1956, 332.

[154] Degree courses were offered by five colleges of the University of London, Birmingham, Cambridge, Durham, Glasgow, Leeds, and Manchester; diploma courses at King's College, London; Imperial College; University College, London; Battersea; West Ham; Universities of Birmingham, Durham and Leeds; Manchester College of Technology; Loughborough College; Birmingham College; and Bradford Institute of Technology.

[155] See Appendix.

[156] The Ramsay chair at UCL was the sole professorship before the war, following Hinchley's death. By 1960, chairs in chemical engineering had been established at Imperial College (1945); Leeds (1945); Cambridge (1946); Birmingham (1947); University College, Swansea (1954); King's College, Newcastle (1954); Edinburgh (1955); King's College, London (1955); Sheffield (1955); Strathclyde (1956); a second at Imperial and at King's College, London (1956); Manchester (1956); Nottingham (1956). See Appendix.

decided that it should stand alone. As a result, the professor sat on the Boards of both the Engineering, and Physics and Chemistry, faculties. After appointment of a professor, lecturer and demonstrator by 1947, teaching at undergraduate level commenced almost immediately, with the tripos being first set in the 1949–50 session.[157] The professor of chemical engineering eventually appointed, Terence C. R. Fox (1912–62), had been trained as a mechanical engineer at Cambridge before working for several years at ICI's high pressure plant at Billingham. He had then returned to Cambridge to teach mechanical engineering.[158] Fox visited American universities with chemical engineering programmes in 1947, and was also advised by John Oriel 'as to how the needs of industry could best be met by their training'.[159]

University of Birmingham

Originally a Department of Oil Engineering and Refining under Professor A. W. Nash, the Birmingham programme was re-organised after 1942 by Professor F. H. Garner. In 1946 the Department of Chemical Engineering was created, with four lecturers plus Garner. According to one of them, Frank Morton, 'a certain amount of consultancy was encouraged'. Garner was 'not a bit impressed with the way that industry used his graduates', although A. W. Pearce of Esso claimed that 'Garner produced the type of chemical engineers that industry wanted'.[160]

Imperial College

Teaching at Imperial had been disrupted by the war. S. G. M. Ure had died in 1941 and was temporarily replaced by G. W. Himus, a fuel technology lecturer, and then by M. B. Donald who had been at UCL and joined the Forces (and who returned to UCL in 1946 when the department was re-opened). Interestingly, the first generation of organised chemical engineers trained during the first world war wielded control after the second: A. C. Egerton (1886–1959), who had worked with Quinan and Macnab at the Ministry of Munitions during WWI, was replaced on his retirement in 1952 by D. M. Newitt (1945–62), Hinchley's student and founder member of the IChemE. Newitt was the first incumbent of a new, permanent, chair provided by a £110 000 endow-

---

[157] John Bridgwater, *Fifty Years Young – Products and Processes – The Future of Chemical Engineering*, inaugural lecture, Univ. Cambridge, 30 Jan 1995 (Cambridge: Cambridge University Press, 1996).

[158] Fox was offered the post after an Imperial College lecturer declined it. M. B. Donald, in his turn at musical chairs after shepherding the IC chemical engineering programme through the war, grudgingly returned to UCL when Dudley Newitt won the new chair at Imperial. Donald succeeded H. E. Watson as Ramsay Professor at UCL some five years later.

[159] Few documentary records of the early Cambridge department survive.

[160] Garner (1893–1964), gaining an MSc in Chemistry from Birmingham University during the first world war, then studied hydrocarbon combustion at the Royal Aircraft Establishment, Farnborough. He received a PhD from MIT in 1921 and was the first refinery chemist at Fawley. A. W. Pearce, 'Prof. F. H. Garner: the teacher', *TCE* Jul/Aug 1970, CE211–CE214; 'Famous men remembered', *TCE* (Feb, 1985); 'Tradition and innovation: The Professor F. H. Garner Memorial Lecture', D. M. Newitt, *TCE* Jul/Aug 1967, CE148-CE153.

ment (£1.5M in 1999 currency) from the polymer manufacturer, Courtaulds, building on links forged in the war.

A second chair, in 'Chemical Engineering Science', was created in 1957 and filled by Peter V. Danckwerts. Danckwerts (1916–84) had obtained a degree in Chemistry from Oxford in 1939, spent the war in the navy and then completed a MSc at MIT. Before the Imperial College appointment, he had spent two years in development work under Hinton at the Risley facility of the UKAEA.[161] Danckwerts tailored his chemical engineering towards a grounding in physical chemistry and the elements of classical physics, with illustrations of practical problems. Chemical engineering science was introduced in the first year of the degree, and given increased emphasis in the second and third. He also emphasised a design project, which made up a significant part of the third year laboratory course along with demonstrations of unit operations.

DESIGN SKILLS VERSUS CHEMICAL ENGINEERING SCIENCE

After the second world war, most graduates in chemical engineering received a training which stressed the importance of design. This was in sharp contrast to the teaching offered on most courses for civil, for mechanical and for electrical engineers, which increasingly emphasised instruction in 'engineering science' at the cost of training in design and in other practical skills.[162]

The distinctive nature of an academic training in chemical engineering was partly the result of the educational policy of the IChemE. Under the chairmanships of John Oriel of Shell (1945–51), Professor M. B. Donald of UCL (1952–53), and A. S. White of the UKAEA (1953–60), the IChemE education committee continued to insist that a training in design should form a central part of a training in chemical engineering. The undergraduate course at the University of Leeds and that taught jointly at the University of Edinburgh and Heriot-Watt College, Edinburgh, were initially refused recognition by the Institution because of failings in this respect. By contrast, the new undergraduate course at King's College, Newcastle, approved in 1958, dedicated most of the final year to design problems similar to those set by the IChemE.[163]

The course at the University of Cambridge, established in 1948, was the sole exception to the IChemE's policy on design. The Department of Engineering emphasised instruction in the engineering sciences, and T. R. C. Fox remained true to this tradition.[164] Moreover, he considered that it was pointless to

---

[161] *Biog. Mem. FRS 32* (1986) 99–114; P. V. Danckwerts, 'The challenges of chemical engineering science', in: B. Atkinson (ed.), *Research and Innovation for the 1990s: The Chemical Engineering Challenge* (Oxford: Pergamon, 1986).

[162] C. Divall, 'Fundamental science versus design: employers and engineering studies in British universities, 1935–1976', *Minerva 29* (1991), 184–94.

[163] MEC (1946–60). The course at the Manchester Municipal College of Technology was initially refused re-recognition, but the problem has not been identified.

[164] Fox never published a research paper during his career, and was little known outside the Cambridge circle. He was for a short time a member of the Council of the IChemE but never took a very active part in their affairs. E. S. Sellers, 'T. R. C. Fox', *Chem. & Indus.* 15 Dec 1962, 2091–2.

attempt to train students in design because it was impossible to reproduce in a university department the kind of economic and managerial constraints under which a designer in industry would have to work.[165] Fox's opinion prevailed despite John Oriel's membership, until 1961, of the Chemical Engineering Syndicate, a small committee of academics and experts from industry which formally governed the running of the department.[166]

The curriculum at Cambridge contained much material of a scientific nature, particularly chemistry. The majority of students spent the first two years of their course studying for Part I of the tripos for an honours degree in the natural sciences; alternatively, they could study the 'mechanical sciences' (that is, engineering science). Until 1968, the course for the tripos in chemical engineering was of a further two years' duration. The subjects studied depended on the student's previous training, but over the two years a set of courses common to all students gave roughly equal emphasis to the study of applied physical chemistry and chemical processes, and to that of unit operations and the physical sciences associated with them. Instead of practical work in design, students undertook a 'modest research project' in their final year, and in the summer vacation they attended a practice school of industrial experience under the tutelage of academics at Shell's refinery at Stanlow; this was modelled on the famous example at MIT.[167] The IChemE was prepared to recognise this syllabus as exempting it from all academic requirements for membership, including the Home Paper, although it maintained pressure on the university to adopt a project in design.[168]

The majority of industrialists on the education committee agreed that design should form a central part of the curriculum. A suggestion in 1954 that the Home Paper should be abandoned found little support. Nevertheless, the problems in design set in most universities were of a lower standard than the Home Paper.

The inability of the universities to teach design skills to a level judged appropriate by the IChemE was caused by several factors. By the mid 1950s, the disparity of salaries between industry and the universities meant that it

[165] P. V. Danckwerts, 'Famous Men Remembered', *TCE* Dec 1984, 44–5; C. Divall, 'A measure of agreement: employers and engineering studies in the universities of England and Wales, 1897–1939', *Soc. Stud. Sci. 20* (1990), 80; Divall, 'Fundamental Science', *op. cit.*, 173–5.

[166] H. Hartley, 'The John Oriel Memorial Lecture' *TCE* Sep 1969, CE138–CE45; J. F. Davidson, 'Chemical engineering in Cambridge: a review' (n.d.) (typescript, Professor Davidson's personal papers): 1–2. Design studies eventually were instituted under Fox's successor, P. V. Danckwerts.

[167] Lists of lectures in *Cambridge University Reporter* (1949–50, 1959–60); 'Department of Chemical Engineering', *op. cit.*, A35–A36; Davidson, 'Chemical Engineering', *op. cit.*, 4–5. In the first sessions, Cambridge and then Birmingham students participated. The scheme was extended the following year to the Shell Haven refinery and then to the Esso Fawley refinery. Letter, D. Freshwater to Johnston, 29 Sep 1998.

[168] It is not clear why the Institution made this decision. Since graduates from Cambridge were easily employed and the department came rapidly to enjoy a good reputation with industry, it probably felt it had no choice. [MEC (1955–6); E. S. Sellers, 'Education for the chemical engineering profession', *TCE* Aug 1957, A45; J. F. Davidson, 'IChemE and I: a personal view', *TCE* Jan 1971, 1; letter, Davidson to Divall 17 May 1991].

was difficult to recruit teachers with sufficient practical experience, although professors of chemical engineering had generally spent longer in industry than their colleagues in other branches of engineering. For example, both the professors appointed at the Manchester College of Science and Technology (the former Municipal College of Technology) had many years' industrial experience; this had not been true of the professors of mechanical and of electrical engineering since before the second world war.[169] The widespread appointment of teachers as industrial consultants would have alleviated the problem, but firms were unwilling to employ the more junior members of staff.[170]

Another problem stemmed from the desire of teachers to introduce into courses the results of their academic research. A similar desire amongst academics in the other principal branches of engineering had become apparent shortly after the second world war.[171] The first theoretical treatment of the behaviour of material subject to the unit operations had been developed between the wars. Such research continued with greater vigour after 1945, particularly via the committee of inquiry established by the Department of Scientific and Industrial Research under Harold Cremer.[172] By the late 1950s, academics had undertaken a considerable amount of research into the transfer of momentum, heat and matter in fluid flows in order to elucidate the 'mechanisms' of the unit operations; they had also started to develop insights into the factors influencing the efficiency and rate of chemical reactions on the industrial scale, through the study of chemical thermodynamics and reaction kinetics.[173] The scale of the expansion of teaching tended to discourage extensive reform of the curriculum until late in the 1950s; this was the case, for instance, at University College.[174] Nevertheless, by 1960, undergraduate courses in several universities were paying considerable attention to subjects in 'chemical engineering science'. For example, the new syllabus for the honours degree introduced in 1956 at the Manchester College of Science and Technology included theoretical courses on the transfer of heat and mass; a similar development was apparent at

---

[169] Personnel Files, UMISTA (1934–67); Divall, 'A measure of agreement', *op. cit.*, 88.

[170] 'A survey of output of chemical engineers in Great Britain', *TCE* Apr 1958, A39; P. V. Danckwerts, 'Chemical Engineering in a University', in *Papers to be presented at a Symposium on Chemical Engineering Education* (London: IChemE, 1957), p. 59; 'Engineering design', *TCE* Jun 1963, CE195; 'Private consulting work' (1957–73), UMISTA.

[171] Divall, 'Fundamental science', *op. cit.*, 172–81.

[172] PRO DSIR17/440–17/447, 'Chemical Engineering Research Committee', Organisation and Meetings 1 to 9.

[173] Report of the Committee on Chemical Engineering Research (London: HMSO, 1951); H. Hartley, 'Chemical engineering at the crossroads', *Trans. IChemE 30* (1952), 13–9; D. M. Newitt, 'The scope of research in a university chemical engineering department', *Quart. Bull. IChemE*, Oct 1955, xliii-xlv; Danckwerts, 'Chemical engineering in a university', *op. cit.*, 65–9; J. M. Coulson, 'The changing nature of the unit operation concept', *Papers to be presented, op. cit.*, pp. 46–50.

[174] S. R. Tailby, 'Chemical Engineering Education Today', *Chem. & Indus.* (20 January 1973): 77–85; P. N. Rowe and A. R. Burgess, 'Chemical engineering at University College London' in N. Peppas (ed.), *One Hundred Years of Chemical Engineering* (Dordrecht: Kluwer, 1989), 223–36.

Imperial College.[175] The insistence of Fox, at Cambridge, on a thorough training in the theoretical articulation of chemical engineering phenomena was
regarded by teachers elsewhere as the norm for courses leading to an honours
degree. By contrast, the standard of scientific understanding required by the
IChemE was said to be about that of an ordinary degree.[176]

Some employers approved of the emphasis on the teaching of the theoretical
science of chemical engineering; an extreme example was Dr D. Clayton of
ICI, who said in 1957 that it should be encouraged at the expense of the design
problem.[177] However, there was a divergence between the views of larger and
smaller firms. The former tended to expect graduates to have a facility only
with process design, and they welcomed the insights that could be expected
from those with a thorough theoretical training. If graduates were to be
employed in the erection, commissioning and operation of plant, any deficiencies in their academic training could be corrected during a period of instruction
in the firm. Smaller firms were far less likely to invest in training. Hence they
tended to employ more graduates with an ordinary degree, since these engineers
were judged to possess a more practical attitude than those with the honours
award.[178] The Manchester College of Science and Technology established, in
1956, a 'practical' course leading to the award of a pass degree; instead of
taking advanced courses in thermodynamics, kinetics and reactor design, flui-
disation techniques and mass transfer, students studied factory management
and layout, structures and other aspects of engineering design. However, this
initiative caused considerable controversy within the IChemE over the desirability of lowering the academic standard of admission to the institution from
that of the honours award, and no other university appears to have developed
a separate course for the pass degree.[179]

Despite these criticisms and difficulties, in the early 1960s the standard of
training in design in chemical engineering was generally higher than that found
in other university departments of engineering. In 1963, such standards for

[175] F. H. Garner, 'Full-time undergraduate training' in *Papers to be presented, op. cit.,* 18–22;
       'Chemical engineering as a science and an art', *TCE* Apr 1962, A73–A74; Danckwerts, 'Chemical
       engineering in a university', *op. cit.,* 62–4; F. Morton, 'The expansion of chemical engineering
       in the Manchester College of Science and Technology', *TCE* Oct 1957, A39–A41; *Calendar,*
       Imperial College (1950–1, 1957–8, 1960–1, 1970–1); Bainbridge, 'The Department of Chemical
       Engineering', *op. cit.,* 16–7.
[176] E. S. Sellers, 'Education for the chemical engineering profession', *op. cit.,* A45–A47.
[177] D. Clayton, 'The requirements of industry: the chemical industry' in *Papers to be presented,
       op. cit.,* 2–5.
[178] C. Buck, T. Hayes and R. Williams, 'Design, erection and operation of pilot plant', *Trans.
       IChemE 24* (1946), 44–53; R. Holroyd, 'Considerations of technical staff requirements in the
       evolution of an industrial chemical project', *Trans. IChemE 32* (1954), 210–5; Kearton, 'The
       chemical engineer', *op. cit.,* pp. 216–7; *Discussion on papers presented at a Symposium on
       Chemical Engineering Education* (London, 1957), A2; R. D. Hayes, 'Production management
       in the chemical industry', *TCE* Oct 1960, A43–A44; M. C. Coe, 'Industrial training for graduate
       chemical engineers', *TCE* Jul/ Aug 1964, CE167–CE73.
[179] Morton, 'The expansion of chemical engineering', *op. cit.,* pp. A39–A41; 'University of
       Manchester, Faculty of Technology, Ordinary Degree in Chemical Engineering', Minutes of
       the Board of Moderators, in MEC (24 October, 1961).

other forms of engineering were severely criticised by the committee of inquiry established by the DSIR on engineering design (the Feilden committee). The committee also censured the professional societies of engineers for their lack of attention to the provision of training in design. The content of courses in chemical engineering and the IChemE were excluded from these criticisms.[180]

Moreover, most teachers of chemical engineering, even those most closely concerned with the theoretical articulation of the subject, continued to believe that undergraduate courses should cultivate design skills, at least in the preparation of thermal and mass flow sheets. In sharp contrast to the practice in the other branches of engineering, teachers who favoured a philosophy of design were being appointed to the more eminent departments. For example, Peter Danckwerts, the first Professor of Chemical Engineering Science at Imperial College, thought that an academic training in chemical engineering should focus on process design; he introduced the design project into the tripos at Cambridge after he succeeded Fox.[181]

Sandwich courses

The mingling of two trends – academic emphasis on 'engineering science', coupled with growing diversity of chemical engineering occupations – promoted more direct interaction between industry and academe. The result was the expansion of programmes teaching chemical engineering students by a combination of course work and experience from industrial placements. The schemes found favour with senior industrial engineers, many of whom had been trained by a combination of university studies and on-the-job training before the war.[182] John Oriel, the Shell industrialist and IChemE official, was a strong proponent. Some academics, on the other hand, initially voiced the argument that work experience would disrupt the habit of study.[183]

Such 'sandwich courses' had long been in operation in a few teaching institutions, notably Glasgow Tech and Salford during the 1920s. At a few universities such as Imperial College and Cambridge, students were encouraged to spend a year in industry before their academic studies, but this was unsupervised and not mandatory. In 1956, however, with the introduction of the newly designated Colleges of Advanced Technology (CATs) and the new award for technological training (denoted the Dip.Tech.), a government initiative set up a new supervisory body. A Committee under Lord Hives designed courses, intended to be equivalent to a university degree, for this new class of award. The courses were required to provide at least twelve months of supervised training in industry as part of the instruction. This was performed either in

---

[180] Divall, 'Fundamental science', *op. cit.*, 189–91; 'Engineering design', *TCE* Jun 1963, CE167.
[181] Danckwerts, 'Chemical engineering in a university', *op. cit.*, pp. 64–5; 'Science in chemical engineering', *TCE* Jul/Aug 1966, CE155–CE59; 'Chemical engineering science', *TCE* Jun 1972, 222–4; K. G. Denbigh, 'Peter Victor Danckwerts', *Biog. Mem. FRS 32* (1986): 102. See also e.g., R. W. H. Sargent, 'Chemical engineering and engineering science', *TCE* May 1963, CE152.
[182] Notable among them was Christopher Hinton; see his *Engineers and Engineering, op. cit.*
[183] R. S. Tailby interview with Divall, 29 Jul 1992.

two industrial 'slices' of six months each, between two or three academic periods of study (known as the 'thin sandwich scheme'), or in a single period of twelve months between the third and final year of study (the 'thick sandwich' scheme). The industrial placements were organised jointly by academics and industrial personnel, and the students were visited regularly by college staff during their working periods. Such courses spanned the engineering disciplines and were vetted by the Hives Committee, later replaced by the National Council for Technological Awards.

Such Dip.Tech. courses were a required format for the CATs, but were also taken up by other technical colleges. In 1956, for example, Glamorgan Technical College instituted a four-year sandwich course, and the Royal Technical College at Salford started a three-year course. The IChemE began to recognise such courses within a year, but did not grant exemption from the Home Paper for diplomates.[184] Despite this hesitant recognition by the Institution, sandwich-course diplomates and, later, graduates were popular with industrial employers.

INSTITUTIONAL INFLUENCE ON ACADEMIC CURRICULA

While academic programmes were expanding satisfyingly after the war, their disparate origins and personnel still led to more independence and diversity than the IChemE Education Committee would have liked.

In 1946 the Committee sought to extend its control to degree courses. The members agreed that the recognition of degree courses should be to a higher standard than other types, and should include examination of the syllabus, standards of teaching, equipment, examinations and marking. The Institution would then grant approval for 5 years (this compared rather strictly with 'accredited' courses, which would merely prepare graduates for the Associate Membership examination). The IChemE recognised the degree programmes of London, Birmingham, and Cambridge in 1949, but rejected that of Manchester until 1955. Similarly, the Leeds course was not exempted from Institution examinations until it incorporated a design element that year. The committee found its authority to accredit challenged, however. Members were warned that universities would not tolerate the monitoring of marked scripts. Nor did all institutions want to be assessed; Manchester Technical College declined such an offer when first advanced in 1952.

The mechanism of recognition was also adapted. By 1957, a condition of recognition of 'certain courses' was that external examiners be approved by the Institution. Two years later a Panel of Moderators was established. At the same time, the Committee decided that there was no need to visit any educational institution in receipt of a University Grants Committee grant, nor would their endorsements be limited to a fixed time period. Other institutions, however, would be visited by at least two moderators and receive a fixed-term approval. By the end of the decade, a dozen UK universities were recognised, all offering honours courses.

---

[184] CM 18 Jul 1956; 20 Nov 1957.

As with the setting of membership criteria in the early 1920s, policies on education were contentious in the 1950s. The design problem, and its potential for deterring membership applications, was at the centre of the debate. Some members, such as John Oriel, were keen that standards not be reduced, but others noted that the insistence that all graduates take the Home Paper would lead to a drop in applications. This appears, indeed, to have been the case for students from sandwich courses: because their Dip. Tech. was not recognised by the IChemE as equivalent to a degree, they were not exempted from its written examinations, as were their counterparts applying to other engineering institutions. Terence Fox's programme at Cambridge was unique in Britain in excluding a design element. G. G. Haselden, at Imperial College, while favouring the design project, thought that flexibility was required. The majority nevertheless agreed that the Home Paper lay at the core of the identity of chemical engineering as an engineering discipline.

The influence of industry on courses was also mixed. There is no evidence of any course having been 'made to order', in the manner suggested by David Noble with reference to the USA.[185] The great majority of employers who were sympathetic to the growth of the new profession did not feel competent to comment in detail on curricular matters, although, by comparison with the other societies of professional engineers, the IChemE was able to achieve a remarkable degree of support and co-operation from industrialists for policies on academic training. The importance attached by teachers and by consultants to the cultivation of design skills found favour with some employers as a test of a student's ability to apply theoretical knowledge to real problems, even though the majority of chemical engineers would not practice as designers.

Thus in sum, governmental and industrial policy in the postwar period encouraged and financed a comparatively high level of provision of academic training. But the development of curricula was left largely to teachers. They pursued their own professional aspirations as academics by developing the science of chemical engineering. The consequential attention paid to the teaching of science met largely with the approval of industrialists, because for the most part, teachers continued to recognise the need to inculcate in students an appreciation of the practical limitations of formal knowledge. Their changing attitudes were exemplified by D. Clayton, head of the Department of Engineering Research at ICI, Billingham, who broke with earlier ICI opinion by observing in 1957 that the position of the chemical engineer was 'as well-defined as those of civil, mechanical and electrical engineers', and was important for large and small firms to promote efficient production. Indeed, he opposed the view of his predecessor, William Rintoul, stating:

> Some students are recommended to take a full first degree in chemistry or engineering and then do chemical engineering in the following year (or two). This has usually been done on the mistaken idea that only in this way can a good university training

---

[185] D. F. Noble, *America By Design, op. cit.*

be secured; this idea is gradually disappearing with the recognition that chemical engineering is a suitable university discipline in its own right.[186]

A growing convergence of the occupation and of the discipline was becoming evident. The establishment of autonomous departments of chemical engineering meant that, from the late 1950s, teachers could respond, without fear of a confusion of professional boundaries, to employers' demands for chemical engineers who were better trained in the chemistry of manufacturing. The postwar years saw sky-rocketing growth in output of chemical engineers: from a mere 31 produced in 1949, to 277 in 1956. The total number practising in the country that year was estimated at 1500, with some 2200 predicted for the end of the decade.[187] Combined with the continuing growth of the processing industry, the prospects for a stable profession appeared excellent.

NEW PROFESSIONAL LIAISONS AND THEIR ORIGINS

While the expansion of industry and the securing of teaching provision were crucial to the occupation and discipline after the war, both were supported by a growing web of organisational links. The extension and nurturing of these disparate affiliational threads came to occupy a growing fraction of the energy of IChemE administrators.

Three varieties of interaction assumed significance: the growth of regionalism within the institution, via local and international branches; the extension of government/industry linkages; and the quest for a more formalised status for the profession. Each activity was a case of crossing boundaries, either regional, national or professional. And in each case, the expansion was accompanied by a dilution of central control.

Shifting balance: regional branches of the IChemE in Britain

The authority represented by the IChemE Council was challenged, to some extent, by change from within. In a series of formations in the two decades after the second world war, local branches began to play an increasing role in members' professional lives. These branches were not the result of a change in administrative policy by the IChemE; nor were they a concerted response by members outside London to any perceived aloofness of Council. Instead, the branch structure within the Institution evolved because of local circumstances and correspondingly received a mixed reception at the London headquarters.

The first platform for other public voices within the Institution was the formation of a Graduates and Students ('G&S') section in late 1928. This was intended to represent junior members and, while not explicitly regional, did

---

[186] D. Clayton, 'The requirements of industry: the chemical industry', in *Papers to be presented, op. cit.*, 2–5. Clayton's son studied chemical engineering at Birmingham.

[187] *Scientific and Technical Manpower in Great Britain* (London: HMSO, 1956).

promote local lectures and activities.[188] This hierarchically-stratified group promoted other forms of division within the Institution. The formation of the G&S Section had been supported by prominent corporate members, and its events began to attract Associate Members and Members, too – so much so that, in 1938, Joseph L. Rosenbaum, the Associate Editor of the *Chemical Trade Journal and Chemical Engineer*, suggested to the Institution that it form a Manchester branch based on this model. A. J. V. Underwood, as Joint Honorary Secretary, noted discouragingly that work by honorary officers would be needed, and that the 20-odd members in Manchester itself were too few. He proposed instead that a North-Western Section be formed from the 70 Lancashire members, if demand warranted.[189]

For Council the principal benefit of such a Section would be to raise the profile of chemical engineering in a region where it was of high potential demand – i.e. lobbying industrialists more than providing an immediate service to members. Julius L. F. Nagel wrote to Underwood that December,

> I deduce that the Council is anxious to promote the usefulness of the Institution as a profession, the value of which is not I believe realised nearly enough among industrialists ... Most large factories with considerable electric plant for power and lighting now have qualified electrical engineers on the staff. Comparatively few, however, engage chemical engineers, whose functions naturally are less definitely obvious, although in many places their value would be apparent if they were given a chance.

Despite similar encouragement from other chemical engineers and industrialists in Lancashire, no individuals offered to take responsibility for the organisational tasks.

The second world war, however, promoted organisation of the Section just as the first had triggered the Institution itself. In early 1944, A. Rees Jones, a Graduate Member, lobbied council for a regional section, noting that wartime relocations of chemical engineers had swelled the local population. Such a branch already existed in practice, if not with official recognition. In 1941, 'due to the fact that a number of young Chemical Engineers formerly in London were taking up new appointments in the North', Rees Jones and others had drafted and sent a circular letter to the Graduates and Students in and near Manchester. Typically fifty to sixty G&S members attended local meetings held through 1942–3, with corporate members also participating. Fourteen such meetings had been held between 1941 and 1944, and Rees Jones proposed continuing these under the aegis of a Branch. A poll of members in the region was positive (some noting, however, that they could give support only after the war), and the IChemE Council approved the branch that Spring.

While members in Lancashire and Cheshire thus established regular regional

---

[188] Gayfere archive box III/3, file 'Formation of Graduates and Students Branch/Section'. By 1970 the Section had eight local centres said to 'frequently field new ideas to council in a provocative and forthright manner' [J. F. Richardson, 'The Institution of Chemical Engineers', *Chem. & Indus.* 4 Jul 1970, 879–82].

[189] Gayfere archive Box III/2 'Branches', folder 'North Western Branch'.

events, the Council also gained what they sought: promotion of the IChemE among employers in the region. The episode again refutes the widespread myth that a monolithic ICI was consistently opposed to chemical engineering. The board of ICI (General Chemicals) offered assistance and active encouragement, and suggested that the development also was viewed positively in the Dyestuffs, Leathercloth, Lime and Alkali Divisions.[190] While this offer appears not to have been taken up by the Institution, at least in the short term, the eleven candidates for the local committee included four ICI employees.[191]

At the same time, a junior group was proposed in Scotland as part of a planned Scottish Engineering Students Association. In late 1944, William Cumming suggested that such a branch would unite the roughly 100 members there and open their horizons to cognate forms of engineering. The idea of a general engineering association was scotched, however, because 'the three main engineering institutions took the view that their own students were well catered for locally and that the inauguration of another students' society would be apt to distract from their own organisations'. A decade later, W. L. Wood of BP Chemicals, Grangemouth, notified the IChemE council that he, Cumming and two ICI employees had been trying to form a local group without the Institution's involvement. Council encouraged the activity, and the group started that June.[192]

The formation of the second successful branch, in the Midlands in 1950, was similar to the first, in that it grew out of an informal group originating around a local Graduates & Students Section. A Scunthorpe branch began the following year as an informal group of members, and again received small grants-in-aid from the Institution. A South Wales group, after tentative moves in 1945 and 1948, was eventually formed in 1956. In each of these cases, the operation of the branch was required to follow organisational rules concerning the election of officers and the frequency of meetings, but was in other respects independent of London.

A North-East branch, while ostensibly paralleling its predecessors, caused more concern in London. In 1958, the Honorary Secretary of the North-East centre of the G&S Section wrote to notify Council that he and another Student Member had transformed the centre into a regional branch for all members.[193] The IChemE General Secretary, J. B. Brennan, concerned at the unilateral declaration and lack of information, urgently sought clarification from members in the region. J. M. Coulson, who had been teaching an undergraduate course for three years at Durham, replied laconically that he, his students and others felt it was time to form a group in the area, and had done so informally. After a visit to the chairman they had elected, Brennan returned somewhat soothed.

---

[190] Letters, E. H. Bramwill to F. A. Greene, 20 Jul and 25 Jul, 1944. Bramwill suggested that he could be of most assistance if he were himself a member, and wondered 'whether there is any means by which I could apply for and be elected to membership in the relatively short time that exists before the inaugural meeting'.

[191] Two of them, from the Huddersfield plant of ICI, were elected.

[192] Gayfere archive box III/2, 'Scottish branch', May–Dec 1944.

[193] Letter, John C. Rose to J. B. Brennan 4 Oct 1958.

He intimated to the IChemE president, however, his misgivings that the chairman was not very senior and that the Honorary Secretaries were mere students.[194]

In each of these cases, and others involving British and foreign branches, local relevance and autonomy were recurring themes. Each branch sought control of funds to support their activities, consisting principally of evening meetings. Some requested grants-in-aid (ranging from £10 per annum for the Scunthorpe branch to £240 for the North-West in the early 1950s, or some £150 to £3000 in 1999 currency); others, a surcharge for members participating or living in the region. Such arrangements were negotiated on a case-by-case basis.[195]

By 1971, branch memberships typically numbered in the hundreds, and twenty years later the most successful branches had some 1000 to 3000 members. The largest of these were subdivided in the early 1990s.[196] By this time, however, the branch structure was to some extent weakening. None of the branches was as successful as the original North-West Branch in providing self-sustaining activities and local enthusiasm. Their success was hampered by the gradual restriction of activities to evening meetings, rather than daytime sessions which had required some support from employers. No longer viewed as an important form of professional development, attendance at such general social gatherings began to decline. The much more popular form of member participation during this later period were the Subject Groups.[197]

The British model of chemical engineering abroad

At least as important as regional branches to the IChemE were the pressures to form foreign branches. Largely instigated by outside pressures rather than internal initiatives – and part of larger political and nationalistic trends away from colonial and Commonwealth ties – these national initiatives nevertheless provided a means of consolidating and extending jurisdiction of both the IChemE and its British variant of chemical engineering training, qualification and practice. The calls for national representation, like the wider movements in national independence, were catalysed by the end of the second world war.

Unlike the American Institute of Chemical Engineers, the IChemE, while equally an organisation firmly rooted in a national context, never explicitly limited itself to a British scope. Nevertheless, its image as 'international' was delimited by national boundaries. The territorial space was bounded largely by the frontiers and possessions of the British Commonwealth. Some of the IChemE members in these countries were Britons working abroad (such as at the petroleum sources in Arabia, or those of saltpetre in Chile); others were foreign nationals who had trained at British colleges and universities; and the

---

[194] Gayfere archive box III/2, file 'NE Branch'.
[195] Gayfere archive box III/2.
[196] See Appendix.
[197] Approved in 1977, these groups attracted some one-quarter of the membership by the late 1990s. See Appendix.

remainder were chemical engineers who had trained at indigenous academic institutions (increasingly accredited by the IChemE) in their own countries.[198] To varying degrees, the three types of member applied the British model of chemical engineering abroad. The opposing model was that developed in the USA, and promulgated through publications, college courses and expatriot American chemical engineers. The British influence, however, had distinct national flavours. The history of foreign branches parallels that of the colonisation of cognate *occupations* discussed earlier in this chapter, and of *disciplines* as discussed in Chapter 4, but along *geographical* lines.

The clearest tension between independence and association was in India. While chemical engineering courses started in 1921, the Indian Institute of Chemical Engineers (IIChE) formed in 1947, the year of India's independence from British rule. Within a year, the new Institute had attracted more members than were affiliated with the IChemE in India.[199]

Into this new nationally-oriented professional environment, the IChemE fitted uncomfortably. While it shared some common members with the IIChE, they appeared to be more receptive to American influences.[200] According to Hugh Griffiths, who toured the country in 1949, chemical engineering training in India benefited from 'lavish amounts of equipment, mainly of American origin'.[201] Indeed, he lamented that 'in every office I entered the usual American textbooks could be seen and the attitude towards chemical engineering is essentially the Gospel according to Walker, Lewis and McAdams'. The founding President of the IIChE cited mainly American examples as the way forward for Indian chemical engineers. His first degree, and that of his associate from the 1920s, were from American universities.[202]

Nor was the discipline centred along precisely the same lines as the British version. Mirroring the British experience with regional influences, courses in

---

[198] The first full membership listing of mid-1929 indicates that 29% of the 563 members lived outside Britain: 1.9% in Europe, 1.2% in Asia, 2.6% in the Americas excluding Canada, and nearly 13% in the Commonwealth, including 4% in India, 3% in S. Africa, 2.1% in Australia and 1.4% in Canada. These fractions did not change significantly for decades: in 1996 one-quarter of the 22 000 members lived abroad and 13% lived in former Empire countries.

[199] Asit K. Mitra, 'Glimpses of History of the IIChE (1947–1997)', (typescript, 1997); R. A. Mashelkar and J. V. Rajan, 'Chemical engineering developments in India', in N. A. Peppas, *One Hundred Years of Chemical Engineering: From Lewis M. Norton (M.I.T. 1888) to Present* (Dordrecht: Kluwer, 1989), 153–222, and D. H. Barker and C. R. Mitra, 'A history of chemical technology and chemical engineering in India', in: Furter, *History of Chemical Engineering, op. cit.*, 227–48.

[200] Letter, IIChE to M. B. Donald, 24 May 1949, Gayfere archive box V/1: file 'Non-IChemE activities'.

[201] The BCPMA also was considering starting a branch of the Association there. N. Neville, BCPMA memo 'India', C.216, 12 Dec 1945, Swindin collection box 44.

[202] Hira Lal Roy, 'Presidential address', Indian Institute of Chemical Engineers, 23 Mar 1949, Gayfere archive box V/1: file 'Non-IChemE activities'. Roy (1889–1965) had a obtained a BSc in chemistry from Harvard in 1913 and a PhD from Charlottenburg in 1923. Roy's successors at the IIChE over the next decade had all received their postgraduate training in chemical engineering at colleges of the University of London. See AF 491 and Mashelkar & Rajan, *op. cit.*, 185.

several Indian regions were shaped by the local economic importance of particular chemical technologies: dye technology in western India, edible oils in the south, coal in the east and sugar technology in several regions.[203]

Linkages between the IChemE and South African practitioners had a similar postwar history. The academic origins of chemical engineering there can be traced to the emergence of the diamond and gold mining industries in the last third of the nineteenth century and the appearance of schools of mining in Kimberley and then Johannesburg, which led to a close association with metallurgy and explosives production.[204]

Local circumstances caused the indigenous organisation to diverge from British interests. The South African branch affiliated with the (South African) Associated Scientific and Technical Societies in 1952, and formed an autonomous South African Institution of Chemical Engineers in 1964, following its government's edict on apartheid.[205] By 1980 the IChemE retained some 400 members in South Africa and recognised four teaching institutions.[206]

These two instances suggest that the factors of regional identity and local circumstances generally worked against the IChemE. The case of Australia, as a geographically distant Commonwealth country, provides further evidence.[207]

The lecturer of the first diploma course in industrial chemistry and chemical engineering, started at the Sydney Technical College in 1915, was American and a member of the AIChE.[208] By the end of the second world war there were a number of programmes in the distinctly different and competitive Australian states. Postwar advisors to the IChemE noted that while Sydney to some extent saw the USA as a model, parts of Queensland, for example, looked to Britain.[209]

As early as the late 1920s, there were several organisations capable of representing chemical engineers in Australia, including the Australian Chemical Institute (ACI), Institution of Engineers, Australia (IEAust), Society of Chemical Industry, Royal Society of New South Wales and Institute of Mining and Metallurgy.[210] Interest by the IChemE increased in 1946, when the

---

[203] Mashelkar & Rajan, *op. cit.*, 196.

[204] At the South African School of Mines (1896) and Transvaal Technical Institute (1903), respectively.

[205] Gayfere archive Box III/2 'Branches', folder 'South African Branch'; CM 17 Apr 1963.

[206] Pretoria, Cape Town, Natal and Witwatersrand. See T. J. Evans, 'South African reflections', *TCE* Jun 1980, 377, 381.

[207] For the general context of Australian industry, see J. A. Brodie, *Australia's Engineering Milestones: The Chemical Industry* (Sandy Bay, 1987).

[208] He first contacted the IChemE in 1927, proposing an Australian branch or associated body which would control acceptance of candidates locally. Letter, R. K. Murphy to Hinchley, 16 May 1927, Gayfere archives box III/2, 'Australian branch'; CM 20 Jul 1927. Interview, Murphy (1887–1972) with N. Whiffen, 1964.

[209] Correspondence from Australian members suggested that 'bitter jealousies' between Australian states, variations between Australian chemical engineering schools and large distances made branch organisation on a State basis necessary. S. Robson, 'Chemical engineering in Australia', Gayfere archive box III/2, folder 'Australian branch', Sep 1952.

[210] There later was overlap of professional territory between the IEAust and IChemE. The IEAust, founded in 1919 and granted a Royal Charter in 1938, had a 'Chemical Engineering Branch' in its Sydney Division as of 1973. Like the IChemE, the IEAust was a qualifying body holding

Australian Chemical Institute was preparing a scheme for the establishment of chemical engineering degree courses. For the IChemE council, this appeared to be a clear case of professional boundary-jumping, but opinions of Australian members were ambivalent.[211] An Australian Advisory Committee consisting of chemical engineering professors recommended a uniform system of professional qualification for chemical engineers, with the Institution administering qualifications with the support of strong governing divisions in each country. Because the Institution was relatively unknown in Australia, the Committee suggested that the only way to increase membership might be to elect members without examination. This delegation of power was resisted by the IChemE council. IChemE representatives reported 'quite a battle' to assert its right to assess courses, and to emphasise its desire for higher standards in training courses, particularly those concerned with plant design.[212]

The expansion of regional Australian Groups accelerated from the 1960s, first with State, then regional, organisations.[213] Yet regional engineering alliances continued to cause conflict. Unlike other professional institutions which had Australian branches involved in learned society activities, the IChemE in Australia dealt with professional matters such as course accreditation, which the IEAust looked on as being legitimately within its province.[214] In 1969, 1973, 1986 and 1992 studies aiming at the merger the IEAust and IChemE in Australia were defeated principally on the grounds of financial viability.[215] In this atmosphere of growing independence, in 1995 an IChemE Board in Australia was set up.[216]

By contrast, Canadian members of the IChemE were almost certainly the least shaped by British connections of those in all the Commonwealth countries.[217] Despite the historical linking of British and Canadian industries, Canadian chemical engineers gradually became more closely associated with American educational and occupational conditions while successfully building

---

its own examinations. See A. H. Corbett, *The Institution of Engineers Australia: A History of Their First Fifty Years 1919–1969* (Institution of Engineers Australia, 1973). The New Zealand Institution of Engineers similarly formed a Chemical Engineering Group.

[211] S. Robson, 'Chemical engineering in Australia', Sep. 1952, Gayfere archive box III/2, folder 'Australian branch'.

[212] *Ibid.*, H. R. C. Pratt, 'Australian notes', *TCE*, Nov 1971, 389.

[213] Victorian (Melbourne) Group: 1961, with Pratt as chairman; NSW (Sydney) Group: 1963; Queensland Group: 1968; S. Australian Group: 1970; W. Australian Group: 1973.

[214] Corbett, *op. cit.* The IChemE accreditation of nine universities was conducted in parallel with the accreditation programme of the IEAust [T. J. Evans, 'Internationalisation', *Diary & News* Sep 1993, 8].

[215] G. Ossipoff, *The Institution of Chemical Engineers in Australia: A Historical Chronology 1960–1994* (typescript, 1997).

[216] 'Institution news', *Chemical Engineering in Australia*, 20 (1995).

[217] For the Canadian context, see T. H. G. Michael, 'The association aspects of chemical engineering in Canada', and L. W. Shemilt, 'A century of chemical engineering education in Canada', both in: Furter, *History of Chemical Engineering, op. cit.*, 199–204 and 167–98; T. H. G. Michael, L. W. Shemilt and H. K. Rae, 'Foundations and evolution of the Canadian Society for Chemical Engineering', in: Leslie W. Shemilt (ed.), *Chemical Engineering in Canada: An Historical Perspective* (Ottawa: Canadian Society for Chemical Engineering, 1991), 49–61.

an indigenous organisation. In 1902, some seven years before a British course, the first chemical engineering programme was established at Queen's University, Ontario.[218] Canadian branches of the SCI were formed, and in 1921 the Canadian Institute of Chemistry (CIC) was established.[219] Like the SCI, the CIC included an undifferentiated collection of practising chemical engineers and chemists. Neither body operated nationally. During the second world war, a new grouping – the Chemical Institute of Canada – was organised to unify chemical interests and bodies, and soon after approved a Chemical Engineering Division.

Foreign bodies of chemical engineering did not influence the Canadian situation directly. The AIChE held regional meetings in Canada in 1920, 1937 and 1949 and formed two student chapters (in Montreal in 1935, and British Columbia in 1942). Nervous of this professional imperialism, the directors of the CIC resisted the American overtures to set up Canadian branches.[220] The IChemE, for its part, saw its own influence waning.[221] The pattern of independence was consolidated in 1965, when the Chemical Institute of Canada restructured to transform its Chemical Engineering Division into the Canadian Society of Chemical Engineers (CSChE).

Into Europe

Canada, under the strong influence of a powerful neighbour, was qualitatively different from the other countries cited, each of which was relatively isolated culturally, geographically, or by regionally-important industries.[222] But because of this, the Canadian case was more typical of foreign reaction to the IChemE. European interactions, in particular, proved difficult to nurture. As noted in Chapter 1, the analogies between professional and national aspirations are very close. The difficulties that the IChemE had experienced with competing professions were to be paralleled by new experiences with European influence. But European interactions were qualitatively different: while British and American chemical engineering were broadly similar in disciplinary and professional scope, European constructions of the specialism differed more fundamentally.

A European branch was first suggested in 1952 by Robert Edgeworth-Johnstone as a means of 'extending the influence of a professional body across national frontiers' and a way of 'breaking down international barriers'. The current environment, he argued, was one in which no European country hosted an independent professional association of chemical engineers, and in which organisations of applied chemistry were seeking to usurp intellectual territory

---

[218] Shemilt, *op. cit.* The first department of chemical engineering was at the University of Toronto in 1908.

[219] See C. J. S. Warrington and R. V.V. Nicholls, *A History of Chemistry in Canada* (Toronto, 1949).

[220] Michael, Shemilt and Rae, *op. cit.*, 53.

[221] The IChemE 'did not display any interest in being active in Canada, nor is there any indication that Canadians sought it out' (*ibid.*).

[222] On the other hand, the British template was more closely traced in some countries that did not form local branches, such as Nigeria and Malaysia.

by adopting the term chemical engineering 'to the concern of genuine chemical engineers'. Edgeworth Johnstone warned that the interest in other countries was due partly to recent lecture tours by American chemical engineers, and that 'the present fluid situation will crystallise into a pattern of little impecunious groups in each country'. A less contentious argument to support what he admitted 'may appear to be a revolutionary proposal' was that it was a logical extension of the ongoing establishment of branches in India, South Africa and Australia.[223] Thus extension of both the discipline and profession in the British style might be accomplished in one fell swoop.

In at least one other respect, too, the proposal was novel. It was an uncharacteristically proactive recommendation, supporting the organisation of chemical engineering outside Britain in the absence of an indigenous demand. Council members noted that if Germany were to form a large chemical engineering body, then the Scandinavian countries 'might look to Britain for affinity'. They decided, however, that with only forty members in Europe (more than half of whom were in the Netherlands) the 'time was not ripe' for pre-emptive tactics.[224]

Co-operative internationalism

If direct intellectual colonisation seemed out of the question for what the IChemE characterised as 'disorganised' European territory, then less persuasive connections were also unattractive. For the IChemE Council, foreign linkages in the first decade after the war were prefigured by prewar models. Thus the 'internationalism' of the Institution was in the form of participation and joint administration of conferences and symposia with cognate organisations. This was more ambitious than the bi-partisan congresses held with the AIChE in 1925 and 1928 and the 'International Congress of Chemical Engineering' held in London in October 1930, which seems to have been a thoroughly British affair.[225] But the Chemical Engineering Congress, a portion of the World Power Conference held in June 1936, was closer to the postwar versions. Planned

---

[223] Letter, R. Edgeworth Johnstone to S. Robson, 7 Jun 1952, Gayfere archive, box III/2, file 'European Branch'. A European branch was formed belatedly in 1975, its presence aided by the route to qualification that the IChemE had provided for Dutch engineers after the war, and by its accreditation of courses at three French universities (Toulouse, Nancy and Compiègne). The later additions of such foreign branches, even more than their earlier Commonwealth counterparts, were concerned pragmatically with securing accreditation of courses, qualifying members and linking local practitioners.

[224] CM 29 Jul 1952. Closer ties with Dutch chemical engineers and the possibility of a Dutch branch or advisory committee were also being explored. The high fraction of postwar Dutch members was attributed to a disciplinary vacuum caused by the absence of chemical engineering programmes at Dutch technical universities. See, for example, 'Note on visit of Dr. F. A. Freeth to Holland Feb 29 – March 3, 1952, exclusive', Gayfere archive box III/1, 'Branches – Holland' and III/2, 'Documents relating to formation of Dutch branch'. A strong professional impact was nevertheless unlikely; with merely a dozen corporate members in Holland and an existing Chemical Engineering Section in the Royal Netherlands Institution of Engineers, an IChemE branch was deemed impracticable.

[225] 'F.L.N. 4' file, Nathan papers, IChemE.

from 1931 and held in London, the Congress included delegates from nearly fifty countries.[226]

The European Federation of Chemical Engineering

Yet Edgeworth Johnstone's evaluation was wrong: there were indigenous sources of chemical engineering in Europe, albeit with no identity closely mapping onto the British template. In 1950, a group of chemical engineers met in Milan to consider a pan-European organisation. These discussions were continued in Frankfurt in late 1952, when several European countries collaborated to consider a European Union of Chemical Engineering at the ACHEMA conference.[227]

To some extent, the IChemE and other established organisations had been caught unawares. The UNESCO Department of Natural Science had been asked by the Swiss Society of Engineers and Architects for its advice on whether they should be represented in the proposed Europaische Union für Chemische Technik. UNESCO representatives had passed the enquiry to the World Power Congress, whose secretary in turn contacted the IChemE for information.[228]

After polling ten European members of the Institution, Council decided not to participate. In support of its decision, it cited the polled opinions: these suggested that the Union would be dominated by the proposed French and German secretariats; that the 'open' memberships of most relevant European societies would dilute the contributions possible by 'qualified' chemical engineers; and that the Institution should take a leading role in any such European organisation to promote the educational aspects. Several felt that chemical engineering as a profession in France and Germany was in its infancy, but that Germany, in the form of DECHEMA, was a natural home for a secretariat.[229] The Netherlands, with its relatively close ties and large IChemE membership, was also proposed as home for the headquarters. Thus the IChemE leadership felt that chemical engineering in Europe was ill-defined, devoid of credentials and too nationally dominated.

The proposed European 'Union' of Chemical Engineering was founded as the European Federation of Chemical Engineering (EFCE) in 1953 by fifteen technical and scientific societies from eight European countries.[230] Britain was

---

[226] Letter, C. H. Gray, 27 June 1932; 'Chemical Engineering Congress Grand Council', Gayfere archive box XIV/1. See also Chapter 5.

[227] Francis Arthur Freeth (1884–1970) represented the IChemE, becoming a member of the first Executive Committee of the EFCE the following year. He had been chief chemist at Brunner Mond before the first world war, developed a process in 1916 for direct production of ammonium nitrate from ammonium sulphate and Chile saltpetre, and was subsequently research leader at ICI Winnington.

[228] CM 3 Dec 1952, 7 Jan 1953.

[229] J. B. Brennan, 'Proposed European Union of Chemical Engineering', 1 May 1953, Gayfere archives, box XIV/1 – EFCE (I).

[230] K. R. Westerterp, 'Forty years – European Federation of Chemical Engineering', ACHEMA Yearbook 1991 (Frankfurt am Main: DECHEMA), pp. A8 – A12 and 'Europäische Föderation für Chemie-Ingenieur-Wesen. Ergebnisse und Probleme', ACHEMA Jarhrbook 1971/1973, Band I, 6–10.

not among them. Four years later, however, the IChemE and SCI joined the
EFCE, which already had 27 member organisations in 14 countries. This
appears to have been due to an expectation of having greater influence, as
much to a concern about being left behind. The Secretariat was shared between
Frankfurt, Paris and London, each carrying responsibility for specific countries.[231] The Federation also partitioned the subject itself more precisely, claiming that while 'chemical engineering has taken over much from the practice of
the physicist, the chemist and the engineer' it nevertheless 'does not itself belong
to any of these disciplines'.[232] It was equally careful not to usurp too much
territory. The EFCE set up a parallel organisation, the European Federation
of Corrosion, in 1955, and a similar Federation of Biotechnology in 1978, citing
these specialisms as closely linked to, but not of sole interest, to the broader
discipline of chemical engineering.

Beyond such professional boundary-setting and policing activities, there were
internal power balances to regulate, with consequences for the discipline. A
Dutch member suggested that the 'German method' of training with a 'sound
knowledge of chemistry and a sight-seeing of several processes' was gaining
ground over the 'Anglo-American method' of providing a 'sound knowledge of
fundamental chemical engineering unit processes'.[233] Here the Federation
deferred to practice in member countries. Its influence was felt through the
programmes agreed for symposia and joint conferences.

The first two congresses were held in Frankfurt (1955) and Brussels (1958).
The third, in London (1962), was the first opportunity for the IChemE, and
British chemical engineering, to be at centre stage. Yet the British version of
the profession and the discipline was not pressed; instead, the IChemE traded
its previous protectionist stance regarding professional identity for the influence
offered by international participation. The chair, C. E. Spearing, noted that
'other European countries have different ideas from our own about the education and training of chemical engineers. These ideas are neither better, nor
worse, than our own; they are just different'.[234] The EFCE stressed communication between the disparate national bodies and sought standards of training of
chemical engineers. At the fourth congress in 1965, again in London, the
metropolis was claimed to be the centre of the world's largest and most powerful
chemical engineering contracting industry.[235] By 1971, 47 techno-scientific societies in 19 European countries were members, and the IChemE Council established a Committee to deal with its proliferating overseas affairs. Its activities

[231] The secretariat offices were at the Societé de Chimie Industrielle in Paris, the DECHEMA
       Deutsche Gesellschaft für chemisches Apparatewesen in Frankfurt and the IChemE in London.
       The London office was made responsible for activities in Britain, Australia, India, Ireland,
       Japan, Norway, South Africa and North America. While non-European countries were to be
       represented, the economic interests of Europe were to be promoted first.
[232] European Federation of Chemical Engineering: Rules and Regulations (1956).
[233] Letter, J. Eekels to IChemE, 15 Jan 1957, Gayfere archive box III/1, 'Branches – Holland'.
[234] 'Third congress of the European Federation of Chemical Engineering', Chem. & Indus. 30 Jun
       1962, 1199; 'Chemical engineering: a unique occasion', Chem. & Indus. 21 Jul 1962, 1314–7.
[235] 'Joint meeting of chemical engineers', Chem. & Indus. 17 Jul 1965, 1277–83.

included direct involvement with the EFCE and the Commonwealth Engineering Conference and indirect links with bodies such as FEANI (Fédération Européenne d'Associations Nationales D'Ingenieurs) and the World Federation of Engineering Organisations (WFEO). Nevertheless, internationalism in chemical engineering had been achieved by splitting the subject largely along national lines.

Exporting the British model

The history of IChemE regional interactions is, then, one of gradual accommodation from direct control, to partnership, through to loose affiliation. There was a continuum of influence shaping professional and institutional identities. These cases – from the disparate British and national branches to the EFCE – underscore the importance of regional variation in the chemical engineering profession. The forms taken by these bodies and the subjects they treated were mutated by local circumstances.

In none of these countries was the British archetype of chemical engineering mapped faithfully. The IChemE, exploring the advantages of foreign branches only after the second world war, discovered its professional territory largely claimed by indigenous organisations. Several of these Commonwealth bodies cut across specialisms, and so showed less rivalry and fragmentation in their native countries than did their British counterparts.[236] Other professions, too, had varying claims to chemical engineering in different national contexts: in Canada, chemical associations had won long-term support; in the antipodes, engineering affiliations proved most fruitful. So, too, was the intellectual domain already staked by American texts and chemical firms. This geographical mutation of the specialism is not surprising. If the profession was to be fluid and malleable enough to adapt successfully to changing circumstances, then it follows that it would assume different forms at different times and places. In a parallel with natural evolution, chemical engineering accommodated itself to the industries, academic institutions and professional organisations that applied the strongest adaptive pressures locally. In such circumstances, the British model could not retain its influence indefinitely.

CORPORATIST TRENDS

These long-distance interactions were matched by a proliferation of connections at home. But there was a difference of scale and perspective. The foreign arrangements were with bodies weaker than, or of comparable strength to, the IChemE. A more influential class of activities, however, and one exerting stronger pressures on the Institution's leadership, was linkage with British government and business. The IChemE had emerged from the war considerably

---

[236] The Institution of Engineers, Australia, the New Zealand Institution of Engineers, the Engineering Institute of Canada, and the Institution of Engineers, India were largely successful in representing a range of engineering professions nationally. For the broader differences between British and Commonwealth engineering professions, see Watson, *op. cit.*

stronger, thanks largely to new relationships forged between government, industry and the chemical engineering profession. The postwar period was characterised by a new interplay between Institutional projections, state patronage and
industrial support. For the first time, IChemE plans became closely coupled
with those of other bodies in these spheres.

As they had been immediately after the first world war, connections between
government, industry and the professions were seen as inevitable, and largely
in positive terms. Moreover, such a network of mutual influence and exchange
promised to be long-lived. Lord Percy noted in 1953 that:

> the impulse which was given to the creation of the profession of chemical engineering
> in the first world war had died out or had faded away between the wars, but the
> impulse given by the last war showed signs of being a good deal more permanent
> and we might perhaps feel now that the profession was fairly launched.[237]

The future of the IChemE appeared to lie squarely with government liaison.

The concerns raised by governmental, industrial and professional representatives at IChemE annual meetings through the war years proved to be a central
issue during the 1950s as well. Chemical plant designers and suppliers, chemical
producers and their employees faced serious international competition in a
country nearly bankrupted by war. But in 1950 Sir Robert Sinclair, President
of the Federation of British Industry, noted already a decline of co-operation
since the war:

> We look back to the diminution, or even disappearance of many of our war-time
> bases where problems, war-time enthusiasm for co-operation have all disappeared
> and our members, if alive, are to-day occupied with private enterprise, if they have
> not been inoculated against it by nationalisation. Concerns for healthy exports, he
> said, 'lead to a government concern for industry and to government planning'.[238]

The supply of new chemical engineers was not adequate for the predicted
demand. All sides were vocal in identifying economic success with more, and
better qualified, technologists. In 1950, newly appointed Professor Dudley
Newitt made 'Technology and the State' the subject of his Presidential address.
He agreed with the 'political creed' that better technical training was imperative,
and urged an organisation of government ministries on a technological basis.
Newitt was vague, though, on how industry could collaborate.[239] Hugh Beaver
focused his own presidential speech in 1958 on the inescapability of the need
for central direction of education, labour supply and technical research.[240] For
research in particular, Beaver was critical of the 'haphazard' independent and
uncoordinated programmes of the Atomic Energy Authority, Ministry of
Defence, Research Councils and civil research in independent departments and

---

[237] '31st IChemE Annual Meeting', *Chem. & Indus.* 9 May 1953, 459–60.
[238] 'British Chemical Plant Manufacturers' Association Annual Luncheon', *Chem. & Indus.* 15 Apr.
1950, 290–1.
[239] D. M. Newitt, 'Technology and the State', *Chem. & Indus.* 14 Oct 1950, 694–6.
[240] H. Beaver, 'Science and the state', *Trans. IChemE 36* (1958), 244–52.

ministries. He called repeatedly for a reorganised national research policy beginning with a revamped DSIR.[241]

Chemical engineering research

The new interplay of government, industry and Institution is manifested clearly by the case of research. Indeed, research was used as a rhetorical tool for garnering power. The DSIR and industrial Research Associations had been criticised earlier, particularly by H. W. Cremer, for their inability to communicate their findings to firms and practising engineers. During the war he, too, had urged the creation of a co-ordinating board or central body to act as a link between industry and government-supported research, and proposed the IChemE as the organiser.[242] Just as it proved to be for educational provision, the postwar government was very responsive concerning sponsored research. Cremer wrote as President of the IChemE in early 1949 to Sir Edward Appleton of the DSIR; within a week the DSIR was polling prominent chemical engineers, and by March a committee had been formed. But the IChemE officers, still conscious of their repeated wartime rebuffs by government departments, took no action other than to propose senior members on Cremer's DSIR committee.[243] His 1951 report recommended expanded research under its supervision, a finding that the DSIR supported despite opposition from the IMechE.[244]

Like its equivocal support for regional representation, this hesitancy of the IChemE in interactions with other bodies was characteristic of its actions in

---

[241] Founded in 1915, the Department of Scientific and Industrial Research set up a number of Research Associations (RAs) Some 31 RAs – groupings of companies in a single industry – were in existence by 1931; by 1957 the DSIR directly controlled 14 government research stations and partially financed 46 research organisations. See Beaver, *op. cit.*, and I. Varcoe, 'Co-operative Research Associations in British industry, 1918–34', *Minerva 19* (1981), 433–63; I. Varcoe, *Organising for Science in Britain: A Case Study* (Oxford, 1974); and, R. MacLeod and E. K. Andrews, 'The origins of the D.S.I.R.: reflections on ideas and men, 1915–1916', *Public Administration 48* (1970), 23–48. Centre for the Study of Industrial Innovation, 'Research Associations: the changing pattern' (1972).

[242] H. W. Cremer, 'The development of new chemical processes', *Trans. IChemE 20* (1942), 38–43, and quoted in 'Interpretation of scientific knowledge', *Chem. Age*, 30 Apr. 1949, 613–4. Cremer had had repeated dealings with the DSIR almost from its origins, beginning with his direct employment between 1918 and 1920 to co-write the Quinan reports.

[243] PRO DSIR 17/440 *Chemical Engineering Research Committee, Formation and Constitution.* The committee members selected were a mixture of academics (T. R. C. Fox, F. H. Garner, D. M. Newitt, W. Cumming), government researchers (K. Gordon, Fuel Research Board; R. P. Linstead, Chemical Research Laboratory; M. F. Courtney Harwood, British Launderers' Research Association), consultants (H. Griffiths), industrialists (J. A. Oriel, Shell; S. Robson, Imperial Smelting Corp; R. Holroyd, ICI) and industrial associations (E. H. T. Hoblyn, BCPMA, J. Davidson Pratt, ABCM). The Ministry of Supply and Board of Trade were also represented.

[244] H. W. Cremer (chairman), *Report of the Committee on Chemical Engineering Research* (London: HMSO 1951); PRO DSIR 17/440 letter 8 Mar 1951 IMechE President to DSIR arguing that the Mechanicals already had 'suitable committees' with 'important representation' by chemical engineers, and denying claims that 'the professional protagonists of chemical engineering are expanding at an abnormal rate to meet an abnormal demand'.

the postwar period. While supporting such activities in principle, the Institution found itself repeatedly side-lined. A. J. V. Underwood complained in 1952, for example, that the DSIR Committee on Chemical Engineering Research set up under Cremer had reached conclusions identical to those he and W. E. Gibbs had presented in 1933.[245] The Committee had recommended a central technical bureau backed by the DSIR that would organise chemical engineering research and communication. Significantly, the IChemE was not proposed as being central to, or even involved with, such a research bureau.

In 1953, a year after the formation of a DSIR committee to guide and establish a chemical engineering 'cell' at the Chemical Research Laboratory at Teddington, the IChemE council convened a committee to take action against it. The members were adamant that such a project was inadequate, would side-track work on major research problems and, in consequence, marginalise chemical engineering itself. They agreed to lobby the Board of Trade, Ministry of Supply and Colonial Secretary to emphasise the importance of chemical engineering research for increasing exports.[246] But Sir Harold Hartley, in particular, put a damper on such lobbying as being 'inopportune' the following year.[247]

By 1955, the IChemE remained ineffectual in influencing research activities. A research station for chemical engineering had not found official favour at the DSIR, but its director, Sir Ben Lockspeiser, wanted to establish a Research Association for chemical engineering. University-funded research no longer seemed the preferred course. A dismayed J. B. Brennan informed President John Oriel that 'chemical engineering research will be absorbed piecemeal by different organisations'. Research activities were dispersed between heat transfer studies directed by the IMechE, chemical engineering projects at the British Hydromechanics Research Association, and a new chemical engineering unit at the Chemical Research Laboratory.[248] The Intelligence Division of the DSIR was beginning to co-ordinate these research projects, but did not employ a qualified chemical engineer and had few resources.[249]

Research from the manufacturers' perspective

While the IChemE took this involuntary distancing from state-sponsored research seriously, its importance was more symbolic than practical. In practice, research was dominated by ICI, which had been publishing the work of its Research Department since the war.[250] The goals of ICI, which sought merely

---

[245] A. J. V. Underwood, *The Times* 17 Feb 1933.

[246] Minutes of special meeting, Research Advisory Committee, 10 Jul 1953.

[247] Letter, J. B. Brennan to J. Oriel, 24 Jun 1955.

[248] The CRL was headed by P. H. Calderbank (1919–88), who had studied at King's College and taught at UCL and the University of Toronto, and was later Professor of Chemical Engineering at the University of Edinburgh.

[249] Brennan, *op. cit.*

[250] J. Taylor, 'The progressing of new projects from the laboratory to the plant', *Chem. & Indus* 4 Jun 1955.

to solve commercial problems encountered by its 'chemists', were not, however, synonymous with those of the IChemE. Nor were the interests of the manufacturers' associations always conducive to continued collective initiatives. The BCPMA took an active role in government liaisons – so much so that the efforts of the IChemE were down-played or left unmentioned.

The BCPMA was unquestionably on the ascendant after the war. From its foundation in 1920, it had been concerned with more than simply representing manufacturers; its initial and repeatedly-voiced concerns included engineering standardisation and chemical engineering education.[251] During the war it had frequent dealings with government departments, particularly the Factories Dept. of the Ministry of Labour. From 45 member firms in 1935, it had grown to 215 by 1954, doubling in the latter six years alone. The BCPMA attributed its growth to the removal of the war-time 'horrors of control' including building licences and permits. Similarly, D. W. Scott, deputy Managing Director of Monsanto Chemicals in Britain, complained that 'controls, allocations, frustrations, lack of incentives seem to have become part of our thinking' and stated that 'solutions would not be found in any concept of central planning'.[252] On the other hand, he saw a need for more communication, particularly between chemical industry bodies. Thus the BCPMA promoted *corporate* liaisons – between the component bodies of the chemical sector – as being more productive than a pan-industry affiliation of employers. More positively, the next Association chairman presented the Ministry of Supply as 'the industry's sponsoring Ministry', and praised the appointment, by the Minister, of Kenneth Gordon, a chemical engineer, as Director-General of Ordnance Factories in 1953.[253] The BCPMA chairman pointed to the 'spirit of mutual respect and co-operation ... built up with the Board of Trade' and its active work in initiating chemical engineering education'.[254]

The ABCM did not associate itself as clearly with chemical engineering as did the BCPMA. It was, however, equally active in promoting the occupation. The ABCM decided in 1955 to respond to the Cremer report by setting up a Chemical Engineering Research and Advisory Committee, with a liaison member from the BCPMA but no IChemE representation. This committee, employing a full-time administrator, was to survey and interchange information between firms. John Oriel, recognising that the IChemE was to be largely cut out of the project, rationalised the situation:

the ABCM have a prior claim to stake their case in view of the fact that it is they who through their membership will eventually have to find the money to operate any scheme of research.

---

[251] *Industrial Trade Associations: Activities and Organisation* (London, 1957), 27.

[252] BCPMA annual dinner, *Chem. & Indus.* 12 Apr. 1952, 329–32.

[253] BCPMA annual dinner, *Chem. & Indus.* 27 Nov. 1954, 1478–9. Ministry officials were prominent guests at annual dinners through the decade. Other frequent guests were representatives of the DSIR and the Royal Institute of Chemistry.

[254] BCPMA annual dinner, *Chem. & Indus.* 26 Nov. 1955, 1548–50.

He urged the Association to sponsor more focused research at universities, but granted that the selection would be in the hands of the ABCM committee alone.

Within a year, the nine-strong committee was collecting information from over 100 member firms. Its resulting report called for research into six areas of technical weakness.[255] The ABCM and BCPMA consequently sponsored research projects at nine British universities, amounting to some £30 000 over three years.[256] Probably more galling to the Institution, the committee 'made observations to the IChemE on the proposed Graduate Training Scheme'. Repeatedly throughout the affair, the Institution found itself either dictated to, side-stepped or ignored.

DECENTRING THE IChemE

As was the case with its limited involvement in BCPMA and ABCM initiatives, the IChemE struggled to remain prominent in other spheres. Visibility of the Institution and its members remained a chronic problem. In 1952 the Organisation for European Economic Co-operation (OEEC) sponsored a Technical Assistance Mission to the US to study the state of chemical manufacturing and engineering there. Its report, 'Chemical Apparatus in the USA', while contrasting the American and British industries, made no mention of the IChemE or the CEG. It depicted chemical engineering outside America as small-scale and poorly organised. An editorial in *Chemistry & Industry*, taking the OEEC findings at face value, was bitterly criticised by the IChemE general secretary, J. B. Brennan. Rehearsing the history and accomplishments of the IChemE, he argued that Britain was in no way typical of the rest of Europe.[257]

The Institution was increasingly marginalised in other ways. A 1958 'Chemical and Petroleum Engineering Exhibition', unlike the earlier 1926, 1931 and 1936 exhibitions, was not centrally organised by the IChemE.[258] Instead the BCPMA joined with the Council of British Manufacturers of Petroleum Equipment to sponsor the exhibition. The IChemE found its role in 'the largest gathering of chemical engineers this country has known' diluted to part of a concurrent meeting of the European Federation of Chemical Engineering, which itself had been instigated by the BCPMA.[259]

Similarly, the ABCM and BCPMA were becoming more successful than the IChemE in other forms of publicity and self-promotion. The ABCM had long published *British Chemicals and Their Manufacturers* and provided it free of charge to a list of 5000 plant users. In the mid-1950s it set up a publicity

---

[255] Subjects included filtration, distillation, ion exchange and drying.

[256] ABCM annual dinners and reports, *Chem. & Indus.* 22 Oct 1955, 1379–81; 28 Apr 1956, 308; 26 Oct 1957, 1413; 1 Nov 1958, 1428–9.

[257] 'Challenge to chemical engineers', 25 Oct 1952, 1039 and J. Brennan, 22 Nov 1952, 1157, both in *Chem. & Indus.*

[258] After the war, the council of the BCPMA had been divided about resuming chemical plant exhibitions, and had suppressed moves for one in the early 1950s on the grounds that order books were over-full and raw materials were still scarce. *Industrial Trade Associations: Activities and Organisation* (London: Political and Economic Planning, 1957), 117.

[259] 'The chemical and petroleum engineering exhibition', *Chem. & Indus.* 10 May 1958, 537.

committee, and published a *Press Guide to British Chemicals*, listing chemical manufacturers, their products and trade names. The BCPMA, in its turn, produced a biennial directory, *British Chemical Plant*, which included an address list of members and their overseas agents. By contrast, the IChemE resisted publication of a members' list through the 1950s. While there are many reasons why professional bodies may not publish membership lists, the grounds stated by the IChemE at the time were that this would lead to a loss of advertising revenue. This desire to control access to information consequently had the effect of reducing the prominence of the Institution.

The IChemE was able to sustain its precarious niche largely through the connections of its presidents. In 1957, for example, Sir Hugh Beaver was President of both the IChemE and Federation of British Industry, which represented both the ABCM and BCMPA. In a replay of attempts by IChemE officials during the early 1920s, he and other industrialists through the decade saw a much broader stage for the activities of the Institution. For the manufacturers' and professional bodies of the late 1950s, the relevant issues were far more than just training and manpower allocation. They included 'the degree of Governmental intervention, the rival theories of nationalization and private enterprise, the fight with international communism, the changing pattern of the Commonwealth and now the impact of the idea of the Free Trade Area'.[260]

An official identity: the Royal Charter

The rising voices of manufacturers' organisations, DSIR Committees, regional branches at home and abroad, and the EFCE, each diminished the centrality of the IChemE leadership in chemical engineering initiatives. But the IChemE gained influence as the recognised representative of the rapidly growing numbers of chemical engineers. Institutional activities focusing on professional status therefore grew in importance during the postwar years.

In 1953, the IChemE Council had one of its first direction-setting meetings to discuss the further development of the Institution. To the special meeting, founding members and former presidents C. S. Garland, D. M. Newitt and F. H. Rogers were invited.[261] Among the new initiatives they favoured were strong support for the HNC programme; expanding the Institution through new premises; and, an application for the Royal Charter.

Such work depended on lobbyists, the most prominent advocate of the profession during this period being Sir Harold Hartley. IChemE President both in 1950–2 and 1954, Hartley actively promoted chemical engineering through his numerous links with government and industry.[262] In a postwar culture transformed by advertising, one of his associates at the IChemE credited him 'with "selling" chemical engineering in the country in a manner in which

[260] G. N. Hodson, BCPMA annual dinner, *Chem. & Indus.* 9 Nov. 1957, 1470–2. In particular, the ABCM supported Britain's entry into the EEC 'unswervingly' from 1958 [*Chem. & Indus.* 20 Oct 1962, 1835].
[261] Minutes, special meeting, 15 Dec 1953.
[262] See, e.g., S. R. Tailby, 'Sir Harold Hartley', *TCE* Nov 1985, 4–5.

perhaps no one had tried before'.[263] The Duke of Edinburgh, too, whom
Hartley advised and convinced to become patron of the Institution, observed
that 'chemical engineers were his passion, and he was the greatest of sales-
men';[264] Hartley's recruiting of Sir Christopher Hinton as a prominent member
has been discussed earlier.[265] Hartley targeted industrialists, too, arguing that
chemical engineering was the basis of the chemical plant industry, and urging
that the BCPMA include 'engineering' in its title. Nevertheless, later that year
Hartley admitted that 'he and his chemical engineering friends had rather failed
to put their case across' concerning co-operation between plant manufacturers
and engineers.[266] Nor were new premises obtained for another three years.[267]

But even if industry was not responsive to such appeals, Hartley and some
of the members of council saw the pursuit of a Royal Charter as a way to
solidify the reputation of the Institution. Even if providing little in the way of
concrete legal benefits, in the British context it was a means of gaining official
recognition and ranking amongst the older professions at the highest level of
government. Not appreciated in the initial attempts, however, was how chal-
lenging the request for a Charter would be to established Institutions, and how
much support would be needed from them. The Charter was a formal means,
then, of probing and developing professional alliances.[268]

Recognition of the Institution by Royal Charter had been an elusive goal
over most of its first thirty-five years of existence. The 1920s had seen registration attempts by many professions, and the subject first appears in IChemE
Council discussions in early 1928. At that time, the Institution's solicitors
investigated the details of the process, which included a formal petition, opportunity for public opposition, and finally recommendation by the Privy Council
to the King. They advised soberly that such Charters were becoming increasingly difficult to obtain, that a body granted one would have to establish its
accomplishments and services of public importance, and that little material
advantage would be obtained in any event. The second President, Sir Frederic
Nathan, advised against premature application in such circumstances, and the
matter was allowed to drop. The subject was raised in Council annually over
the succeeding nine years, but repeatedly deferred.[269]

[263] Ralph Harris and Arthur Seldon, *Advertising in Action* (London: Andre Deutsch, 1962).
[264] He wrote to the Duke of Edinburgh's secretary 'Chemical engineering has until now been the
junior partner; the Royal Charter gives it the same status as the other three institutions. It
seems to me that it would be very fitting if the Prince Philip should be the Royal Patron of
this latest branch of engineering which is becoming of such enormous importance to the
industrial future of the U.K ... we are up against considerable competition from the U.S. and
Germany.' 22 Mar 1957 letter Hartley to Sir Frederick Browning, Hartley papers box 144.
[265] IChemE annual meeting, *Chem. & Indus.* 7 May 1955, 529–32. See also H. Hartley, 'Chemical
engineering at the cross-roads', *Trans. IChemE 30* (1952), 13–9 and 'Chemical engineering: the
way ahead', *Chem. & Indus.* 14 May 1955, 544–9.
[266] BCPMA annual dinner, *Chem. & Indus.* 25 Apr. 1953, 404–5.
[267] In 1956 the IChemE moved into the new headquarters in Belgrave Square, London, purchased
by the SCI on a long-term lease.
[268] The pursuit of the charter was only the second such formal testing of status in the Institution's
existence, the first being the case of the wartime Central Register.
[269] CM 1928–36.

The subject was studied again in 1937, when William Cullen as President recommended waiting until the membership exceeded 1000 and the Institution had a more established position in the chemical industry to convince the Privy Council of 'not only the importance of our Institution but what it is attempting to do'.[270] Perhaps in an attempt to establish the technological lineage and expertise embodied by the IChemE, Cullen also instituted a programme to record the history and biographies of founding members of the Institution.[271]

Postwar prospects for a Royal Charter appeared more promising. Other seemingly comparable bodies, such as the Institution of Gas Engineers in 1929 and the Institute of Energy in 1946, had been granted charters. Treating the process as an uncontentious if arduous legal exercise, the Council and its solicitors honed the application, presenting it to the Privy Council in 1949 without informing the membership.

Responses from the other professions were not long in coming. The following month, Herbert Cremer and Hugh Griffiths, the current and immediate past Presidents of the IChemE, respectively, were invited to lunch with officers of the IMechE. There they were told that the IMechE would not oppose the petition, but they saw no evidence of enthusiastic support.[272] In August, Council learned that the Institution of Civil Engineers and the Institution of Electrical Engineers had both lodged opposing Petitions. The Civils had decided in March to 'co-operate with the Institutions of Mechanical and Electrical Engineers in opposing the Petition of the Institution of Chemical Engineers for the grant of a Charter'.[273] They opposed the Charter for the IChemE on the grounds that Associate Membership could 'be satisfied by the holding of other than an engineering qualification and that a degree in chemistry confers complete exemption from this requirement';[274] civil engineers, by contrast, were all required to pass exams and undergo training to gain corporate membership. The Electricals employed almost identical language in their petition against the IChemE. In an internal discussion paper, they also marshalled a numerical argument based on inter-Institutional membership, claiming that 'the persons comprised in The Institution of Chemical Engineers are primarily chemists', because

> only 10% of these members are Corporate Members of one or other of the three Engineering Institutions, whereas some 35% are Corporate Members of the Royal Institute of Chemistry.[275]

---

[270] Gayfere archive, box VII/2; *CM* 16 Jun 1937

[271] Gayfere archive, box VII/1, file 7. Of 31 suggested 'founders', seven had died by 1939. The project, under H. Talbot and S. G. M. Ure, was never completed, being diverted instead into capsule biographies of IChemE presidents in the 'Presidents' book', which was updated until the retirement of the first General Secretary in 1969, a period of recession during which the past was downplayed.

[272] Gayfere archive, box VII/2, 'Petition for a Royal Charter', Apr. 1951, p. 4. See also Watson, *op. cit.*, 88.

[273] ICE CM 26 Apr 1949, minute 7227.

[274] Not true. An indication of the difficulty of the IChemE exams is that 45% of candidates passed in 1953 [Q.B. No 115 Jan 1954].

[275] IEE CM 21 Apr 1949; Document No. 3780/3.

The IChemE submitted formal 'Observations' to the protests, which were summarily rejected by the Privy Council. The special committee appointed to deal with the Charter ascribed their failure to a lack of influence. They suggested that the Ministry of Education, Ministry of Supply, Board of Trade and DSIR had supported the Petition, but that the Royal Society and Royal Institute of Chemistry had not been consulted. The Electricals were 'strongly opposed to the grant of a Royal Charter, and ... the Institution of Mechanical Engineers was at least unsympathetic'. The committee suggested more informal contact with officials of the Privy Council to counteract their suspected 'belief that royal recognition of sub-division of the corpus of professional engineering should not proceed further'.[276]

Two years later opinions were mixed about the advisability of seeking a Charter. Charles Garland suggested that building educational programmes and membership should take priority. Oriel and others felt that 'we must wipe out the stain of refusal' by lobbying industry and the Privy Council for at least 15 months. Newitt urged immediate preparation of a new petition.[277] In 1955, with John Oriel as President, another special committee was formed to draft a Petition and Charter. There is evidence that the desire for a charter was determining how the Institution sought to position itself within the ecology of professions. P. K. Standring of ICI, a Vice-President of the IChemE, noted that 'In the Petition, it is essential that the Institution provide convincing evidence that it is not a 'splinter group' and that, on the contrary, Chemical Engineering is a 'new', successful and helpful technology and not in any way simply a part of some well recognised form of engineering'.[278] Support, this time, was more encouraging. The advice of ICI and the employment of new solicitors with experience in similar work also strengthened the application.[279] Oriel reported that the Institutions of Electrical and Mechanical Engineers would support the Petition, and that Sir Hugh Beaver, Sir Harold Hartley, Sir Christopher Hinton and Sir Ewart Smith had consented to act as private sponsors for the Petition.[280] Indeed, the latter strongly supported closer ties with the IChemE and pressed support for the its Royal Charter in the IMechE Council. The Civils and Electricals 'recognised that the Chemicals were nearer our own particular field than theirs and they, therefore, stood aside while we were "making up our minds",' wrote the IMechE secretary.[281]

The Charter, granted in February 1957, effected a change in legal status of

---

[276] Gayfere archive, box VII/2, 'Petition for a Royal Charter', Apr. 1951, p. 5.

[277] Special CM 15 Dec 1953.

[278] Phrasing nearly identical to that employed by the INucE in their own struggle for autonomy the following year.

[279] J. B. Brennan, *The First Fifty Years: A History of the Institution of Chemical Engineers 1922–1972* (unpublished typescript), p. 26.

[280] Beaver (1890–1967) was the current President. Hartley had been responsible for gaining Hinton's continued support for chemical engineering since his second term as President in 1954. Hinton, in turn, was instrumental in marshalling Sir Ewart Smith, Chief Engineer at Billingham during the war; both were vocal members of the IMechE council. Minutes, 6 Oct 1955, Gayfere archive box III/5, 'Petition for a Royal Charter'.

[281] B. G. Robbins, 25 Sep 1958, IMechE COP11/7. See also ICE CM 16 Sep 1958.

the Institution from a limited company to a body incorporated by Royal Charter, and subtly promoted it with respect to the Big Three. The ICE, IEE and IMechE had for at least half a decade held 'Three Presidents' Meetings' to discuss matters of common interest. The idea of adding the IChemE President to the group was pressed on the IMechE council by Sir Christopher Hinton, who argued that chemical engineering was a 'primary technology' like their own. The ICE and IEE presidents were positive, but the Mechanicals kept the arrangement informal.[282]

Thus, in parallel with the dramatic expansion of industry and the successful negotiating with government and universities for educational provision, the IChemE was transformed after the war from a small, centrally organised body to an expanding organisation with proliferating branches. This growth brought it stronger links with other bodies, and a new effectiveness. A measure of decentralisation arguably strengthened the overall claim of the IChemE to representativeness, for example. But it also shifted the distribution of power towards the periphery. Paradoxically, the Institution was finding aspects of its influence both augmented and threatened as it entered the new decade of the 1960s.

---

[282] Letters, S. E. Goodall (IEE), 23 Oct 1958, F. A. Whitaker (ICE), 28 Oct 1958, both IMechE COP 11/7. Nearly all sixteen past-Presidents and Vice-Presidents of the IMechE supported Sir Owen Jones in his decision to make only an informal invitation 'on appropriate occasions', agreeing that admission of the IChemE into the fold would cause awkward requests for participation from other institutions.

CHAPTER 7

UNSTABLE EQUILIBRIUM

If the 1950s charted the rise of a new profession, the following dozen years threatened to see it fall. According to most indicators, chemical engineering in Britain did slow its rate of growth after the heady 1950s: the number of jobs, journals, courses and professorial appointments, the rate of students graduating and book publishing were all to reach saturation by the early 1970s.[1] This was largely inevitable: rapid growth must diminish as territory becomes more densely populated and resources become scarce. On a wider scale, science and technology in the western world mirrored these events.[2] But alternately charac-terised as a 'mature profession and discipline' or 'stagnant occupation' by protagonists and opponents, respectively, chemical engineers saw themselves again struggling in a highly competitive environment made more complex by the now explicit and continuous involvement of the state. Set amidst a backdrop of retrenchment in employment, two aspects of the chemical engineer's identity nevertheless showed modest gains during this period: occupational visibility and the establishment of disciplinary status.

THE PENETRATION OF CHEMICAL ENGINEERS INTO INDUSTRY

By the end of the 1950s, the acknowledgement and acceptance of chemical engineers in British industry had attained a plateau, but their role remained equivocal. Two examples will illustrate their changing perception. At ICI Billingham, the occupational profile of the technical staff – predominantly chemists and mechanical engineers since its formation after the first world war – had been abruptly overturned. One employee described the period 1926–40 as 'the era of the civil and mechanical engineer at Billingham when new buildings and plant were constructed', 1944–55 as 'the era of the chemist when new products were developed', and 1955 and beyond 'the era of the chemical

---

1   See Appendix.
2   On the inevitability of the end to such indicators of 'progress', see Derek de Solla Price, *Little Science, Big Science* (New York: Columbia University Press, 1963).

engineer'.[3] From scarcely ten chemical engineers before 1958, six years later the number employed in two sections had risen tenfold, many of them replacing their occupational rivals, chemists.[4] They comprised a sixth of all engineers, and one-third of all 'chemists' in Agricultural R&D. The sudden change was motivated by two factors: bad commercial performance in 1958 and 1961 because of heightened international competition, and the consequent decision to reorganise the Division and modernise its technology by converting from coal-coke to oil as a production feedstock between 1962–5.[5] The Director of R&D at Billingham saw the future use of chemical engineers as plant managers because of economies over employing engineers and a chemist. But was this really a reversal of the dominant ICI stance? It did not, in fact, constitute a rejection of the corporate culture espoused by Lord McGowan during the second world war, when he suggested that chemical engineers were most suited to smaller manufacturing units; it was merely an admission that not even ICI could always afford 'the ideal arrangement' of 'the employment of an engineer along with a chemist'.[6]

Courtaulds, the synthetic fibre manufacturer, was a similar case. Its chairman, C. F. Kearton, observed that chemical engineers 'became fashionable' around 1955. A decade later, the company was recruiting one to ten per year, mostly from the universities.[7] The Central Chemical Engineering Dept (founded by Kearton in 1946) employing some 100 chemical engineers, functioned as an R&D division dealing with 'trouble shooting' at plants, but its personnel came to influence the entire company's operations. Chemical engineers were used for the management of chemical processes, and all recruits typically spent two to three years on operations before they were 30 years old. Nevertheless, the formation of a new Research Division in 1964 was seen by some as having a structure that once again subordinated the chemical engineer to the scientist, with the accusation that there was a consequent limitation of their career prospects.[8] What could go up could as easily come down.

Despite these ambiguous examples, chemical engineers were unquestionably better established than a decade earlier. A 1963 survey carried out by the Institution, itemising how its members were distributed in British companies,

---

[3]   Andrew Pettigrew, *The Awakening Giant: Continuity and Change in ICI* (Oxford: Basil Blackwell, 1985), pp. 123–4. The changes of the late 1950s were part of a wider shift of the corporate culture at Billingham and at other ICI divisions.

[4]   Heavy Organic and Agricultural Divisions. The distribution of tasks, among 65 identified chemical engineers, was: R&D, 30; Project Design & Engineering, 20; Production, 13; Long term techno-business planning; 2. P. W. Reynolds, 'Chemical engineering with ICI at Billingham', *TCE* Mar 1965, CE55–CE59.

[5]   Pettigrew, *op. cit.*, 132–3.

[6]   Lord McGowan, *Proc. CEG* 25 (1943), 27.

[7]   C. F. Kearton, 'What industry wants in its chemical engineers', *TCE* Sep 1965, CE209-CE212. Christopher Frank (later Lord) Kearton (1911–92) was with ICI from 1933–40, worked on the Manhattan project through the war, and joined Courtaulds thereafter until 1975, successively as Department Manager, Director and Chairman. Obituary, *The Independent* 29 Jul 1992, p. 15.

[8]   Evans archive R8: Rules for professional conduct.

Table 7-1 New corporate membership by occupation, 1963–4 and 1973–4[9]

Self description	1963–4	1973–4
Chemical engineer	24%	17%
Process engineer	10%	8%
Chemist	9%	6%
Engineer	15%	19%
Manager	30%	22%
Educator	6%	7%
Consultant	2%	0%
Technologist/assistant	4%	19%
Metallurgist	0%	0%
Sales representative	1%	2%

showed that about one-quarter of corporate members worked in some 125 companies employing moderate-sized groups of chemical engineers. The largest employers – ICI, UKAEA, Shell, DCL, Esso and Monsanto – in multiple divisions together employed nearly 12% of the membership. The employment of Members, Associate Members and Graduates varied distinctly with employer, presumably because of differences in pay scales and the nature of the business. At ICI, most of the IChemE members were employed by the Dyestuffs and General Chemistry Divisions; the fewest were found at the Fibres and Heavy Organic Chemistry Divisions.[10] The highest-status chemical engineers, as earlier in the history of the Institution, were to be found in positions of research, management and consulting.[11]

The 1960s: aspirations reined in

Despite predicted shortages of chemical engineers as late as 1958, the emerging reality through the following decade was of too many graduates for a job market increasingly threatened by international competition. The fraction of students in the Institution reached a peak in 1959 (comprising a third of the total membership) and began a twenty-year decline thereafter.[12] While the Institution had devoted great effort to improving teaching provision through

---

9   Sample: 94 of a population of approximately 350 new corporate members (1963–4) and 96 of approximately 270 new corporate members (1973–4). For other periods, see Tables 4-2, 5-2, 6-1 and 8-1.

10 Pettigrew (*op. cit.*, 125–32, 328–31) describes the very distinctive organisation and management cultures at the ICI Billingham, Alkali and General Chemicals Divisions, all of which, until the late 1950s, paid little attention to chemical engineers as an occupational group. Alkali Division was a direct descendent of the Brunner, Mond Company and its paternalistic labour policies and ethos of superior management. General Chemicals Div. inherited the 'weaker business lineage, lower prestige, and indistinct traditions' of the United Alkali Company.

11 Making up only about 1% of the membership, some one-fifth of consultants had PhDs.

12 Students comprised 12% of members in 1980, the lowest fraction since the second world war, but stabilised at about 20% during the 1990s.

the Fifties, it turned to face employment questions only diffidently a quarter-century after abandoning its Appointments Bureau.

For the first time since the war, the prognosis for future graduates was not good, and appeared endemic because of the definition of chemical engineering being pursued by the Institution and universities. In 1962 Professor G. G. Haselden at the University of Leeds wrote to the IChemE Council to say that, exceptionally, two men who had graduated in chemical engineering in Leeds months earlier were still unemployed:

> It was evident that this arose primarily from the present industrial recession. If chemical engineering was regarded as a narrow specialism, centred in chemical plant design, then obviously the scope for employment was very limited.[13]

Moreover, the changing definition of 'responsible chemical engineering experience', in Prof. Haselden's opinion, exerted a restricting influence. 'The position could not have been foreseen by the Institution', argued Council member F. E. Warner, 'when it encouraged the expansion of chemical engineering training'.[14] The following March Professor John Coulson wrote, saying that, 'until chemical engineers were spread over industry, as works chemists used to be spread, there would not be a substantial increase in the vacancies for chemical engineers'.[15] While being more visible than a decade earlier, chemical engineers had an occupational identity that was deemed inadequately broad and fragile.

The chemical industry had always been cyclic, and the demand for chemical engineers by the early 1960s was made even more unstable by waves of investment. Capital expenditure in the chemical industry followed a four year cycle in the 1960s, with sharp peaks in 1962, 1966 and 1970. This economic cycle was exacerbated by delayed feedback: new investment in the increasingly large-scale chemical plants was followed by over-capacity, a reduction in profits and cuts in investment. Employment of chemical engineers soared as plants were designed, constructed and started-up, declined as a few operators were hired to man them, and plummeted as further plant design was deferred. High unemployment among chemical engineers followed the economic swings, peaking in 1962–3, 1966–8 and 1971–2 (Figure 7-1).

The increasing scale of new plants also introduced new problems for contractors. The capacity of new ethylene plants, for example, had risen from some 10 000 tons per annum in 1950 to 100 000 t/pa in 1959, 300 000 in 1966 to 450 000 t/pa in 1970.[16] New classes of process plant such as petrochemical and pharmaceutical plant, had widely varying situations of process design, perfor-

---

[13]   CM 19 Dec 1962.

[14]   *Ibid.*

[15]   CM 20 Mar 1963. J. M. Coulson (1911–90) was educated at Cambridge and took the Imperial College postgraduate course in chemical engineering. An assistant lecturer at Imperial College before the war, he did wartime service at Royal Ordnance Factories and then resumed as lecturer and then reader at Imperial. From 1954 to 1975 he held the chair of chemical engineering at Newcastle-upon-Tyne.

[16]   *Chem. Age* 10 Jul 1970, 4.

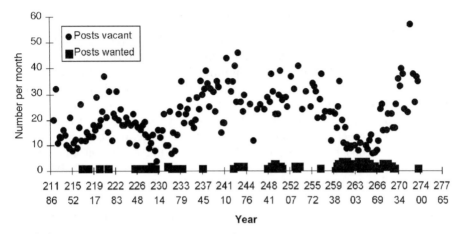

Figure 7-1   Chemical engineering posts, vacant and sought, advertised in *The Diary*.

mance criteria, start-up and so on. Many such plants were designed, constructed, operated and/or maintained via lump-sum contracts. Yet the contract conditions, guided by model forms of contract established by other engineering institutions, were increasingly inadequate. An IChemE committee thus began examining the problem in 1964 and published its model form four years later. After its thorough evaluation by trade associations, engineering institutions and other professional bodies and legal counsel, the contract stood for a decade without revision.[17] Given the ceding of research initiative to the BCPMA in the early 1950s, this responsibility for contract models was something of an about-face.

Other economic trends were not so readily addressed. Companies were increasingly assailed by the mixed effects of foreign investment and mergers. The case of Balfour Scott is typical. Surviving through its promotion of glass-lined vessel technology in the 1930s, the firm was bought outright by the American company Pfaudler, which had provided the technology, in 1962. The American firm had little interest in the original process plant and chemical engineering side of the business, which continued to decline through the 1960s. A major chemical engineering research and development laboratory launched by Balfour Scott in 1959 as a contract facility for Scottish industry, for example, was reduced to scarcely a testing station. After further mergers with water-treatment and medical equipment companies, the combine changed its name and ended chemical engineering and process design activities.[18]

---

[17]   IChemE, *Model Form of Conditions of Contract for Process Plants Suitable for Lump-Sum Contracts in the United Kingdom* (Rugby, 1968). This was followed by revisions and variants, e.g. for minor works contracts,

[18]   'Sybron' sold its glass-lining operations to Sohio Corporation in 1982 and ceased chemical engineering. W. E. Bryden, *Reminiscences of a 5/8 Ruddy Plumber, or into the Caprolactam* (unpublished manuscript, 1994).

The 1970s: economic and professional slump

The mixed employment opportunities through the 1960s were capped by ominous trends by the time the fiftieth anniversary celebrations of the Institution rolled round. Teetering toward what appeared increasingly to be the downward slope from their peak of prosperity, Council decided to celebrate the Jubilee of the Institution in 1972 only if 'the content should be of a forward looking nature rather than a retrospective'.[19] And three years later, amidst plans for the move of the headquarters out of London, the General Secretary wrote of being 'surrounded by gloomy despondency'.[20] If the profession entered the 1960s characterised alternately as 'mature' or 'stagnant', by the early 1970s the industries on which it depended were more uniformly cast as 'contracting' or 'surviving'.

The chemical plant industry entered a recession from about 1970 owing to world-wide overproduction of chemicals and man-made fibres. New chemical engineering processes were no longer appearing at the same heady pace as during the 1950s. While investment in the chemical industry had almost doubled during the 1960s, the rapid growth had been perceived by investors as signalling further opportunities which could not, in fact, be sustained.[21] Advertising revenue for the *The Chemical Engineer* – mirroring prospects for the process plant industry – fell by half. The consequent lack of demand for engineers following what the IChemE called 'a period of acute depression' was

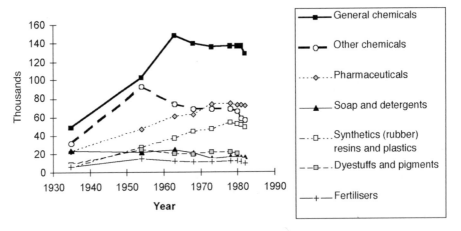

Figure 7-2 General employment in the chemical industry. Adapted from J. Clark, *Technological Trends and Employment 2: Basic Process Industries* (Aldershot: Gower, 1985) Table 2.9.

---

[19] CM 26 Oct 1966.
[20] A. M. McKay, 'The move to Rugby', *TCE* Sep 1975, 492.
[21] 'PPA annual chemical plant dinner', *Chem. & Indus.* 4 Nov. 1972, 819; W. B. Duncan, 'Lessons from the past, challenges and opportunity', in: D. H. Sharp and T. F. West (eds.), *The Chemical Industry* (Chichester: Ellis Horwood, 1982), 15–30; J. Clark, 'Chemicals', in: J. Clark (ed.), *Technological Trends and Employment 2: Basic Process Industries* (Aldershot: Gower, 1985), 70–115.

exacerbated by two other factors. First, the price increases and embargo of the Organisation of Petroleum Exporting Countries (OPEC) in 1973–4 led to a general energy crisis in Britain and other western countries. The price of crude oil quadrupled, and the petrochemicals based on it by almost as much.[22] By the end of 1973, the economic situation was further complicated by industrial action in the mines and railways and the imposition of the 'Three-Day Week' to deal with the crises. Its authority eroded, Edward Heath's Conservative government was replaced by Labour under Harold Wilson the next Spring. More subtly, the Flixborough chemical plant explosion of 1974 damaged the reputation of chemical engineers and the chemical industry much more directly and locally than previous concerns over environmental pollution had done.[23]

EVOLUTION OF EDUCATION AND INDUSTRIAL TRAINING PROGRAMMES

With the decline of employment and the threatened retreat of an occupational identity, the profession nevertheless maintained and extended its disciplinary identity after 1960. Teachers of the subject at degree level came to new arrangements with industry; extended the chemistry and management studies components in courses, thereby altering the identity of the graduates they generated; and indeed gained a near monopoly on producing new chemical engineers.

The warnings of the worsening job situation that had come from educators such as Haselden and Coulson were typical of the reactions of the IChemE, decoupled from industry itself. While industrialists within the Institution were largely accustomed to swings in chemical engineering employment, teachers were soon threatened by dropping enrolments. The health of their departments, created or expanded over the previous decade, mirrored the health of the profession as a whole.

As with employment, the rapid evolution of chemical engineering teaching through the 1950s decelerated during the next decade. There was a weakening of the comparatively close connection between chemical firms and teachers of chemical engineering. Teachers could no longer rely on chemical firms to employ all of their graduates, and it became necessary to broaden the range of subjects taught on undergraduate courses to include those intended to develop management skills. This met with the approval of many employers, although for some the process was not sufficiently developed.

Until the mid-1950s, the IChemE's views regarding training in industry had also coincided with those of most employers: the Institution said that structured courses of training were not generally appropriate for would-be professional engineers, and that periods of supervised, practical experience in 'responsible' posts were more valuable. Only the largest firms, such as those in the petroleum sector, had thought the cost of structured courses worthwhile, particularly in

---

[22]   IChemE Annual Report, 1973; Peter H. Spitz, *Petrochemicals: the Rise of an Industry* (New York: John Wiley, 1988), pp. 476–81.

[23]   For more on Flixborough and its consequences, see Chapter 8. Chemical engineers were, of course, working at anti-pollution bodies such as the Warren Spring Laboratories for Air Pollution from the 1960s and the Alkali Inspectorate from the time of G. E. Davis.

the context of the high salaries they had to pay to attract graduates in the highly competitive labour market.[24] The general climate was not yet amenable to the establishment of separate state-sponsored organisations dedicated to the development of industrial training. Consequently, the MNLS had little option but to utilise the IChemE's resources.

Industrialists at the time were naturally keen to co-operate with the IChemE in order to gain some of the scarce personnel they wanted. The Institution worked with the ABCM and the BCPMA to devise a set of guidelines for training schemes; this helped to reduce any feeling among industrialists that they were being forced to accept training without any relevance to their needs. The presence of senior engineers on the IChemE's education committee also helped, because it was this committee which assessed the appropriateness of each company's method of working.[25]

The success of this bipartite arrangement depended on the continuance of compulsory military service, from which trainees were exempt. By June 1960, the IChemE had considered 48 training schemes, and by July 1963 it had approved 45. But the ending of national service in 1960 marked the start of a sharp decline in provision. In 1963, probably less than half of newly-qualified chemical engineers were receiving a formal training in industry, and two years later the proportion was judged to be as low as a fifth. Moreover, many of the firms that continued to provide training did not like the Institution's general scheme. Training now tended to relate closely to the specific nature of a firm's own business.[26] As a consequence, the training of the resulting specialists tended to revert more towards chemical technology than chemical engineering, threatening to re-open the jurisdictional problems of the interwar period.

From the mid-Sixties, the state became more directly involved in matters of training. This further undermined the IChemE's position, weakening the co-operation between the Institution and employers. In 1967, the Labour government enacted provisions in the 1964 Industrial Training Act which established the Chemical and Allied Industries Industrial Training Board (CAPITB) and the Petroleum Industry Training Board (PITB). These two training boards covered the majority of chemical engineers employed in private industry. A number of other boards, such as the Gas ITB and the Water Supply ITB (both founded in 1965), catered for many of those in the 'fringe' industries.[27]

The boards were tripartite organisations. Representatives of employers and the trades unions sat in approximately equal measure alongside ministerial

---

24 'The price of engineers', *Quart. Bull.* Jan 1954, xxvii–xxviii; 'Practical training for graduates and undergraduates', *Quart. Bull.* Apr 1954, xxxv–xxxvi; M. C. Coe, 'Industrial training for graduate chemical engineers', *TCE* Jul/Aug 1964, CE 167–C173.

25 'Symposium on staff training', *J. Inst. Petroleum* 43 (1957), 131–51; 'Graduate training in chemical engineering', *TCE* Aug 1958, A43–A44; A. S. White, 'Graduate training in industry', *TCE* Feb 1961, A44–A49; 'Graduate training in industry – discussion', *ibid.*, A53–A59.

26 MEC 5 Jul 1963; J. B. Gardner, 'Survey of graduate training schemes for chemical engineers', *TCE* Feb 1961, A49–A52; 'Under consideration', *TCE* Jan/Feb 1965, CE2; Annual Report, *Trans. IChemE 42* (1964), T167; Coe, *op. cit.*, CE167–CE173.

27 MEC 8 Jan 1967.

appointees. By contrast, the IChemE's influence on the boards was slight, partly because most of the work was concerned with employees below the professional level. Moreover, the boards preferred to work with organisations that could deal with all the issues pertinent to a particular industry – for example, in 1969–75 the CAPITB chose to work with the new Chemical Industries Association (CIA) in trying to predict levels of demand for trained personnel. The IChemE could not offer this degree of co-operation. Finally, since the training boards enjoyed powers of compulsory levy over firms which did not meet agreed standards of training, industrialists naturally tended to turn to the boards for guidance on major matters of policy.[28]

These changes did, however, encourage the IChemE to develop a more market-oriented relationship with firms. In association with institutions of higher education, it developed and provided programmes of continuing education, partly intended to help employers meet their training responsibilities. In the 1970s, short courses, usually of no more than a few days' duration, were popular with companies wishing either to update or extend their employees' technical knowledge or, more rarely, to improve their managerial skills.[29]

Demand for greater academic attention to chemistry

In comparison with the other principal engineering disciplines, employers were, throughout the 1960s, broadly satisfied with the technical and scientific instruction given to chemical engineers in the universities. In 1960, a survey of 1 865 chemical engineers suggested that the syllabus recommended by the IChemE covered the majority of skills used by employees in a wide variety of posts in private industry and government service; a similar exercise undertaken five years later produced the same result.[30] Nevertheless, by the early 1960s, there were significant discrepancies between the IChemE's recommended syllabus and what was being taught in the more progressive universities. The IChemE was becoming less relevant to the articulation of employers' wants, because once a university course had been recognised it was not subject to any kind of periodic examination. Moreover, the IChemE Education Committee was increasingly populated with older members less aware of prevailing industrial concerns.[31] The first generation of chemical engineers still held sway, notably

---

[28] The training boards' powers of levy were weakened from 1973, but they did not disappear. Annual reports of the CAPITB and PITB (1968–78); H. W. Ashton, 'Taking stock', *TCE* Jul/Aug 1973, 353–9; G. Terry Page, *The Industrial Training Act and After* (London, 1967), 257–71; P. J. C. Perry, *The Evolution of British Manpower Policy* (London, 1976), p. 248.

[29] MEC (1974–5); D. C. Freshwater, 'Symposium on continuing education in the process industries', *TCE* Nov 1972, 429–30.

[30] R. Edgeworth Johnstone, 'A survey of chemical engineering education and practice', *Trans. IChemE 39* (1961), 263–72; Edgeworth Johnstone and C. B. Lax, 'A second survey of chemical engineering practice', *TCE* Jan/Feb 1966, CE7–CE15; O. G. Weller, 'The demand for chemical engineers', *TCE* Dec 1961, A73–A78; Kearton, *op. cit.*

[31] 'Academic requirements for Associate Members', MEC 20 Feb 1959. The committee reserved the right – apparently never exercised until the policy towards accreditation was altered – to reconsider any course at any time.

through Hugh Griffiths' influence over the content of the Home Paper. Nevertheless, new departments and courses were still being founded, and so the Institution sought to identify those advances in teaching which it considered most promising in the light of the current and future skills likely to be required of industrial chemical engineers.

By 1963 – in the midst of the deepest postwar recession for the industry – a committee chaired by P. V. Danckwerts had identified a trend within many of the chemical industries towards a greater concern for the manufacture of products, both existing and novel, by means of new, more sophisticated chemical reactions and new catalysts. This implied that more attention than hitherto should be given within process design to the efficiency of the chemical reactor, and hence to the detailed study of chemical reactions on the practical scale; this 'chemical reaction engineering' drew upon the theoretical insights of applied chemical thermodynamics and chemical kinetics that were already being developed through academic research.[32] The committee also considered that some knowledge of subjects which had traditionally been dealt with by courses in industrial and applied chemistry, and in chemical technology, would be useful. These included studies of raw materials, of industrial processes and products, and of the reactions used in industrial practice. Such courses might treat in a qualitative and descriptive manner those aspects of industrial practice which were not amenable to mathematical and theoretical analysis.[33]

A sub-committee of the Education Committee, chaired by K. G. Denbigh, professor of chemical technology at the University of Edinburgh and an expert on the design of chemical reactors, developed these suggestions into a new set of recommendations for university courses. The new syllabus emphasised the teaching of chemical thermodynamics, reactor theory, and chemical processes and catalysts. It was complemented by another syllabus oriented towards the design of the physical side of processes and the construction of process plant; it placed a greater emphasis on materials, power thermodynamics and fuel technology. The IChemE finally published, in 1964, a single set of recommendations giving university departments sufficient flexibility to enable them to teach either, or both, of these kinds of courses. This course of action enabled the Institution to emphasise that the cultivation of design skills – be they of the design of plant or of chemical processes – should remain central to the philosophy of all courses.

A good example of the kind of syllabus that was now approved by the institution was that at the University of Birmingham. From the early 1960s, students could choose at the end of their first year of study to follow an honours course in 'chemical engineering science' or one in 'chemical process engineering'. Both included the study of unit operations, the problems of scale,

[32] The principal American text of the period was Byron Bird, Warren E. Stewart and Edwin N. Lightfoot, *Transport Phenomena* (1960). See G. Astarita and J. M. Ottino, 'Thirty-five years of BSL', *Industrial and Engineering Chemistry Research 34* (1995), 3177–84.

[33] 'Second report of the Exploratory Committee', *TCE* Jul/Aug 1964, CE174–CE176; S. P. Waldram, 'What is chemical reaction engineering?', *TCE* Jul 1983, 28–31.

thermodynamics, mathematical analysis, organic reactions, reaction kinetics and reactor design. However, only students following the 'chemical' stream were taught the details of several specific chemical processes and catalysts, and chemical process design and choice; those following the 'physical' stream acquired a more detailed knowledge of fluid flow and physical transport mechanisms. Finally, both streams required the successful completion of a design problem.[34]

Training of chemical engineers in management studies

Despite the emphasis of the founders of the IChemE on the potential of the chemical engineer for management, by 1960 many employers (and many of their employees) thought that university courses were an inadequate preparation for such responsibilities. Not enough time was being spent on subjects such as industrial economics and management science, and in developing communicative skills.[35] This criticism was also being made about courses in other branches of engineering; it was partly the result of a general change of opinion among industrialists on the practicability of cultivating management skills through an academic training.[36] However, the problem was felt particularly keenly by chemical engineers. First, because the profession had expanded so greatly since the late 1940s, there was a comparatively high proportion of individuals moving from the technical posts in which they had been first employed and on into management. Between 1953 and 1960, this 'bubble' in managerial positions increased from about 20 percent to over 30 percent of members of the IChemE.[37] Secondly, the trend in the 1960s towards much larger scale production of chemicals presented novel problems of management, and hence required better training. Finally, the temporary saturation, in 1962, of the traditional job market in the heavy chemical industries suggested to some academics that there was a need to broaden the range of skills covered by an academic training.[38]

Under the last of these circumstances, there was a risk that the priorities attached by teachers to the subjects taught in a course of academic training in chemical engineering would diverge from those of the traditional employers. In order to defend their own position and to increase the prospects for employment of their students in novel industries and occupations, some teachers suggested that the subjects to be included in a course on chemical engineering

---

[34] MEC (1963–4); *Scheme for A Degree Course in Chemical Engineering* (London, 1964); J. T. Davies, 'Chemical engineering education for the future', *TCE* Sep 1965, CE232–CE233; 'Discussion of papers presented at the second session', *TCE* Sep 1965, CE239–CE240.

[35] F. H. H. Valentin, 'Chemical engineering education', *TCE* Dec 1957, A37–A41; Coe, *op. cit.*, CE172; Weller, *op. cit.*, A78; Edgeworth Johnstone, *op. cit.*, 274–8; Edgeworth Johnstone & Lax, *op. cit.*, CE12; T. R. Bott, 'Does industry get the chemical engineers it needs?', *TCE* Mar 1964, CE38–CE50.

[36] Divall, 'Fundamental science versus design: employers and engineering studies in British universities, 1935–1976', *Minerva* 29 (1991), 167–94, esp. 181–3.

[37] Coe, *op. cit.*, CE169.

[38] 'The chemical engineer – profile for the future', *TCE* Sep 1968, CE273–CE278.

should be chosen on the basis of their intellectual rigour and ability to encourage logical and creative thinking, rather than according to the functional requirements (including those relating to management) of employers in the chemical industries. Training within the firm could correct any discrepancy between the two sets of requirements.[39]

The revision in 1964 of the IChemE's recommended syllabus had not adequately respected the opinions of employers on training for management. There is no obvious reason for this, although the need to influence the educational policy of the newly founded Engineering Institutions Joint Council, founded in 1963, might have affected the judgement of the Education Committee.[40]

In 1966 the IChemE established a committee of academics and industrialists, chaired by E. S. Sellers of British Petroleum, to review the requirements for the training of chemical engineers over the next 10 to 15 years. Sellers had been a member of the original teaching staff at Cambridge before moving, in 1955, to head the new department at University College, Swansea; but he was also a practical engineer with a considerable reputation.[41] His committee agreed that an undergraduate course could no longer be expected to provide a complete academic training for a chemical engineer. The teaching of technical material should be 'more related to the scientific fundamentals', leaving the training of graduates in the details of particular processes and technologies, or non-technical specialisms such as economics or marketing, to the postgraduate level or to training in industry. A 'core' of recommended subjects, defining the minimum that should be required academically of a chemical engineer, included studies of the physical properties of materials; descriptive chemistry; the transfer of heat, mass and momentum; classical and chemical thermodynamics; rate processes; process control; mathematical and computer techniques; and techniques of economic evaluation. By integrating these core subjects in a manner which demonstrated their practical relevance, usually by means of a design problem, undergraduates would develop their 'managerial, entrepreneurial, human and innovative abilities'.[42]

The pressure on the IChemE to reform its recommended syllabus was reduced as the prospects of employment in the traditional chemical industries improved

---

[39] E.g., Edgeworth Johnstone, *op. cit.*, 273, 275. Some teachers had always stressed the non-vocational aspects of an academic training in chemical engineering. See, for instance, D. M. Newitt, 'Technology and the state', *Trans. IChemE 28* (1950), 1–3.

[40] MEC (1963–6). Academics claimed that managerial issues were covered in seminars; there was a certain lack of precision concerning the teaching requirements from employers.

[41] Ernest Stanley Sellers (1912–88), educated at Manchester Municipal College of Technology in Applied Chemistry, managed oil refineries before the war. After a twelve year academic career, he became IChemE President in 1964, managing the BP Research Centre at Sunbury-upon-Thames before becoming Assistant General manager of its Refineries Department. Obituary, *Diary & News* May 1988, 4.

[42] 'Profile for the future', *op. cit.*, CE273–CE278; MEC 13 May 1968. The first computer in the UK used to control chemical plant was installed at Winnington in 1959. Through the 1960s, computers also were increasingly used for optimising plant design, via critical path analysis and automated flow charting, for example. The process control and design functions entrained both a skilling of chemical engineers in computer methods and a deskilling of operatives.

markedly from the mid 1960s. The universities, however, continued to introduce new courses for degrees, as well as modifying existing ones, partly in response to the changing qualifications and expectations of school-leavers. A noticeable innovation was the introduction of courses leading to the award of a degree with joint honours including a subject other than chemical engineering; such as that at the University of Exeter. A survey in 1972 of all 25 colleges offering undergraduate courses in chemical engineering showed that sixteen included teaching in economics, and fifteen in management, although attendance at courses in these subjects was not always compulsory. There was some criticism that the teaching in these subjects was not relevant to the specific needs of engineers, and that it was not properly integrated with the scientific and technical work. There was a small number of structured courses in which the teaching of management was particularly prominent; for example, at the Universities of Leeds, Aston, Surrey and Loughborough. All the courses in chemical engineering included a compulsory project in design, and in this respect they escaped the criticism of the lack of design still being levelled at many courses in the other principal branches of engineering.[43]

New dominance of degree courses

As we saw in the last chapter, chemical engineering teaching during the 1950s was transformed from postgraduate and other diplomas to degree courses. This process accelerated during the 1960s. At the beginning of the decade there were flourishing undergraduate courses at Birmingham, Imperial College, UMIST, Leeds, Bradford, Battersea and Loughborough. By 1968, the Education Committee of the Institution recognised eleven postgraduate 'conversion' courses, but these were not greatly popular; Bradford and Leeds, for example, abandoned theirs by 1969.[44]

A synopsis of the postgraduate course at King's College, London is worthy of note because its closure, in 1966, was symptomatic of the problems facing 'conversion' courses. The altered climate in favour of chemical engineering after 1945 meant that there was little doubt that chemical engineering would re-emerge there after the vicissitudes of the war. An informal advisory committee was established in 1946; by 1948, this numbered J. A. Oriel (Shell), J. Davidson Pratt (ABCM), F. A. Greene (a King's trained engineer, consulting engineer in the paper, gas, explosives and chemical industries, associated with the IChemE from the days of the Chemical Engineering Group and a long-standing member of the Education Committee), E. V. Evans, Keith Fraser and Hugh Griffiths. The course (for which the award was now a Diploma) re-opened in 1947 after the wartime hiatus. Although the syllabus and general requirements had changed little from those of the prewar era (H. W. Cremer, President

---

[43]   R. Edgeworth Johnstone, 'From practice to theory and back', *TCE* Apr 1967, CE68; W. Smith, 'A university view', *TCE* Jan 1973, 26–9; I. Berkovitch, 'Professional skill, adaptation and development', *TCE* Oct 1974, 647–52; S. R. Tailby, 'Chemical engineering education today', *Chem & Indus.* Jan 1973, 77–85; MEC 2 Jun 1975.

[44]   MEC 1969.

of the IChemE in 1947, was still Director of Studies), the sub-department now fell within the Faculty of Engineering as part of the joint Department of Civil, Mechanical and Chemical Engineering. The question of revising the course was considered by the advisory committee in 1948, but nothing came of it. However, recognition by the IChemE was still judged important; the accreditation of 1929 having been deemed by the IChemE to have lapsed, the course was granted fresh recognition, for five years, by Professor Garner (of the University of Birmingham) and F. E. Warner in 1950. Griffiths raised the question of the curriculum again in 1950. He considered that students with a background in engineering needed to study mechanics of fluids, that both sets of students should be instructed in heat transfer, and that the theory of vapour pressure and other 'physical topics of wide application' should perhaps be included. But the advisory committee considered the syllabus full and again recommended no changes.[45]

The course was still regarded as a small one; the capacity in 1948–9 (determined partly by the number of students that the engineering departments could accommodate) was six, although an increase to nine was judged 'desirable'. The decision was also reached to encourage the recruitment of engineers by reserving a proportion of the places for such individuals. The first engineer of the postwar era took the course in 1950–1, although the overwhelming bulk of entrants until the demise of the course remained chemists; a substantial waiting list existed for such candidates (over 40 applicants for six places in 1951). By 1955, fifty diplomas had been awarded.

Industrial firms continued to support the course by way of industrial placements, and by 1953 were looking for a greater output from the college. In 1954, Shell Petroleum, through Oriel, offered the annual sum of £1 100 for studentships; the firm expressed the hope that when more accommodation was available, the money would be used to attract industrially experienced students. Such accommodation was completed in 1956, permitting an annual intake of 16–18 students (although this figure was revised downwards shortly after, probably because of the pressure on the, by then, sole member of the teaching staff, S. B. Watkins). In the 1956–7 session, the department operated for the first time as a fully independent entity. The future looked rosy: yet within a decade, the department was to be closed.

An indication of problems to come was the withdrawal of no less than half of the sixteen students accepted for entry in the 1956–7 session. Another blot on the horizon was the difficulty experienced, in 1955–6, in recruiting an additional member of staff to relieve the pressure on Watkins. The explanation for the latter lay solely in the salary on offer, a problem not unique to KCL.[46] But the loss of students was serious; although the surviving records do not offer a definitive explanation, it is more than likely that the problem lay in the increasingly obsolete nature of the curriculum, combined with the increase in facilities for training elsewhere.

---

[45]   King's College, London Chemical Engineering Committee, archive KDCM/M1.

[46]   D. R. Morris, a former research student of the department, was eventually appointed in 1959.

By 1961, a fall in quality of applicants was being experienced, at least among those for the Shell awards, and the company was threatening to withdraw all support. The difficulties continued as Watkins suffered a serious illness; numbers were maintained increasingly by recruitment from overseas, and the failure rate increased sharply as a result. Although there was something of a rally in the session 1964–5, the death of Watkins in November 1965 and the impending departure of Morris brought matters to a head. Morris argued that although there was a demand for the diploma, the department was too small in its present form; an undergraduate course was needed to secure its future. Lacking support from the advisory committee, the decision was taken, in November 1966, to close the department. Although the difficulties experienced at King's were not duplicated exactly elsewhere, by 1972 the number of conversion courses in the rest of the country had fallen to five; these appealed largely to students from overseas.[47]

Despite this evidence of an increasing trend towards degree studies, calls for a wider spectrum of training in chemical engineering persisted. In 1960, there had been discussion in Council that more graduates having a pass degree were needed. Council saw the greatest employment opportunities for chemical engineers as process operators in the numerous chemical plants. Moreover, pass degrees were seen as a natural progression from the HNC programme which would lead to standardisation on degree-trained chemical engineers. The following year, Manchester applied for recognition of its pass degree. The Education Committee decided in July 1962 to retain the requirement of honours degree, but 'after long and detailed discussion', the Manchester pass degree was accepted that December. The matter was effectively sealed when the Engineering Industries Joint Council decided to accept the pass degree as a standard for engineers in 1964.[48]

Supporting the divergence of chemical engineering training options, the Higher National Certificate in chemical engineering was re-examined. The Ministry of Education began a reconstruction of course programmes for technicians, craftsmen and operatives in English and Welsh technical colleges in 1961, and the Institution decided that the time was ripe to revise the HNC scheme. In 1963, the Education Committee decided that the HNC was a 'technician qualification', although it could lead to a professional qualification. The course was consequently revised along these lines.[49] By contrast, five years later the need for chemical engineering education was projected as 'an education rather than just a training for a job'.[50]

---

[47]   Minutes of the Advisory Committee, King's College, London; Gordon Huelin, *King's College London 1828–1978* (London, 1978), 140; Tailby, *op. cit.*, 80. The teaching of the science of process manufacturing continued at King's in the Department of Powder Technology. J. E. English, 'Research and teaching in the Department of Powder Technology, King's College, London', *TCE* May 1970, CE113-CE114.

[48]   MEC 1960–3.

[49]   [S. R. Tailby], *Scheme for a part-time course in chemical engineering* (London: IChemE, 1964); 'Chemical engineering technicians', *TCE* Apr 1968, CE62–CE63.

[50]   MEC 1963–8.

In 1965, the Higher National Diploma (HND) was instituted. A lower-standard certification than a degree, the HND was slow to gain popularity and never reached the yearly output of the HNC, partly because it was usually followed full-time instead of part-time. Between 1952 and 1970, over 1200 HNCs were awarded. By 1970, however, total admissions were declining. The drop in entrants was a result of the decision of the new Council of Engineering Institutions (successor to the EIJC) to close the part-time route to professional qualification, making full-time degrees the standard, a response to the changing pattern of education introduced by the establishment of more universities in the mid-1960s.[51] An IChemE working party under Professor S. R. Tailby of the University of Surrey correspondingly retailored the HNC and HND programmes to suit the training of chemical engineering technicians.[52] Courses were reduced in theoretical and mathematical content to fit them between the Advanced Chemical Technician certificate and chemical engineering degree courses. By 1971, as a result of these altered goals, the number of HNDs awarded exceeded HNCs for the first time.[53]

But further soundings of industrial opinion indicated that there was little strong support for the HND. While a few organisations employed chemical engineering technicians with success, most met their needs with a combination of university graduates and relatively unskilled process workers. The six colleges running HND courses increasingly attracted their students from overseas. Most of the local students obtained HNDs as a stepping stone to further qualifications, and few went directly into industry. The Institution recommended that no further courses be encouraged, and that their number, in fact, be reduced.[54]

The IChemE continued to administer the HNC and HND programmes until 1974 in England and Wales (Scotland, with a separate educational administration, was administered longer). The Institution's influence was eroded by the introduction of the Technician Education Council (TEC) and Scottish Technical Education Council (SCOTEC). While these bodies sought the support of the IChemE in their training plans, they provided no direct representation for it.

The problems of defining suitable education programmes continued into the 1970s. Two major problems arose in trying to match IChemE standards with those of universities and the increasingly important Council of Engineering Institutions.[55] First, the CEI scheme meant that a course might meet their requirements, but not those of the IChemE. Second, new courses were being defined by universities on academic grounds, but the Institution had to be concerned with professional implications. These conflicting motivations were

---

[51]   IChemE Working Party, 'The future of the HNC/HND in chemical engineering', *TCE*, June 1975, 378–9.

[52]   IChemE, *The Education of Chemical Engineering Technicians*, Dec. 1968.

[53]   See Fig. 6-1.

[54]   IChemE, 'The future of HNC/HND', *op. cit.*, 378–9. Of 25 companies responding, all but one deemed the HND unnecessary, and 14 never employed chemical engineering technicians.

[55]   The CEI, formed in 1965, was established and nurtured by the new Ministry of Technology of the Labour government, and planned as one means of revitalising the competitiveness of British science and technology. Relations with the CEI are discussed further in the next chapter.

exacerbated by the absence of an agreed definition of both the modern chemical engineer and the minimum chemical engineering content of education. One industrialist on the Education Committee noted that 'the Institution did not occupy a key position between employer and employee'.[56]

In 1971, a severe reduction in opportunities for new graduates within the chemical industries encouraged the IChemE again to examine its role in relation to the qualification of chemical engineers. Although prospects of employment improved rapidly during 1973, the heavy chemical industry was affected for much longer. Graduates turned to the so-called 'fringe industries', such as food processing, plastics and ferrous metals, and to jobs unrelated to chemical engineering, particularly accountancy.[57] The Education Committee, under P. P. King, an industrialist from ICI, established a working party early in 1972 to examine the possibility of broadening the syllabus in chemical engineering so that it could better serve as an education for a variety of posts. This committee revived the notion that there should be a 'core' of subjects defining the minimum content of an academic training in chemical engineering, and argued that only 55 percent of the time allocated for formal study need be spent on these technical, scientific and economic subjects. The balance could be allocated to the study of 'allied' subjects, such as environmental engineering, biochemical engineering, or to the further study of subjects relevant to management; however, both of the industrial members of the working party, and some other industrialists, were opposed to the teaching of management studies at the undergraduate level. After consultations with the heads of departments, these proposals were accepted by the IChemE in 1974.[58]

But these measures seemed insufficient to re-establish the credibility of the IChemE with either the universities or with employers. The greatest potential was thought to lie in involvement with continuing education. The Institution would sponsor and encourage the development of short courses in those areas not thought particularly suitable for an undergraduate training; for example, safety studies. Nevertheless, the IChemE considered other measures with regard to undergraduate training. In 1975 it was decided, with the support of the CEI, periodically to monitor existing courses in the universities. Lengthy discussions took place over the question of once again requiring an honours degree for admission to the Institution. Some members felt that this would enhance the significance of membership of the Institution both for employers and for chemical engineers. However, the IChemE was divided on the merits of such a move, and since the CEI was not enthusiastic, no progress was made until a decade later.[59]

---

[56] Hamm, MEC, 1975.

[57] J. Medley and M. J. Knight, 'A student viewpoint on chemical engineering employment', *TCE* Jun 1974, 395; 'First employment of graduands in chemical engineering', *TCE* Sep 1974, 570; Berkovitch, 'Professional skill', *op. cit.*, 647; 'Future requirements of university graduates in chemical engineering', *TCE* Mar 1975, 118.

[58] MEC (1971–1975); P. G. Caudle, 'IChemE and I', *TCE* May 1971, 173; 'A scheme for a degree course in chemical engineering', MEC 3 Dec 1973, Appendix I.

[59] R. Edgeworth Johnstone *Report on Continuing Education in Engineering with Special Reference to Chemical Engineering* (London: IChemE, 1969); MEC (1972–6); Report of the Education Committee, *TCE* Mar 1975, 163.

Thus, up to the mid 1970s, the teaching of chemical engineering in the universities did not meet all of employers' wishes. But there was no systematic expression of discontent similar to that felt by employers of mechanical, of electrical and, to some extent, of civil engineers, over the failure of graduates to apply the insights of engineering science to practical problems.

TOWARDS NEW PROFESSIONAL AND INSTITUTIONAL IDENTITIES

If the 1960s provided relative stability for practising chemical engineers, the early and mid 1970s were difficult times for the chemical industry and the IChemE, when many employers no longer relied on the Institution to articulate their wants. The institution's reputation with the universities had also suffered, partly because the Council of Engineering Institutions had set the level of academic training at that of an ordinary degree. The ability of graduates to obtain employment without becoming members of the IChemE weakened the Institution's influence over syllabuses.[60]

These factors combined with growing financial constraints for the Institution. The IChemE Annual Report of 1973 noted that 'at the beginning of the year the chemical industry was still in a period of acute depression'. Advertisements in 'The Diary' for vacant positions had plummeted since 1971, while those seeking posts there had for the first time become significant (see Figure 7-1).

These events were mirrored in a downturn in the intake of new members of the IChemE, and some 600 existing members – a record number – were deleted for non-payment of dues in a single year. Recessions typically affected chemical engineers unequally. Those employed as design detailers were culled more quickly than chemical engineers engaged in conceptual design. The younger component of the profession thus turned towards more promising occupations, decimating prospects for subsequent growth.

The production of new chemical engineers, too, was declining. Degree courses in chemical engineering were scarcely half full, and admissions of new Graduate Members were also seriously truncated.[61] And, for those chemical engineers who did graduate, a decreasing fraction found employment in the profession.[62]

---

[60] MEC 2 Oct 1972, 7 Apr 1975; 'G & S News', *TCE* Sep 1972, 360–1; R. W. H. Sargent, 'Future policy of the Institution', *TCE* Jul/Aug 1974, 438–9; R. J. Kingsley, 'The future of the Institution', CM, document F.115, 2 Sep 1974.

[61] Course admissions dropped by over 40% between 1970 and 1974. Home students graduating from UK courses fell from 680 in 1973 to only 365 in 1977. The IChemE total membership hovered at the 10 000 mark from 1969 to 1976; Student and Graduate membership, in particular, declined during this period: admissions of new graduates to the IChemE fell from about 60% to 34%. The same trend was evident at the larger AIChE, where membership remained nearly static at 38 000 through the early 1970s. See IChemE Annual Report, 1973; 'Who joins the Institution?', *TCE* May 1978, 329; R. T. W. Hall, 'Supply and demand for chemical engineers', *TCE* May 1978, 388–90.

[62] Of the 1968 graduates of the Ramsay school of chemical engineering at UCL, 79% found employment in chemical engineering, 2% in other subjects and 19% returned for further university studies; for the 1972 class, the figures were 49% in chemical engineering, 23% in other subjects and 28% in further study. See J. Medley and M. J. Knight, 'A student viewpoint on chemical engineering employment' *TCE* Jun 1974, 394–5.

The IChemE Council blamed the fall on a combination of factors: (1) poor teaching in secondary schools and high and demanding entry requirements; (2) general disaffection amongst school leavers with the advantages of a techno-logically based society; and (3) recession in job opportunities and severe redun-dancy problems.[63]

The crisis for the profession was variously interpreted. Some presented it positively as an opportunity for entering new disciplinary territory. Heat recov-ery and efficiency, for instance, were trumpeted as new goals for chemical engineering; soon, claimed one correspondent, 'cooling towers and flare stacks will be considered monuments to an engineer's failure'.[64] But others saw the events of the new decade as heralding the end of a temporary and market-driven profession. W. S. Norman, teaching at UCL, described three periods of employment: the 1930s (poor), the 1950s (good), and the late 1960s and beyond (poor again).[65] Chemical engineers may have been valuable in increasing the efficiency of processes in industries 'built up over generations by a loose, often uncomfortable association of chemists and mechanical engineers', went the argument, but declining industries no longer needed as many engineers as the academic world produced. As a result, there were 'no longer any grey-haired practitioners, because they have gone to more profitable posts'.[66]

The president for 1973–4, R. W. H. Sargent, noting the feeling by many members that the Institution had 'run out of steam', attributed this more directly to weaknesses of the organisation: that energies were being dissipated in relations with the CEI while doing little for chemical engineering itself.[67] What future, indeed, did chemical engineers or their institution have?

---

[63]   CM 2 Apr 1974.
[64]   M. Kneale, 'The energy crisis means work for us', *TCE* Mar 1974, 129; P. V. L. Barratt, 'A crisis of our own creation', *TCE* Apr 1974, 197.
[65]   W. S. Norman, 'Chemical engineering outlook', *J. Ramsay Society* 20–21 (1973), 60–5.
[66]   M. Stacey, *TCE* Dec 1987, 4.
[67]   R. W. H. Sargent, 'Future policy for the Institution', *TCE* Jul/Aug 1974, 438–9. For a member's view, see M. C. Singer, letter to the editor, *TCE* Oct 1971, 386.

CHAPTER 8

INSTITUTION VERSUS PROFESSION

With a half-century of significant battles of professional recognition behind it, the identities of the Institution and its members were gradually transformed from the mid 1970s. Part of this was attributable to external contingencies of the economy, the oil crisis and increasingly prominent public concerns surrounding pollution and nuclear issues. Other, deeper, changes concerning the membership can be ascribed to wider social changes affecting engineering practice. The self-identity of practising chemical engineers changed more rapidly during the 1970s than ever before because of factors such as the new overwhelming dominance of degree-educated chemical engineers – mirroring the CEI standard for the engineering professions established during the 1960s, but having earlier origins – and the rise of women in the profession.[1] And besides the impetus for change that came from within the IChemE – and which brought a new administrative structure, membership categories, premises and General Secretaries – were new liaisons. The most important of these was a growing affiliation with other bodies which came to influence professional orientation and to dominate the activities of the IChemE Council, if not the interests of its members. The unavoidable result of these factors is a divergent narrative: the history of the Institution and profession, rather than the occupational specialism, becomes crucial for understanding this period, despite the tepid co-operation of rank and file member chemical engineers themselves. In the midst of attempts to unify the technical professions was a curious rending of occupational and professional identities.

NEW BODIES, NEW ALLIANCES

The imperfect match between the aspirations of the IChemE, as understood by its London administration, and that of practising chemical engineers was an undercurrent to the narrative of Chapter 6. This disjuncture of imagined and pragmatic identities became more evident through the 1960s. The manufac-

---

[1] By the end of the decade, the IChemE led all other British engineering institutions in its fraction of 'academic' members (with 69% having a degree only, rather than a degree and HNC). Policy Studies Institute, *Professional Engineers and Their Careers* (1979), p. 6.

turers' associations, which had increasingly become involved in chemical engineering education and research, reduced their direct involvement from the early
1960s and concentrated on government lobbying. The relationships between
the Associations, the IChemE and the government became perceptibly more
formalised. The establishment of a Ministry of Technology in 1963, for example,
appeared initially to be precisely what Herbert Cremer had called for a decade
earlier, namely a revitalised DSIR. The new Ministry proved within a few
months, however, to be undertaking a much broader role. It soon incorporated
the UK Atomic Energy Authority and other departments that had engaged in
substantial research and development.

Also in 1963 the government established the National Economic
Development Council (NEDC), which was welcomed by the ABCM and
BCPMA. Both organisations protested, however, when suggestions were made
to set up a state-sponsored chemical engineering consulting organisation to
promote international trade.[2] The ABCM, in particular, sought new arrangements to influence government. But, as in the immediate postwar period, the
organisation wanted a particular form of interaction. The 1964 chairman said:

> we realise that government intervention in industry is inevitable, and intervention of
> the right sort should be welcome. Equally important is the need for industry's views
> to be clearly spelled out in Whitehall and Westminster. Let us not call this interven-
> tion, but rather mutual co-operation and an extension of the traditional boundaries
> of Government and industry.[3]

But the ABCM found earlier distinctions increasingly difficult to maintain,
separating the responsibility for wages and working conditions within industry,
on the one hand, from commercial and economic questions on the other. The
ABCM thus decided to merge with the Association of Chemical and Allied
Employers at the end of 1964. The new Chemical Industries Association (CIA)
took responsibility not only for commercial questions but also labour and
government negotiations.[4] Similarly, in 1965 the major contractors formed
their own trade association, the British Chemical Engineering Contractors
Association (BCECA), to represent their interests.[5]

As we saw in the last chapter, the IChemE during this period was becoming
more closely involved with the CEI. Its 1968 annual report, for example, noted
a considerable amount of work concerning assessment of academic qualifications, administering examinations and providing information on a range of
issues concerning professional engineering.[6]

The CEI was an attempt to foster professional unity of the engineering

[2]   BCPMA annual dinner, *Chem. & Indus.* 9 Nov 1963, 1800; ABCM annual dinner, *Chem. &
      Indus.* 24 Oct 1964, 1809–10.
[3]   M. J. C. Hutton-Wilson, *Chem. & Indus.* 24 Oct 1964, 1809.
[4]   *Chem. & Indus.* 6 Nov. 1965, 1865.
[5]   This was restricted to British companies until 1990, when the changing nature of the industry
      (particularly the mergers with, and subsequent control by, American firms) forced the removal
      of this limitation on membership.
[6]   *Chem. & Indus.* 25 May 1968, 662.

professions and to increase their status in British society. Rather than 'borrow status from science', the CEI aimed to 'raise engineering's status by emphasizing its *own* identity and importance'.[7] It was an outgrowth of postwar trends starting before even the Percy Report on higher technological education of 1945, which had inextricably mixed the interests of government, industry and academia in producing and accrediting engineers. The immediate origin of the CEI was a consequence, from the late 1950s, of the desire among the councils of the Big Three institutions (Civils, Mechanicals and Electricals) to have closer contact. The institutions had found themselves collaborating ever more closely as a result of government initiatives to boost the number of engineering graduates through expanded university, technical college and HNC courses. In 1961 the Engineering Institutions Joint Council (EIJC) was founded to explore opportunities for professional restructuring and promoting the status of engineering.[8] Its chief accomplishment was the creation of the CEI as an expanded body. This federation of thirteen chartered engineering institutions represented, by 1965, nearly a quarter million engineers, some 130 000 having corporate status, and operating under a Royal Charter.[9] In its early days, its promotion became linked with the efforts of the Labour government to revitalise science and technology in Britain.

The CEI had a profound effect on the long-term goals of its member institutions and on the status of their engineers. It introduced a new qualification, the Chartered Engineer, and set the accepted educational standard as no lower than a university degree. In effect, the part-time HNC route to professional status opened up after the war was being closed off. This redefinition of standards was contentious, particularly among the smaller institutions having a large fraction of non-graduates, such as the Institution of Plant Engineers.[10] Within the IChemE, however, this definition attracted the opposite criticism: that chemical engineers might now become corporate members by obtaining merely a pass degree while bypassing the Institution's Associate Membership examination.

From the late 1960s trade and professional organisations became rather more polarised in their sphere of activities and interactions. The chairman of the BCPMA, at its fiftieth anniversary in 1970, naturally enough derided government 'interference' with the plant industry via the increase of interest rates for export. On the other hand, he praised the sort of 'interference which the industry needed', namely the creation of a working party to support associations. The trade associations, consisting of engineering designers and contractors, equipment manufacturers, tank and pipe fabricators and erectors

---

[7]   Ian A. Glover and Michael P. Kelly, *Engineers in Britain: A Sociological Study of the Engineering Dimension* (London: Allen & Unwin, 1987), p. 78.

[8]   Watson, H. B. *Organizational Bases of Professional Status: A Comparative Study of the Engineering Profession* (Ph.D. thesis, U. London, 1976), pp. 279–88.

[9]   *TCE* Dec 1965, CE308.

[10]   The CEI invited some thirty to forty excluded bodies to participate in a confederation paralleling its own, leading to the Standing Conference of Nationally Qualified Technicians.

were, he argued, all 'part of a team'.[11] The BCPMA used this forum to combine
with two other plant associations, the Food Machinery Association and the
Tank and Industrial Plant Association, to form the Process Plant Association
(PPA). The new association, a commercial mutual-help body, placed new
emphasis on the promotion of international trade. Unlike its predecessor, the
BCPMA, the PPA avoided questions of education and training. It did, however,
sponsor research, some of it in chemical engineering.[12]

Moves towards merger

The new combinations of trade organisations were mirrored by similar attempts
among professional institutions. In common with many British professions, the
late 1960s and early 1970s were decisive for the IChemE, as successive Councils
considered a question both foundational and pragmatic: its identity as an
independent organisation. With the questioning of institutional identity came
continued probing of its relationship with others. The 'other' of constant
importance was the Society of Chemical Industry. The episode is worth
expounding in some detail, because it represented the first extended evaluation
by the Institution of its prospects and continued autonomy, and set the course
for similar negotiations through succeeding decades.

The IChemE and the SCI Chemical Engineering Group had collaborated
more closely since 1937, allowing the CEG to 'form a bridge between the
Society of Chemical Industry and the Institution accessible from both ends'.[13]
At that time, the CEG had a membership of 495 (about half that of the
IChemE), of whom 120 were members of the Institution. Then, and later, the
CEG membership was largely London-based and more senior than the average
of the farther-flung IChemE membership. During the war the two bodies had
held all ordinary meetings jointly and issued a joint volume. The relationship
was relaxed at the end of 1945 at the Group's request, but both bodies agreed
to share use of the Institution's library, to confer on the publication of papers,
and to accord special purchase arrangements for journals.[14] The distancing of
the two organisations left more time for the permanent staff, which grew in the
1950s, to attend to Institution business.[15]

Since 1966, there had been discussion regarding the future of the two organisations, and the initial impetus concerned accommodations. In the early postwar
years, the SCI had obtained grants from industry and purchased long leases
on a London property for their headquarters, a portion going to the IChemE
for only the cost of occupation charges.[16] Within a decade, financial realities

---

[11] 'BCPMA Golden Jubilee dinner', *Chem. & Indus.*, 7 Nov 1970, 1418.
[12] 'Process Plant Association and the IChemE', *TCE* Mar 1972, 89.
[13] Memorandum, 11 Dec 1937, Gayfere archive box XIV/1.
[14] IChemE Annual Report for 1945.
[15] Letter, R. Mason to Johnston, 6 Jan 1998.
[16] 14–15–16 Belgrave Square, occupied by the SCI and IChemE in 1956.

led first to rental, and then wider, negotiations.[17] The prevailing mood was for reorganisation of chemical and engineering organisations in the face of inefficiencies and duplication.[18] But there were distinct priorities on each side. While premises dominated the concerns of the IChemE, rationalisation of chemical organisations was at the top of the agenda for the relatively prosperous SCI. The IChemE Council had mixed reactions to the proposal for merger, in which the SCI would become responsible for all learned society activities while the Institution dealt with professional activities and maintained its current restrictions on membership.[19] The majority favoured a close association provided that professional autonomy and a clear occupational identity could be retained. A few months later, the Council decided to follow a 'two-stream approach' by starting separate discussions with the Institution of Gas Engineers.

By early 1967 the SCI was beginning to become embroiled in negotiations of its own with the Royal Institute of Chemistry and the Chemical Society for amalgamation of the three organisations. The three chemical bodies had jointly appointed Sir James Taylor to examine the possibility of their coming together, although the SCI found itself almost immediately on the defensive.[20] The SCI consequently sought options in its alliances, exploring collaboration with the Chemical Industries Association as well as the IChemE.

Taylor, ex ICI and a Council member of the SCI, RIC and CS nevertheless kept these chartered chemical organisations separate from the IChemE, which was classed among the Chartered Engineering Institutions.[21] Despite the mediocre terms offered to the SCI, there was a strong body of opinion within it to join the new body.[22] This was countered by the CEG and other influential members who were in favour of a closer association with the IChemE.

---

[17] C. S. Windebank, 'The Institution of Chemical Engineers and the Society of Chemical Industry', *TCE*, Jul/Aug 1968, CE245. The IChemE agreed to pay rent beginning in Sept 1967 but this was more than doubled the following year, with the Institution urged to seek industrial sponsorship for larger premises SCI CM 11 Aug 1967 and Document 35/68 (1968).

[18] Sir Harry Melville, President of the Chemical Society, made rationalisation of British chemical bodies the subject of his Presidential address in 1967. G. A. Dummett, IChemE VP that year and President in 1968, having polled members of his and other companies, recalled a widespread desire to reduce the number of organisations and journals. Letter, Dummett to Johnston, 2 Apr 1997.

[19] IChemE CM, 19 Jan 1966.

[20] The SCI secretary protested that 'the Society's "position of weakness with regard to possible amalgamation with the Chemical Society or the Royal Institute of Chemistry" referred to a numerical weakness of membership only'. SCI CM, 10 Mar 1967. The memberships in 1966 of the SCI and IChemE were similar at about 8 000 each, compared to 14 000 and over 17 000 for the Chemical Society and RIC, respectively.

[21] J. Taylor, *Examination of possible methods of merging the three chemical bodies* (1968). For a more detailed account of merger negotiations from the perspective of the chemical bodies, see C. A. Russell, N. G. Coley and G. K. Roberts, *Chemists by Profession* (Milton Keynes: Open University Press, 1977), pp. 301–24.

[22] The 'final' terms as outlined by Taylor were that the SCI should lose its name, its journal, and its premises which were to be sold, the proceedings to go to the new body. D. H. Sharp, 'Note on SCI/IChemE relationship' (unpublished typescript), p. 4, 6 Feb 1997.

Nevertheless, the SCI council decided that the other two bodies faced fewer problems merging than did the SCI with either.

In early 1969 the Presidents of the SCI and IChemE met optimistically to discuss the merger negotiations.[23] The advantages of a merger for the IChemE were deemed to be the gaining of good premises, good financial returns and a good journal, and for the SCI the gaining of a cachet of professional and research status.[24] They decided to appoint a joint General Secretary and to merge their libraries. The SCI Council agreed that it was an opportunity that should not be missed, since it was becoming clear that the SCI would not be joining the RIC and CS in a unified body.[25]

The General Secretaries of the SCI and IChemE, F. J. Griffen and J. B. Brennan, respectively, correspondingly retired in 1969 and were replaced by a Joint Secretary, David Sharp.[26] Sharp, who had been Assistant Director of the Federation of British Industry (FBI), was also uniquely a member of Council of both the RIC and SCI during their 1968 negotiations.[27] His objective was to fuse the IChemE and SCI into a new body organised much like the new Royal Society of Chemistry, which was now forming from the amalgamation of the CS and RIC.[28] The RSC had successfully overcome the difficulties of merging a professional and qualifying body (the RIC) with an 'open' organisation (the CS), and he believed the IChemE and SCI could do the same.[29]

The new SCI president, G. H. Beeby, favoured continued independence.[30] Only two of his council supported him: of the remainder, thirteen voted to join a new chemical society, and 28 to form a joint learned society with the IChemE.[31] The 2:1 preference of the SCI Council for merger with the IChemE was not polled among the membership. Following a postal referendum of members in late 1970 in which merger with the IChemE was excluded as an

[23] Neil Iliff of the SCI was with Shell; Tony Dummett (b. 1907) of the IChemE spent most of his career with APV.

[24] Communication, G. A. Dummett with Johnston, 2 Apr 1997.

[25] SCI CM 10 Jan 1969. See also 'Closer association between the Chemical Bodies: a personal note from the immediate past President and Chairman of Council, Mr Neil Iliff' *Chem & Indus* (4 Oct 1969), 1431–2 and SCI CM 7 Oct 1969; N. Illiff, 'Progress of negotiations between SCI, IChemE and the chemical bodies', *Chem. & Indus.* 7 Feb 1970, 163.

[26] John Basil Brennan (b. 1910), a mechanical engineer and former Principal of an Irish college, and Francis J. Griffin (b. 1904), each devoted to extending the identities of their respective organisations, wrote unpublished histories on their retirement: J. B. Brennan, *The First Fifty Years: A History of the Institution of Chemical Engineers 1922–1972* (1972) and F. J. Griffin, *The History of the Society of Chemical Industry* (c1974).

[27] *Chem. & Indus.* 15 Mar 1969, 331. Sharp had done research before and during the second world war at the Chemical Defence Research Establishment at Porton Down, at two companies and the British Ceramic Research Association. He became the technical director of the Confederation of British Industry upon its formation in 1965.

[28] See D. H. Whiffen and D. H. Hey, *The Royal Society of Chemistry: The First 150 Years* (London, 1991). The RSC was formed in 1971.

[29] D. H. Sharp, 'Note on SCI/IChemE relationship', (unpublished typescript), 6 Feb 1997.

[30] SCI CM 1967–70, B.59 document 22 (March 1970) 'Relationship between SCI, Chemical Society, RIC and Institution of Chemical Engineers'.

[31] SCI CM 13 Mar 1970.

option, about half responded very strongly in favour of merging with CS/RIC, but this represented less than half the membership. Beeby had effectively scuttled the debate by narrowing the choices and by taking a public stand against amalgamation with the CS/RIC.[32] The IChemE, for its part, was decidedly luke-warm to the idea of losing its own institutional identity by dilution. The IChemE Council informed their counterparts that they would discuss the possibility of forming a joint learned society, but 'not if the Society occupied a subservient place in the CS/RIC structure and substantially lost its identity'. They did not, however, rule out 'a fusion of two pairs of two if the CS/RIC amalgamation went through and the IChemE/SCI association was also put into effect'.[33]

In 1972 Beeby was succeeded as president by Lord Kearton, then Chairman of Courtaulds. A notable chemical engineer, he was thought, among the IChemE Council, to be amenable to a merger. The then-President of the Institution, Herbert Ashton of BP Chemicals, made a semi-formal approach but Lord Kearton quickly rejected the idea.[34] Reasons seem to have centred on the original question of sharing finances and premises.[35]   Under Lord Kearton, the support for a Joint General Secretary died, and Sharp opted to remain as secretary of the SCI alone, being replaced as General Secretary for a 17 month tenure by Alex M. McKay.[36]

Under McKay, the modernisation of the institutional structure continued. Then-President Roger J. Kingsley prepared a major paper on the future of institution, drawing on 15 previous papers. His 'Future of the Institution' said 'a weakness of the Institution in the past had been that issues were submitted to the Council in isolation and the Institution perforce proceeded by a series of isolated ad hoc decisions, rather than by conforming to a clear development pattern'.[37] A follow-on steering group under a former President, H. W. Ashton,

---

[32] G. H. Beeby, 'Amalgamation with the new Chemical Society', SCI CM 1967–70, document B.206, Nov 1970.

[33] SCI CM 8 May 1970.

[34] D. H. Sharp, 'Note on SCI/IChemE relationship', unpublished typescript, 6 Feb 1997, p. 3. On Ashton (b. 1911), see AF 10289.

[35] Hugh Anderson, then a member of the committee seeking closer collaboration, blamed funds, and Windebank ascribed the problems primarily to the issue of premises. Communications to Johnston by H. D. Anderson 25 Jan 1997; D. H. Sharp 6 Feb 1997; G. A. Dummett 2 Apr 1997. Under IChemE President Roger Sargent (b. 1926) in 1973, the decision was made to move from Belgrave Square, with Rugby eventually chosen. The proposal of Leatherhead, in particular, provoked vocal opposition from the membership (see, for example, *TCE* Sep 1974, 571 and *TCE* Oct 1974, 656), and alternative locations in the industrially important north-west were suggested. As to the merger plans, as early as 1976 the new Presidents were discussing a 'structured relationship'. By 1978, despite the diffident support of the two General Secretaries (Sharp and Evans), plans for closer co-operation had been approved by both councils. Evans archive, R3: Relationships with Kindred Bodies.

[36] CM 14 Mar 1974. For a brief biography of Maj. Gen. A. M. McKay (d. 1999), see *TCE* Dec 1974, 796.

[37] CM 13 Aug 1974; R. J. Kingsley, 'Future of the Institution' (letter/memorandum to membership, 1974). Kingsley (b. 1922) was a 1949 chemical engineering graduate of Manchester, and was MD of Lankro chemicals by 1972. Presidents' book; *TCE* Jan 1976, 12.

found 'apathy and strong dissatisfaction' by many members towards Institution activities. The committee called for the IChemE to become more representative of all members, to streamline decision-making, to give members better value for membership and to make chemical engineers better known to the outside world.[38] Members themselves had been redefined under David Sharp's period as General Secretary.[39]

The forced independence for the IChemE in the mid 1970s that resulted from an involuntary decoupling with industry, a rise of state involvement, and the failure of merger negotiations, was cathartic and signalled a profound transition for the Institution. So, too, did the move of some thirty staff to new premises in Rugby in 1976 and restructuring of the workings of the Institution.[40] The act of moving contributed greatly to internal change. Several staff, junior and senior, were unwilling to move from London. The General Secretary, Alex McKay, delegated administration of day-to-day operations at Rugby to his assistant, Trevor J. Evans, while tending to the scaled-down London operations.[41] The mirroring of activities in two centres and the exigencies of selection and storage led to a discontinuity in the documentary record and a dimming of the latter history of the London administration. The organisation was amputated from its past.

When McKay left to become Secretary of the IMechE in 1976, Evans was suitably situated to pass from de facto to official General Secretary. The relocation also strengthened links with industry and other professions. The support of the boards of many companies was crucial for funding the move and in orienting the Institution by providing council members. And the departures of David Sharp and Alex McKay to become General Secretaries of the SCI and IMechE, respectively, helped to give those organisations a more intimate understanding of the IChemE during subsequent interactions.

REPOSITIONING CHEMICAL ENGINEERING

The SCI/IChemE affair was a preview of negotiations to come. From the 1970s, the difficult financial positions of many institutions pressed them towards greater affiliation. Pressure was also applied by the state, which had taken such an active part in increasing the numbers of qualified engineers and conferring

---

[38] A. M. McKay, 'From the General Secretary', *TCE* May 1976, 305–6.

[39] In 1971, the former 'corporate', or voting classes of 'Associate Member' and 'Member' were renamed 'Member' and 'Fellow', respectively. The new 'Associate Member' class was designed for chemical engineering technicians. This brought the Institution's categories into accord with the practice of other institutions and CEI recommendations. New classes of 'Affiliate' and 'Companion' members were added for persons without sufficient qualifications for regular membership. 'Student' and 'Graduate' classes remained unchanged.

[40] The Institution left Belgrave Square in early 1976, and the library services moved at the end of that year. To retain a London presence, the Institution opened an office in mid-1978 at 12 Gayfere St, SW1.

[41] Evans obtained degrees in chemical engineering from UCL, and had worked for the South African Research Council and Ford, UK before being appointed Technical Secretary of the Institution. See 'Our new General Secretary', *TCE* Oct 1976, 637 and *Who's Who*.

status on them since the war. The eventual outcome of three wide-ranging processes – the Finniston Committee, the formation of the Engineering Council and the revisionist Fairclough proposals for unification of the engineering professions – forced an explicit declaration of allegiances from the IChemE and 'neighbouring' professions, a process that had begun fitfully with the second world war's Central Register and continued with the tensions over the Institution's postwar pursuit of the Royal Charter. In so doing it pushed the IChemE towards cognate engineering professions and, temporarily at least, away from its historically diffident chemistry alliances. The period was crucial in repositioning the Institution as one of the 'Big Four' engineering institutions, along with the Civils, Mechanicals and Electricals.[42] Each of these influences will now be considered in detail.

The Finniston Committee

In 1978 Sir Monty Finniston was appointed by the Callaghan Labour government to examine the status of British engineering institutions. Despite the occasional acknowledgement of potential problems, until the early 1970s neither politicians nor high officials in Whitehall had been greatly troubled about the way in which professional engineers were trained and certified. This was regarded as the joint responsibility of the professional associations and the institutions of higher education.[43] Grant Jordan has argued, however, the Institution of Electrical Engineers, in particular, sought to raise the standard of the 'Chartered Engineer' in the mid 1970s.[44] Its Secretary, George Gainsborough, proposed that all engineers should be registered and licensed by a statutory body. The support of the Labour government in instigating the Finniston inquiry therefore represented a major shift in official attitudes. By attempting to redefine the boundaries of entrenched interests, the Finniston Committee provided the Institution of Chemical Engineers with a new opportunity to negotiate a position for itself among the larger professions.

The IChemE devoted considerable energy to providing evidence to the Finniston Committee, and in early January 1980, under the new Conservative government of Margaret Thatcher, the Finniston report was published.[45] The

---

[42] This identification as one of the 'Big Four' had been started by Harold Hartley in the postwar years, e.g.: 'The development of our modern civilization has seen the successive rise of the four engineering professions', in 'The place of chemical engineering in modern industry', *School Science Review* No. 126, Mar 1954, 199–202. Nevertheless, this inclusive definition was more an unfulfilled wish as late as the early Eighties, when a negotiator for the IEE generously referred to the four as 'the big 3-½'. As late as the 1970s, the 'Big Four' were taken as the ICE, IMechE, IEE and Institution of Naval Architects (Watson, *op. cit.*, 3).

[43] C. Divall, 'Fundamental science versus design: employers and engineering studies at British universities, 1935–76', *Minerva* 29 (1991), 167–94.

[44] Grant Jordan, *Engineers and Professional Self-Regulation: From the Finniston Committee to the Engineering Council* (Oxford: Clarendon Press, 1992).

[45] Sir Monty Finniston, *Engineering Our Future: Report of the Committee of Inquiry into the Engineering Profession* (London: HMSO, 1980). pp. 108–11. See also Glover and Kelly, *Engineers in Britain, op. cit.*

Finniston Committee's chief recommendation – to establish a British Engineering Authority (BEA) – initially was welcomed by the IChemE Council. However, a further suggestion to merge engineering institutions to form groups with broadly similar interests was more cautiously received, because of pragmatic concerns for how to handle such aggregations. On the whole, the Institution seemed proud that it appeared to be on a parallel course with Finniston. It had maintained good relations between academia and industry, and it championed a model degree scheme with 'core curriculum' as called for by the committee.[46]

But the official response of the Institution was muted. Finniston's vision of an engineering identity seemed poorly matched to the IChemE's own self-image. The Institution claimed that the first two chapters of Finniston, outlining weaknesses with British competitiveness and markets in the postwar period, did not apply to process industries. Chemical engineering fit awkwardly with the Finniston concept of engineering: the 'design content' for the industry was 'to be found in the manufacturing facilities for the product rather than the product itself'.[47] The proposals appeared unlikely to make a marked contribution to industrial performance in themselves. And while the Institution officers supported the idea of a British Engineering Authority, they wanted it to work with existing institutions and remain small. Other aspects were more contentious. Finniston's proposal to have several new qualifications and titles were criticised, as the term 'Chartered Engineer' had scarcely achieved wide recognition. The Institution supported Finniston's recommendation of retaining the 'valuable and well-earned HND' modified to become a Higher Engineering Diploma. This proved less popular with academics, however, because it would re-open the part-time route to professional status.

Professors of chemical engineering were critical of the Finniston report.[48] The majority view was that its educational recommendations were ill-considered, and that the IChemE should have a central role in such matters. 'The IChemE has established a rapport between the academic profession and industry which we believe is unique among the institutions of the CEI', noted one.[49] They called for the Institution to continue to be responsible for accreditation, and for it to co-ordinate continuing education on a national basis. The other engineering institutions reached similar conclusions.

Other commentators have analysed the complex negotiations within Whitehall that in the early 1980s led to the dilution of Finniston's recommendations.[50] In the context of the history of the close relationship between the

[46]   T. J. Evans, *TCE*, Feb 1980, 73.
[47]   'The IChemE response to the Finniston report', *TCE* May 1980, 389–96.
[48]   'Finniston report: commentary by professors of chemical engineering', *TCE* Jan 1981, 9–10; minutes of Accreditation Board 1981.
[49]   Minutes of Accreditation Board, 1981, p. 364.
[50]   K. J. McCormick, 'Engineering a consensus: unanswered questions and questionable answers after the Finniston report', *Public Administration 63* (1985), 360–3; A. G. Jordan and J. J. Richardson, 'Engineering a consensus: From the Finniston report to the Engineering Council', *Public Administration 62* (1984), 383–400; Grant Jordan, *op. cit.*

IChemE and employers, it was not surprising that the state – via the Department of Industry – failed to secure the wholehearted support of industrialists for some of Finniston's more radical proposals.

Finniston saw the establishment of a statutory Engineering Authority as a critical component of his recommendations. Yet it was not quite the case – contrary to the remarks of two eminent analysts – that, in the discussion of this proposal at the time, the issue of *professional* self-regulation supplanted that of the nation's economic well-being.[51] At least in the chemical sector, employers were as concerned as the IChemE to minimise any loss of autonomy to a statutory body.

The IChemE and employers alike were faced with difficult choices. For many industrialists, Sir Monty's well-known desire to increase their influence by way of a statutory organisation responsible for the certification of professional engineers had to be balanced against the enhanced powers of intervention in industry's affairs that the state might take in return. The chemical employers' experience of the Chemical and Allied Products Industrial Training Board (CAPITB) since the mid-Sixties inclined a significant proportion of them to oppose involvement in any statutory body that they did not control.[52] For this reason, many industrialists wanted a minimum of change as far as the existing professional organisation of chemical engineering was concerned. In its evidence to the Finniston inquiry, the CIA, for example, indicated that although it was not satisfied with the educational and training policies of some of the other professional institutions, it had no complaints regarding the IChemE. In particular, the association said that university courses were appropriate to the needs of the chemical employers.[53]

The IChemE faced a similarly nice judgement. Many senior engineers within the institution agreed that, taken as a group, the professional associations did not enjoy the full confidence of employers in the manufacturing and engineering industries. Hence there might be advantages to the establishment of a new authority with some statutory powers, even if this meant that the IChemE would have to take a subordinate role in determining policy concerning certification. Yet there was also considerable concern over the possibility of increased political intervention in the affairs of the chemical engineering profession.[54]

The precise structure, powers and duties of any statutory authority responsible for the registration of engineers were therefore of the first importance. The IChemE put forward a plan which appealed to many industrialists in the process sector because it was intended to minimise political intervention in the affairs of the engineering professions and hence, indirectly, in those of industry.

51 Jordan and Richardson, *op. cit.*, pp. 384–6 but cf. *op. cit.*, pp. 397–8.
52 'Uneasy start for new training body', *Chem. & Indus.* 17 Jan 1983, 46; 'Controversial first steps for Association's training body', *Chem. & Indus.* 18 Apr 1983, 293; 'Chemical broadside', *TCE* Jul 1984, 5.
53 'CIA submission', *Chem. & Indus.* 4 Mar 1978, 149–51.
54 IChemE, 'A submission to the Committee of Inquiry in to the Engineering Profession', supplement to *TCE* Feb 1978; 'Finniston report: commentary by professors of chemical engineering', *TCE* Jan 1981, 9–14.

The Institution said that the new engineering authority should be constituted in a manner similar to the General Medical Council. The profession would thus be self-regulating on a *de facto* if not a *de jure* basis; the authority was to be answerable to government and parliament only indirectly, through the Privy Council, not directly to a minister, and ministerial appointees and lay persons on the authority were to be kept to a small minority.[55]

Moreover, the IChemE sought to minimise the authority's power of intervention in the affairs of small firms. The institution refused to countenance the general licensing of engineers, that is, the restriction of certain kinds of employment to those holding statutorily designated qualifications. If the Institution had been intent on upholding the model of chemical engineers as 'autonomous' professionals, it presumably would have favoured such a measure as a way of removing employers' ability to recruit unqualified personnel. As it was, the institution proposed that individual firms need only accept on a voluntary basis whatever standards of education and training the Engineering Authority might set. The IChemE was to continue monitoring courses of higher education under the authority of the new body. Hence employers could continue to influence educational policy through the part played by senior engineers within the Institution.[56]

Beyond Finniston

The Engineering Council, Finniston's new unifying body, was formed in 1982. The CEI consequently was wound up that November. Of the 41% of 200 000 British engineers who voted, more than 9 out of 10 wanted to transfer regulation of the profession to the Engineering Council (EngC). As finally constituted, the EngC possessed a structure and powers remarkably similar to those envisaged by the IChemE. The institution's role in 'manpower' policy therefore did not change radically. As both a 'nominated' and 'authorised' body of the council, the IChemE took on responsibility for the formal accreditation of academic courses of chemical engineering. Throughout the 1980s, it built on the fundamentally sound relationship that it had enjoyed with many industrialists and firms in the 1970s. The Institution revised its procedures for canvassing the views of employers and for seeking consensus between them and academics, and a number of comparatively minor changes to the structure and workings of the Education Committee enhanced the role played by senior industrialists in the detailed formulation and implementation of educational policy. For example, the accreditation of university courses now required the participation of senior engineers from industry.[57] This did not, however, lead to any funda-

---

[55]   IChemE, 'Submission', *op. cit.*, para. 6.2; 'The IChemE response to the Finniston report', *TCE* May 1980, 291.

[56]   IChemE, 'Submission', *op. cit.*, para. 6.1; IChemE, 'Response', *op. cit.*, 290.

[57]   MEC (1980–3); 'Joint industry/academic colloquium', *TCE* Apr 1978, 220.

mental revision of the Institution's views on what should be taught in the universities.[58]

By contrast, government policy continued to affect the relationship between the IChemE and firms with regard to industrial training. In 1982, the Conservative government abolished the CAPITB, and companies regained the entire responsibility for training chemical engineers at the professional level. The IChemE found little scope for reasserting any influence over the standard and content of courses provided by firms, partly because the CIA founded a body intended to co-ordinate the efforts of the smaller employers. The situation changed, however, in the late 1980s, when the EngC became concerned with the quality of the training being received by graduates in all branches of engineering. In 1990, it implemented a pilot scheme for the accreditation of programmes of integrated training and practical experience. This scheme operated through the council's 'nominated' and 'authorised' members. Hence the IChemE once again became involved in approving programmes of industrial training devised by firms, although its influence was, of course, restricted to those companies which chose to offer training.[59]

Yet even if the retained autonomy of the Institution was satisfactory, the overall outcome of the EngC was a disappointment. The maze of special interest groups proved impossible to cut through, degenerating, said an IEE representative, into 'petty and obscure wrangling in the institutions'.[60] Moreover, there was a declining mandate for these aggressive policy changes: 'all along', claimed an editorial in The Chemical Engineer, 'there has been a lack of enthusiasm – even interest – from the average engineer'.[61] The same seems to have been true of the state. The honeymoon period for the Engineering Council was as short as that for its predecessor, the CEI. As widely anticipated, the Thatcher government early in 1985 withdrew future financial support for Engineering Council. The EngC responded by launching an unsuccessful appeal to industry to help meet its £2.8M annual budget.

MAINTAINING AN INSTITUTIONAL IDENTITY

The Finniston inquiry stressed the professional context in which the IChemE acted, and encouraged its ongoing self-examination of identity. In his presidential address in 1985, Gordon Beveridge emphasised that the profession was moving gradually away from the chemical industry, but still staked its claim on influencing the process dimension of many other industries. The possibility

58 The IChemE notably decided that the EngC's requirement that all engineering undergraduates should be trained in the application of engineering principles to the solution of practical problems (Finniston's 'EA2') was adequately met by the Institution's long-standing provisions concerning a training in design. Minutes of the Accreditation Committee, IChemE, 15 Nov 1988; R. Aird, 'Teaching engineering applications', TCE Mar 1989, 50–1.
59 'Training – how important is it?', Processing, Oct 1978, 5; 'Uneasy start', op. cit., 46; 'Controversial first steps', op. cit., 293; 'chemical broadside', TCE Jul 1984, 5; Minutes of the Accreditation Board, IChemE 15 Nov 1990.
60 Editorial, Electrical Review 208 (1981), 16–7.
61 'The end of the beginning?', TCE Feb 1983, 1.

of changing the name of the Institution – that most explicit of identifications – was mooted.[62] This wider and vaguer territory was not uncontested, however. The IChemE General Secretary, indeed, later mused that, had a chemical engineering institution been founded in the 1980s, it would have had to give more attention to those qualifying at the boundaries and frontiers of the discipline to compete with other institutions.[63] The IMechE Process Industries Divisional Board, for example, stated that it intended to represent all types of engineering within the process sector. After several meetings, the two bodies agreed to share, and interact on, this common ground.[64] Nevertheless, in the plethora of chartered engineering institutions and bodies for technician engineers and engineering technicians, said Beveridge, 'the IChemE is but one cog'.[65]

But the Institution threatened to be an expendable cog. It was still positioned, in the mid Eighties, precariously between the Council of Science and Technology Institutions (CSTI) and the Engineering Council. Beveridge also

Table 8-1   New corporate membership by occupation, 1983–4 and 1993–4[66]

Self description	1983–4	1993–4
Chemical engineer	7%	8%
Process engineer	26%	19%
Chemist	4%	2%
Engineer	19%	19%
Manager	34%	41%
Educator	2%	0%
Consultant	1%	8%
Technologist/assistant	6%	3%
Metallurgist	0%	0%
Sales representative	1%	1%

---

[62]   In 1983, council had discussed changing the name to the Institution of Chemical and Process Engineers or something similar. As indicated by Table 8–1, the term 'process engineer' was considerably more popular amongst new members. Council discussion eventually led to a consensus that the historical title should be retained. Tradition (and inertia) were important, however; the American and other national organisations employed the term 'chemical engineer', and the profession had struggled to become known under that name. In Scotland, moreover, Process Engineering was sometimes used to describe new technology or electronics. CM 12 Jul 1983. Two years later, a majority of Council members favoured 'Institution of Chemical and Process Engineers' but not 'Institution of Process Engineers'. Some felt that "'chemical' really was seen as bad news amongst the school population". General Secretary Evans suggested the latter name to attract a broader range of new members. CM 30 Jul 1985.

[63]   T. Evans, 'Common interest', *Diary & News* Oct 1989, 24.

[64]   CM 7 Feb 1984.There was consensus on the view that the IChemE should emphasise its learned society roles, which were thought to be achievable now that many issues relating to the Engineering Council had been settled. CM 12 Jul 1983. However, there were continued calls to 'keep the Mechanicals off the IChemE patch' (e.g. CM 8 Oct 1985).

[65]   G. S. G. Beveridge, 'The politics of change, or the moulding of today's Institution', *TCE*, May 1985.

[66]   Sample: 89 of approximately 180 new corporate members (1983–4) and 92 of about 285 (1993–4). For earlier periods, see Tables 4-2, 5-2, 6-1 and 7-1.

saw the distribution of chemical engineers as unhealthily atypical in the engineering industry. Instead of a pyramid topped by professional engineers and backed by large numbers of technician engineers, chemical engineering management had a pattern of an unstable inverted pyramid: 'graduates backed by negligible numbers of other levels' within the profession, a situation that had been highlighted in the evidence to the Finniston Committee.[67] The 'technical support' was, in fact, provided by others in chemistry, mechanical, control or instrumentation engineering – an arrangement that had been useful between the wars to promote higher status for chemical engineers, but which now prevented an internal control of occupational and professional tasks. Professional visibility was further hampered by the fact that the 'engineering work tends not to end up in a visible item of hardware', the result instead being the production of system specifications. There was other evidence of a narrowing and inward-looking membership. The number of Fellows had been rising merely linearly, not exponentially, for a quarter century, and fewer than 10% of members were now affiliated with other institutions.[68]

Constructing affiliations

In this environment of self-questioning, the IChemE was conditioned to seek new alliances. With the new Engineering Council came new groups of Institutions. Five groups were to form; the chemical industry was in the fourth, called the 'Process and Extraction Group', along with Energy, Gas, Metals, Mining, Minerals & Metallurgy. Groups were allocated representation on the EngC by the number of corporate members they had.[69]

Moves towards merger were flexible, if problematic, and appeared to be proceeding well. In 1985 the General Secretary, Trevor Evans, presented a variety of options to the IChemE council for rationalising the chemical engineering profession:

(1) complete merger of Chartered Engineer interests in all Institutions, probably on the basis of the Institution of Civil Engineers, the oldest body, 'welcoming everyone back';
(2) complete merger of all institutions at all levels;
(3) part merger horizontally (across specialisms): e.g. the IChemE with other Chartered Group 4 institutions;
(4) part merger vertically (across professional ranks), e.g. by the IPlantE joining the IChemE as an Engineering Technicians group;
(5) merger with a learned society, e.g. SCI;

---

[67] See also Jordan, *op. cit.*, 109. The lack of growth of Associate membership, and the resulting deviation from Engineering Council recommendations, continued to concern council. *Diary & News* Jul 1990, 8.

[68] Affiliations with other institutions had been held by 72% of new members in the 1920s, 60% in the 1930s, 47% in the 1940s and 20% in the 1950s.

[69] Vulcan, 'Goodbye, CEI', *TCE* Mar 1983, 5.

(6)   go it alone.[70]

Such considerations were on the minds of the councils of many institutions. That April, the IMechE and Institution of Production Engineers (IProdE) circulated a proposal to merge, and by October announced plans to combine within 18 months under the name the Chartered Institution of Mechanical and Production Engineers.[71] IMechE members, however, did not agree with their Council. At about the same time, the Institution of Plant Engineers and Bureau of Engineer Surveyors (BES) amalgamated under an umbrella formula, allowing each to retain its name and identity, and the IProdE later began negotiations for merger with the Institution of Electrical Engineers (IEE).[72] Similarly, the IEE and the Institution of Electronic and Radio Engineers (IERE) edged closer together.

While affiliations, associations and mergers appeared to be in vogue, there were problems from the start for the IChemE. The Institution had gained full acceptance as one of 'Big Four' engineering institutions, but it had a much smaller membership, and so had a problem leading 'Group 4' when there was another primary discipline – metallurgy – present.[73] In fact, other, smaller institutions in the Group could combine to outvote the IChemE. The President therefore suggested examining the list of institutions associated with the Engineering Council to see which had an affinity with the IChemE. Among the chartered bodies, the Institute of Energy and Institution of Gas Engineers looked favourable, and among other institutions, the Institution of Plant Engineers, the Institute of Measurement and Control, the Biological Engineering Society (Group 2) and others not yet associated with the Engineering Council, such as the Institute of Physics and Institute of Petroleum, were promising. But the Gas and Energy Institutes had members from the same industry orientation, although not the same academic discipline, as those in IChemE. Its Presidents were IChemE Fellows, but actual membership overlaps were small, amounting to no more than 150 from the Institution of Gas Engineers and 200 from the Institute of Energy out of some 7000 IChemE corporate members. The consensus of the IChemE council was that the Institution should not become a multi-disciplinary organisation nor reduce its standards of qualification.[74] In 1985, after the suggestion of a merger with the Institution of Plant Engineers, the General Secretary reported that the IPlantE was 'simply not interested and, indeed, did not want to discuss the issue',

---

[70]   T. Evans, Council Paper T15: 'Relationship between IChemE and other societies – rationalisation of the profession'. 6 Feb 1985. Evans archive, R8: Freshwater Working Party on Relationships with Other Institutions.

[71]   The Institution of Production Engineers had been founded in 1921, two years before the IChemE, and organised largely on an American model stressing scientific management, but focusing on the automotive industry. Watson, *op. cit.*, 215–7.

[72]   *TCE* Oct 1985, 8; Mar 1986, 5; Jan 1987, 9. The IPlantE had formed in 1946.

[73]   CM 8 May 1984.

[74]   On the other hand, the CEI had reduced the academic standards for member engineers; industry itself, complained one academic, paid little attention to grades, so raising standards would cause little increase in membership levels.

limiting discussion to issues relating to journals, other publications, meetings and conferences. Council agreed that until a greater level of mutual confidence could be developed with other bodies, it would be inappropriate to move to develop mergers. Some were concerned that rebuffs would tarnish the credibility of the IChemE.[75]

But agreements were difficult to achieve. Later extensive negotiations with the smaller and struggling IEnergy revolving around a shared secretariat (an arrangement reminiscent of that between the SCI and IChemE in the early 1970s) via a management contract, but this promising form of 'strategic alliance' failed to reach their respective councils owing largely to mutual mistrust and conflicting personalities.[76] A working relationship with the Metallurgists, which members saw as being a strong science-based institution, appeared the most promising. The IChemE council also agreed with the Institute of Metals (IMet) in 1987 to set up a standing body of the two Institutions to develop strategies of joint action on education, training, publications, meetings, specialist subject groups and regional activities.[77] In 1990, the Association of Cost Engineers (ACostE) became an Institution-affiliate of the IChemE as defined in the bye-laws of the Engineering Council, to register individuals as Chartered Engineers (CEng). That year, the IMechE tried again to affiliate, this time with the IEE. The IChemE Council, feeling threatened by this potential giant comprising some 200 000 members, investigated the 'horizontal merger' option: participation in a pan-European process engineering federation.[78] In the event, however, the IMechE again rejected their proposed amalgamation and the process engineering federation did not develop. There may have been tentative discussions with the Institute of Metals that year to form a new Institute of Materials (IMat),[79] but the Institute of Ceramics, Institute of Metals and the Plastics and Rubber Institute agreed to combine instead from 1 Jan 1992.[80] By contrast, growth of the IChemE continued to be organic rather than acquisitive.

Finniston rides again: the Fairclough Commission

Sir John Fairclough, becoming chairman of the EngC after 4½ years as scientific advisor to the Prime Minister, re-evaluated the direction of the Engineering Council beginning in 1991. On his steering committee was IChemE President Ted Bavister.[81]

The mood of the IChemE was becoming pessimistic. While there had been 'some success in lobbying along the corridors of power and in improving public perception of engineering', it was apparent that 'individual institutions have

[75] CM 30 Jul 1985.
[76] Evans archive, EAL 7/97: Institute of Energy –1995.
[77] 'At the council table', *Diary & News* Mar 1987, 2; 'Two year old Institute with 120 years of history', *Diary & News* Apr 1987, 2.
[78] 'At the council table', *Diary & News* Jun 1990, 8.
[79] CM 27 Jun 1989; interview, Sir Hugh Ford with Johnston, 2 Dec 1997.
[80] *TCE* 13 Sep 1990, 58.
[81] Fairclough report, *Engineering into the Millennium*, 1993.

held on firmly to standards-setting and regulating within their own branches of engineering'. As a result, the relationship between the institutions and the EngC was going nowhere.[82]

With engineering unification seeming increasingly unlikely, the Institution officers turned back towards science and an international perspective as defining aspects of professional identity. 'One view is ... that there is no such monolith as the engineering profession. Chemical engineers have more in common with each other world-wide – and perhaps with physicists and chemists – than with civil, mechanical or electrical engineers'.[83] Chemical engineering, it was argued, differed in occupational as well as disciplinary emphases from other forms of engineer. Chemical engineers worked more frequently than the others for multinationals and contractors around the world, making national grounding of the profession less imperative, and leading the IChemE leadership to promote its international qualification and learned society roles.[84]

Fairclough proposed a single body to take on the qualifying role for all engineers, while institutions concentrated on 'learned society' activities. Implicit in this proposal was the prospect of a single engineering degree for all engineers, with specialisation in chemical engineering coming later. This two-year general engineering degree, as proposed, had no specifically chemical engineering content, and indeed made no calls for chemistry courses in secondary school. It was criticised by the IChemE Council and academics as threatening to marginalise chemical engineering, defining engineering 'in a way that excludes us'.[85]

The Fairclough proposals had revolutionary consequences for all the Institutions and their officers, and had a uniformly rocky reception. The major institutions rejected the initial proposals outright at the end of 1992. They objected to the loss of control of their 'subordinate' bodies to the a qualifying organisation, which would put the unified professions before distinct disciplines.[86] The IMechE and IEE participated, signing a memorandum of understanding to collaborate in several areas of professional activity. There was nevertheless some distance between the Fairclough proposals and what institutions believed was the way forward.

Stage II, according to Fairclough, would be a negotiation and implementation phase. Its objective was to be the establishment of the 'new relationship' and an examination (rather than imposition) of the concept of a single organisation (and avoidance of the term 'institution').

Representatives of the 'Big 4' met Fairclough in August 1993. With their common experience of the Finniston inquiry behind them, they had agreed beforehand to adopt a tougher line and to 'take over' negotiations from Fairclough. An IEE representative argued that Fairclough's proposals 'had not

---

[82] 'Finniston rides again ... on the horns of a Marxist dilemma', *TCE* 28 Nov 1991, 2; See also 'IChemE response to the Fairclough initiative', 29 Nov 1991, 34.

[83] *Op. cit.*

[84] T. Evans, 'A unified engineering profession?', *Diary & News* Apr 1992, 8.

[85] 'At the council table', *Diary & News* Mar 1993, 8.

[86] 'Unification blues', *TCE* 14 Jan 1993, 22.

received strong support'; that institutions had not supported the Single Institution or Colleges; regional meetings had not supported the Single Institution and had been lukewarm to Colleges; that industry did not support a centralist approach; that government involvement and support 'does not exist' ('they want the institutions to sort themselves out', he argued); and that members did not want assets or activities of the Institutions to be handed to a central body, nor see their own institutions disappear and surrender their charters. 'Therefore Sir John Fairclough has got nowhere to go except to accept the lifeline offered by the four presidents'. For Stage II, then, he recommended that they not discuss a Single Institution; recognise that a federal approach to unity was preferred; that Colleges would be assumed to be 'electoral colleges' only; and that terms of reference of this stage would include the appointment of a project director.[87] The IChemE line, as voiced by Evans, was to support 'any option which maintains our organisational and financial independence' and a council member noted that 'the principle of maximum devolution *must* remain'. In a letter to the *Financial Times*, Evans expanded on this general perspective:

> We want a real partnership with a central body which would then act primarily as a public relations focus for the profession as a whole ... Much has been made of the need for more 'interdisciplinary' abilities in engineering. Success in interdisciplinary areas is not about training each individual with some kind of multidisciplinary tool kit, but rather about a high level of disciplinary specialisation. What we need are first-rate groups of specialists able to appreciate each other's contributions and make effective contributions as a team.[88]

There is a certain irony here, in that chemical engineers in the early days of the Institution had sought precisely to portray themselves as generalists rather than specialists. Sir John Fairclough added his own rhetoric to the debate:

> We seek to build an organisational network involving the whole profession and bond the Institutions to it in a way that will allow us to speak and act with unity when appropriate, but allow each individual institution to develop its own particular role and specialisation.[89]

A major argument of the Fairclough proposals was that existing institutions did not represent either their members views or interests.[90] The IChemE Council decided to poll members to refute the claims. The survey seemed to show both a rejection of Fairclough and a certain lack of interest. The questionnaire yielded responses from about 40%, showing remarkable unanimity. Two-thirds backed the 'Four Presidents' Proposal' for a partnership with a central organisation having less power than that proposed by the Engineering Council. A similar fraction rejected an omnipotent EngC or single institution to represent

[87] Evans archive, EC1 (F) Apr 1993 – Dec 1995.
[88] Evans archive EC1 (F) Unification of the profession 1991–3 and 1993–4.
[89] Sir John Fairclough, 'A Lasting Relationship: The Way Forward' (*Ibid.*, 1 Jun 1994).
[90] 'Unification rumbles on', *TCE* 11 Mar 1993, 13.

engineers.[91] An editorial summed up the Council response to the Fairclough report: 'the idea that these bodies are somehow at war with one another is false – it conjures up a picture more fitting to the bad old days of trade union rivalries ... The world has moved on since Finniston'.[92] The Report of the Council of Presidents, *Engineering into the Millennium* (Apr 1993) recommended a new relationship between the EngC and the Institutions to improve performance and the promotion of the engineering profession.[93]

A month after the release of the Fairclough report, the Presidents of what was by now the 'Big Four' stated that a case for single engineering institution had not been made, and decried the EngC for failing to give the profession the leadership it needed.[94] The EngC responded that central powers were needed to ensure that engineers saw themselves as a class of technical professionals first and as members of an Institution and discipline second. It rejected the IChemE proposal for a discipline-based structure, arguing that unity of the engineering profession would be impeded because disciplines were diverging.[95] Bavister concluded that 'you can change anything in the UK provided that you do not change appearances'.[96]

The separate approach by the 'Big Four' was criticised by the smaller institutions and members of the EngC, as it had been thirteen years earlier by the chairman of the threatened, and soon defunct, CEI.[97] One IChemE Council member noted that 'the Unification debate is becoming about saving the Engineering Council as an organisation in something like its present form, size and status', partly because of fears of EngC staff of loss of status, power or jobs.

Even if the structure of the unifying body was to be revamped, by the following year, the name 'Confederation' had been rejected for continued use of 'Engineering Council'. The new body would have a senate and two executive boards, the Board for Engineering Regulation (BER) – responsible for maintaining register of engineers and technicians who wished to be registered at any of three phases of education, training and experience – and the Board for the Engineering Profession (BEP) – to promote the profession via unity. The two bodies would have different constituencies and mandates: one to deal with training standards and regulation, and the other to deal with engineering affairs.

At the end of the century, matters were eventually coming to a head. The Big Four agreed, said IChemE President Gordon Campbell in the language of

91 T. Evans, 'Your views on unification', *Diary & News* Apr 1993, 1; 'Collision course set on unification', *TCE* 20 Apr 1993, 5.
92 'Please leave the furniture where it is', *TCE* 29 Apr 1993, 3.
93 Evans archive, EAL 7/97: Unification 1992–4.
94 'Big four blow to Fairclough', *TCE* 13 May 1993, 6.
95 'Those whom the gods wish to destroy ...', *Diary & News* Jan 1993, 8.
96 T. Bavister, 'From evolution to revolution: pressure on the profession', 1993 IChemE Presidential address.
97 'It seems to have been the policy of the Department of Industry from the outset to divide and conquer the engineering profession, and I think the activities of certain institutions has encouraged them in the belief that they could succeed in this aim and ignore CEI.' Letter, 6 Mar 1981, P. A. Allaway to P. N. Rowe, Evans archive, Finniston correspondence. See also *The Times* 15 Sep 1980, p. 15.

the times, that the EngC had 'failed to define its customer base, the requirements of its customers, and the way in which those customers are serviced'.[98]

The Big Four institutions did, in fact, agree with Fairclough on the need for a restructuring of the professions around a central focus, and appointed representatives for a Stage II Policy working group in 1994. Its outcome was a proposal that required no changes to the IChemE Charter, but significant changes to that of the EngC. The Institution would retain its accreditation arrangements, but would have control of the new body's policy rather than vice versa. Nevertheless, the initial results suggested that administration and costs were excessive in the new EngC, and further consensus appeared elusive.[99]

PROFESSIONAL IDENTITY FOR THE INDIVIDUAL

The chapter thus far has focused on 'external' relations that shaped institutional identity from the 1960s. The negotiations of the IChemE with other institutions and with the state affected the identity of practising chemical engineers only on the most abstract of levels, principally by defining how they were affiliated with other occupational groups. Like the classification of fauna, institutional affiliations were important to the classifiers, not the classified; they had no effect on the viability of the species, or their individual members, in a complex ecology. Thus we turn now successively to more individually pertinent expressions of the professional image.

Broadening the membership

Related to institutional affiliations was the need for greater representation within them. With the memberships of the merging bodies growing (although generally less quickly than the IChemE – see Figure 8-1), the Institution was threatened with a diminishing voice in a unified profession. Membership levels in the Institution, and means to increase them (possibly at the expense of other institutions' memberships) were a repeated topic of council discussion.

In 1985, the first draft paper on the subject was presented to Council.[100] It considered how the IChemE could achieve a significant increase in its corporate membership through the inclusion within its corporate register of other categories of engineers, scientists or persons employed within the process industries who are excluded from membership by virtue of the Charter, By-laws and Regulations or by their current interpretation. This represented the most dramatic change to date for membership definition, and hence self-defined identity for chemical engineers.

Six categories of person were identified:

(1) qualified engineers and scientists in the process industries;
(2) those who had presented papers at IChemE events, and who might work them up as a thesis in lieu of design project;

---

[98] *TCE* 14 May 1998, 5.
[99] 'At the council table', *Diary & News* Jun 1996, 8; *Diary & News* Nov. 1996, 1.
[100] *Broadening the Basis of Institution Membership Draft paper U111*, 1987.

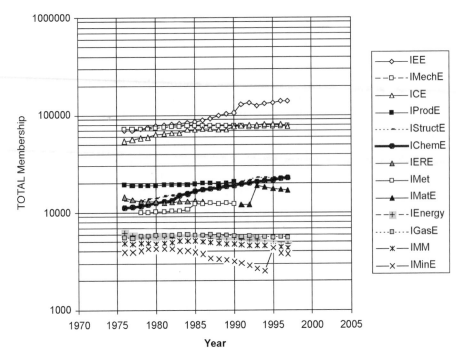

Figure 8-1   Growth of chartered engineering institutions. The relative growth rate of the IChemE during the last quarter of the century exceeded that of all other institutions, except during times of their merger. See acronym list for full names.   Note that this logarithmic plot de-emphasises actual numbers: while the IChemE membership rose by some 11 000 during this period, the IEE membership rose by over 70 000. Source: CEI and EngC, collated by T. Evans.

(3)   those in disciplines employed on the fringe of traditional process industries.

More radical expansion of the membership net was considered, for:

(4)   professional people employed in process sector;
(5)   engineers in other industrial sectors;
(6)   mature graduates – who obtained their academic qualifications before the 1970s, when a HND in chemical engineering with a Home Paper were sufficient for corporate status.

The report suggested creating a new class of Corporate membership widely drawn and thus permissive. When discussed at local branches, the first three categories of persons drew cautious approval, but the latter three were firmly rejected. The re-annealing image fractured along several dimensions. The Northern Branch, for example, rejected biotechnologists as suitable for membership (the Biotechnology Subject Group, in turn, was alone in supporting the admission of professional people employed in the process industries). There was a clear conflict between Institutional and professional identity. The rank-and-file members saw no reason to dilute their own status in order to raise that of the Institution itself. Yet Council members were more conscious of the need for interacting with other groups. One argued that flexibility on qualifications would raise the international profile of the Institution by attracting more Europeans, for whom the IChemE did not recognise many degrees. Another warned that offending an academic by rejection might lead to a generation of alienated students; yet another, that chemical engineers working in other industrial sectors be welcomed in case they should eventually reach positions of power.[101]

Exerting identity: member participation in the IChemE

Member comments on these proposed changes were a belated expression of individual identity which surfaced only in the 1970s. From its origins, the Institution had been administered by an enthusiastic, if atypical, minority of its members. Over half of the first Council were, or had been, consultants, and all had held management or supervisory posts. By contrast, fewer than 5% of the original members were consultants, and about a third managers. This disparity in occupation, and the outlook that followed from it, endured over the succeeding decades.

The Articles of Association of the IChemE did not specify several practical details of elections, notably the selection of candidates. Unsurprisingly, the organisers were the first to control the new Institution but, more unusually, they instigated a direct line of succession. The first Provisional Committee was appointed from those persons attending an organising meeting who had risen to speak.[102] The council elected at the first Corporate Meeting consisted mainly of the Provisional Council which, significantly, proposed the candidates.[103] While nominations could be made by a group of five corporate members, the internal selection of successors had become an informal tradition within a decade. This inward-looking tendency could be carried further: some forty years later, the first General Secretary, J. B. Brennan, complained of the 'nepotistic tendency of some council members to suggest only members in their own company' for nominations.[104]

While Council candidates occasionally were nominated by others, Presidents invariably were proposed by their successors; Council approval ensured that only one candidate was offered at the Annual Corporate Meetings.[105] Such a procedure was not uncommon in institutions; the Civils had operated in this

---

[101] 'At the council table', *Diary & News* Jan 1987, 2.

[102] CM 9 Nov 1921.

[103] In practice, candidates were approached by the Provisional Council and asked to put their names forward. See, for example, letter J. Hinchley to F. L. Nathan, 27 Feb 1923, 'F.L.N. 4' folder, Nathan papers.

[104] Brennan, *op. cit.*, 28.

[105] W. M. Cullen, dissatisfied with this seemingly furtive procedure and complaining that 'our officials have sometimes no record of what was actually done', attempted to standardise it. When he failed to convince his colleagues, he settled for placing his own actions on record. See 'Notes on procedure adopted by William Cullen with regard to the selection of his successor', May 1938, Gayfere box III/5: Miscellaneous sub-committees.

way from their origins, although it led to member complaint as early as 1841.[106]
Council nominations in many British institutions have been 'typically shrouded
in the utmost secrecy and confidentiality, leaving the impression of rule by self-
perpetuating oligarchies'.[107] By the early 1960s, with the term of office having
been reduced to one year owing to the increased work-load, the usual practice
was for the President to have served as a Vice President in the term before his
election so that he was sufficiently knowledgeable during his period of residency.
Past Presidents also consciously selected successors with varying backgrounds,
a typical ratio being two from industry for every one from academe.[108] A
similar practice, active from the 1950s, was for at least one of the multiple vice-
presidents to be chosen from abroad.

The absence of a constitutional need for nominations from the membership
may explain the gradual decline in attendance at Annual Corporate Meetings.
The absolute numbers attending remained relatively constant over at least four
decades, while the corporate membership expanded twenty-fold. As a result,
member representation at the meetings fell from over 20% in 1923 to scarcely
3% by 1960. Indeed, only in 1971 were the by-laws revised to allow non-
corporate members to attend. The weak participation of members at corporate
meetings was underlined by the absence of a postal vote (although proxy voting
had always been possible, if little used).[109] The foreign membership was thus
effectively disenfranchised. Nor did the ACMs often elicit discussion not initiated by the Council itself.[110] Few members took this level of interest in
Institution business. An attempt to update a list of members' occupational and
address details in 1954 yielded answers from merely two-thirds; in the early
1990s some one-fifth of members were unaccounted for.

Outside the Annual Meetings, there were limited opportunities for members'
comment. For the first forty years of the Institution, member suggestions,
questions and grievances appeared only in the form of a trickle of letters to
Council members. The absence of a public forum for chemical engineers was
exacerbated by the unaccommodating publications of the Institution and
others. Neither the *Transactions* nor *Quarterly Bulletin* provided space for
letters to the editor.

This lack of a public voice cannot, however, be blamed wholly on the culture
of the Institution itself. Even after the launch of an informal magazine, *The
Chemical Engineer*, in 1956 with editorial requests for 'articles and comments

[106] Garth Watson, *The Civils: The Story of the Institution of Civil Engineers* (London: Thomas
Telford, 1988), pp. 25–6.
[107] Watson, *op. cit.*, 200–1.
[108] This was, for example, the case in 1991, when 21 past-presidents met and in 1992, when the
President and Past Presidents planned their next three successors. CM 12 Feb 1991, 18 Feb
1992.
[109] The 1989 election of council had a record 1644 votes, representing 21% of corporate members.
[110] Notable exceptions were meetings through the 1950s, at which a sole Member, Ralph John
Low, was a frequent questioner and critic of Council actions. *Minutes of Annual Corporate
Meetings and AGMs 1923–1964*; Gayfere archive boxes II/1 and II/2. Low (1902–67) had
obtained an engineering diploma from King's College in the early Twenties and was subsequently employed by chemical manufacturers. AF 1429, 1982.

by members', a trickle of letters began to appear regularly only a decade later, and these initially concentrated on technical issues. Perhaps unsurprisingly, the 'learned society' functions of the Institution summoned far greater member interest than did its professional functions. Once admitted to the Institution, members had less personal concern for accreditation issues, but an increased interest in continued education, occupational advancement through publications, and the social interactions afforded by technical meetings.[111] Only in the 1970s did yearly publication of letters in *TCE* reach double figures. In that increasingly difficult period for the profession, professional issues of employment, salaries, education and status began to dominate. Nevertheless, editors of *TCE* and of its Graduates & Students section still struggled to promote member participation, and complained of the low rate of correspondence.[112] Whether from lack of relevance to their daily professional life or general satisfaction with the Council's representation, member interaction concerning the broad activities of the Institution proved chronically difficult to excite.

A new constituency: the rise of the female chemical engineer

To some extent, the role of the regular member in the Institution put into relief the situation of a growing fraction: women. And, as with the case of member participation, the rapid evolution of the gender mix from the 1970s was a consequence of a longer history.

The first female members of the Institution were neither numerous nor prominent. Probably the first among them were Hilda Derrick in 1942, and Joyce Holden in 1951, both joining as Student members, and Graduate Alina Borucka, employed as a sales desk chemical engineer while completing her PhD at Battersea Polytechnic in 1952.[113] A handful of women reached Associate Membership in the IChemE in the 1950s, but only Derrick hesitantly voiced the problems she found in following a chemical engineering career.[114] The acceptance of women as equals in traditional institutions was, indeed, a new

---

[111] Technical meetings, whether supported by regional branches (from 1945) or Subject Groups (from 1977), proved popular and, as special meetings, an important and rising source of Institutional income from the 1960s. See Chapter 6.

[112] On the lack of member participation, see also D. G. Bagg, 'Correspondence', *TCE* Nov 1979, 794 and editorial reply.

[113] Derrick (b. 1920 née Stroud, and joining the IChemE with her husband) studied chemical engineering by correspondence course during the war while working at the Fuel Research Station, and afterwards became sales and development engineer for a plastics firm. Holden had studied for a BSc (Hon) in Chemistry and Physics at the University of Bristol during the war, and was then undertaking a diploma in chemical engineering under S. R. Tailby at Battersea Polytechnic. Unlike her male contemporaries, her marital status was clearly signalled in the IChemE records. Borucka had obtained her BSc in chemical engineering at West Ham in 1952. The first public mention of a female chemical engineer appears to have been the Australian Valeria Blakley, employed as a draughtswoman in 1951. See AFs 2472, 2473, 5345, 6042, 6831, *Quart. Bull.* Apr 1951.

[114] Note that 'Associates' were renamed 'Members' in 1972. H. M. Derrick, 'Women in engineering', *TCE*, Oct. 1958, A53. Derrick had represented the IChemE at a conference on 'Careers for girls in chemical engineering' for the Women's Engineering Society at Coventry in Jul 1957.

phenomenon for the civil service, Oxford and Cambridge Universities, and the House of Lords in the 1950s. One estimate put the number of women training in all types of engineering at scarcely fifty per year by the late Fifties.[115] A decade later, the government found the fraction of women averaged over all engineering professions in Britain to be merely 0.2%, well below that in other European countries.[116] The Department of Education and Science consequently designated 1969 'Women in Engineering Year' to encourage recruitment.

The members of the IChemE Council, unlike those of other engineering institutions, had always been encouraging, showing little evidence of institutionalised prejudice.[117] Nevertheless, pronouns on printed application forms were altered from 'him' to 'candidate' only in 1962, an indication of the beginnings of a perceived change in member identity. The Institution also implicitly accepted employers' reticence to employ female chemical engineers. In 1958, an IChemE sub-committee surveyed employers regarding their requirements for hiring graduates. Of thirty firms responding, over half stated that they would definitely not employ women chemical engineers; another four said they were unlikely to do so; and only seven (i.e. less than one-quarter) thought they might employ female chemical engineers for research and development work, 'whilst considering women unacceptable for any post or career involving plant management'. Among those definitely refusing entry were the UKAEA establishments, the Steel Company of Wales and Babcock & Wilcox. Firms blamed women for 'faults in personality and management ability' (Thomas Hedley & Co.), indecision about their career plans (Kellogg International Ltd) or 'uncertainty about the duration of employment of the attractive women scientists' (Mars Ltd). Most argued that women employees would marry and leave industry before the investment in their training was repaid. Shell Refining Co. and Esso Research Ltd were the most positive, although Shell would employ women only as technologists, and Esso had merely one female chemical engineer, a graduate from an American university. Even those willing to accept women owing to the labour shortage saw chemical engineering as demanding more managerial skill than did chemistry posts, supporting the Institution's long-standing identification of the chemical engineer as intrinsically management-centred. One employer sought applicants having chemistry as a primary subject to 'ensure her a job at the laboratory bench if she should be unable to find employment as a chemical engineer'. P. V. Danckwerts, the IChemE committee chairman, correspondingly advised that university departments not fill many of their places with women, 'with their rather lower expectation of useful industrial careers', unless 'a girl ... feels strongly that she wishes to enter industry as a chemical engineer' and would have 'sufficient ability and determi-

---

[115] Harry Hopkins, *The New Look: A Social History of the Forties and Fifties in Britain* (London: Secker & Warburg, 1963), pp. 320–37, esp. 321–2.

[116] E.g. 3.6% in France and 10% in Norway.

[117] See, for example, J. A. Oriel, 'Women in chemical engineering', *TCE*, Dec. 1958, A43.

nation'.[118] The strong cultural bias against female engineers proved largely impassable.

The rise of female membership in the Institution was a phenomenon of the late 1970s and 1980s. Annual female entrants for university chemical engineering courses in Britain hovered in the twenties through the early 1970s.[119] In 1976, when 'lady members' of the Institution were first separately identified in the membership statistics, they made up scarcely 1% of all members, and most remained junior engineers at the Student and Graduate level. By 1984, the fraction of women in the Institution had tripled, but only 20 of nearly 5000 UK-based corporate members were women, and none was a Fellow of the Institution. This was less than half the frequency of UK chartered engineers generally, itself still a very low 1%.[120]

The status of women in engineering remained hindered by entrenched attitudes. With the rise of more permissive sexual mores but before feminism made inroads into the legal domain, advertisements portraying semi-clad women with process equipment appeared regularly in *The Chemical Engineer* from the early 1970s. Such imagery began to attract charges of sexism only a decade later, when the advertisements began gradually to be withdrawn by their firms.[121] The masculine representation of the chemical engineer infused the trade press: the IChemE *Diary* presumed 'members and their ladies' to be the appropriate phrase, and *Pipeline*, a publication of the IChemE distributed to all full-time students of chemical engineering, made little mention of women as chemical engineers until the mid Eighties.[122]

The WISE initiative (Women Into Science and Engineering) in 1984 publicised the problem of chemical engineering employment for women and promoted attempts at solutions.[123] The University of Bradford, for example, combated the pessimistic statistics by providing three-day introductory courses in chemical engineering to sixth form girls. The gradual blurring of the gender line is reflected in the fraction of graduates from departments such as that at the University of Nottingham (Figure 8-2). For UK chemical engineering courses overall, female enrolment rose from 8% in 1980 to 19% in 1987, and to 27% by 1990. And in industry by the late 1980s, 'career breaks' were being promoted as a pragmatic solution in a society within which women were still expected to bear the brunt of raising a family.[124]

---

[118] 'Careers for women in chemical engineering', Gayfere archive box III/5, Miscellaneous sub-committees, file 'Careers for women', 1958.

[119] 'Supply of new graduates in chemical engineering', *TCE* May 1978, 391.

[120] C. Hanson, 'Chemical engineering: a career for women', *TCE* May 1983, 109.

[121] One series of advertisements, showing a nude woman in an industrial safety shower, was defended by its firm as promoting safety and attracting attention to the proper method of use. See letters to the editor, *TCE* Jun 1983; Aug/Sep 1983; Nov 1983 and advertisements Nov 1974 – May 1984.

[122] See, for example, *Diary* May 1977, 348 and 'Wine, women and exams!', *Pipeline* no. 1, Apr 1979, 3 for chemical engineering as a male pursuit. For the subsequent change of attitude, see L. Shelenko, 'Engineers born with a handicap – their sex', *Pipeline* Mar 1986, 4.

[123] S. Bullivant and C. Onions, 'Engineering a career for women', *TCE*, May 1984, 47–9. For the flurry of readers' comments, see letters to the editor, e.g. Aug and Sep 1986.

[124] A. Meldrum, 'Female engineers and the career break', *TCE* Sep 1986, 64.

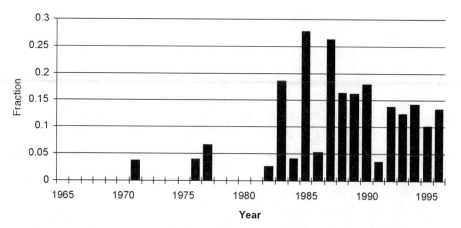

Figure 8-2   Female chemical engineering graduates, University of Nottingham. Source: B. Azzopardi.

The first female Fellow and member of Council of the IChemE were elected in 1989, and a female Technical Director and Deputy Chief Executive shortly afterward.[125] By 1996, the fraction of women had risen steadily to over 13% of all IChemE members, while the increased student fraction hovered at about one in four entrants.[126]

The most difficult hurdle for women remained the workplace. In academe, a mere 4% of university-funded teaching places were filled by women in chemical engineering departments in the early 1990s.[127] A flurry of letters in *The Chemical Engineer* in 1990 criticised the tendency of employment interviewers to question female applicants who were mothers about their ability to cope with a chemical engineering post. Robert Edgeworth Johnstone, first professor of chemical engineering at the University of Nottingham, represented what was, by now, the minority view in public if not practice:

> Female emancipation was a good and necessary thing but there are some women who want to have their cake and eat it ... a married woman with a baby already has a job which she cannot and should not entirely delegate.[128]

Edgeworth Johnstone's view that men had additional responsibilities of 'cover-

---

[125]   There were more than ten female Fellows by the late 1990s. Nevertheless, the age distribution of corporate members remained skewed, presumably owing to the as-yet indistinguishable combination of the recent entry of women to the chemical engineering workforce and their earlier retirement from it: numbers begin declining for women from their early thirties but for men from their late fifties (1997 data collated by T. Evans).

[126]   *TCE* 25 Nov 1993, 6.

[127]   Women were distributed in chemical engineering teaching as one of 42 professors, 3 of 120 readers/senior lecturers and 8 of 135 lecturers.

[128]   *TCE* 13 Sep 1990, 6. Edgeworth Johnstone (1900–94) had been Assistant Director of Ordnance Factories after the war, and conducted widely cited surveys of chemical engineering education and practice in 1961 and 1966, two of the earliest sociological studies of British engineers. See also IChemE AF 1427; R. Edgeworth Johnstone, 'The dark days', *TCE* Jan 1987, 38–39; 'Engineer, musician, philosopher and patriarch', *Diary & News* Jan 1995, 2.

ing expenses' and 'fighting wars' were roundly criticised by other correspondents – but by a factor of merely three to one. Other residual biases against women included such mundane factors as the lack of provision of separate toilet facilities on job sites, and prevention of their working three consecutive shifts. Such practical issues, often tracing back to economic considerations, obstructed the occupational gender shift. Nevertheless, according to official census figures, the number of women in the profession rose five-fold between 1971 and 1991, from about 1.2% to 6.4% of self-declared chemical engineers.[129] Even if women were more prevalent in chemical engineering than in other British engineering professions, at the end of the century the rise of female participation was a trend that had not yet run its course.

Ethics

The institutional reconstruction of professional identity relied on other attributes besides an acknowledgement of gender and member democracy. The 1970s also saw the development of explicit standards for the quality and responsibilities of chemical engineering work.

For the layperson, the image of a 'professional' is generally imbued with attributes of exemplary conduct in the performance of work. It may therefore seem odd that this aspect of 'professionalism' did not often attain overt attention in the IChemE, or indeed in other British technical professions. Yet in common with member democracy the issue of ethics remained largely unaddressed. The subject of an Institutional Committee only from the 1970s, the issue of ethics, or standards of practice, was a late-developing component of professional life in the IChemE.

While professional ethics have been associated with the definition and activities of the older professions throughout their existence, British engineering institutions, including the IChemE, have not been seriously shaped by such concerns. The absence of a policy, at least in the early Institution, contrasts with its American counterpart, which adopted a code of ethics in 1912, four years after its formation.[130] But early ethical codes for engineering organis-

---

[129] *Census 1971 Great Britain: Qualified Manpower Tables* (London: HMSO, 1976); *Census 1991: Qualified Manpower Great Britain* (London: HMSO, 1994).

[130] See 'Ethics statement', *Trans. AIChE 10* (1917), 472–5; G. W. Thompson, 'How the Institute Committees have served the profession: Committee on Ethics', *Trans. AIChE 28* (1932), 314–5; Edwin T. Layton, Jr., *The Revolt of the Engineers: Social Responsibility and the American Engineering Profession* (Cleveland: Press of Case Western Reserve University, 1971), pp. 84–5; Terry S. Reynolds, *75 Years of Progress: A History of the American Institute of Chemical Engineers 1908–1983* (New York: AIChE, 1983), p. 18; and F. J. Van Antwerpen and S. Fourdrinier, *High Lights: The First Fifty Years of the American Institute of Chemical Engineers* (New York: AIChE, 1958), pp. 77–9. The early adoption of a code of ethics by the AIChE, however, must be seen in context. It was adopted at a time when such codes were widely discussed, for example by the American Society of Civil Engineers (1893, 1902) and the American Institute of Electrical Engineers (1907). The first code of ethics was adopted by the new American Institute of Consulting Engineers in 1911, and the AIChE agreed on its own some months after the AIEE. By being one of the first and ahead of most of the 'founder' organisations (ASCE, ASME, AIEE and AIMME), the AIChE council likely saw the embracing of a code of ethics as a means of constructing professional status and demonstrating social responsibility for their new Institute.

ations, in America at least, were vague and usually unenforceable affairs. In practice, the AIChE used its guidelines to screen out unsuitable candidates for membership, rather than to censure the behaviour of existing members.[131]

The role of a code of ethics

Early sociologists and historians of professionalisation argued that the adoption of a code of ethics was a natural stage in the emergence of a profession, and suggested that such codes serve to bind together members of the profession in altruistic motives.[132] But explicit ethical standards were demonstrably lacking through the first half century of the IChemE; the members and Council either saw no particular need for such a code or identified it as a hindrance to be resisted.

As discussed above, standards of professional practice appear to have served little purpose in the internal activities of the IChemE. Jeffrey Berlant, on the other hand, has argued that ethics codes are not a culmination of natural growth but the result of a drive for monopoly, and perform the function of excluding outsiders.[133] This is clearly not the case for the IChemE. Such codes were not used to filter out unsuitable membership candidates as apparently they were for the early AIChE; they did not act as a template against which to judge, and correct, straying members; they were not used as a model of suitable moral behaviour in work-a-day practice.

When codes were finally introduced, they were largely a result of specific outside pressures.[134] Consultants' codes of practice were motivated by external calls for the Institution to recommend specialists, and by the Institution's need to ensure a clear relationship of trust between itself and the client (and secondarily, between the client and consultant). The general code of ethics followed the merging of interests of the IChemE with other engineering bodies. Promotion of the code as a symbol of professional responsibility, and the taking of an active part in ethical issues such as safety, followed the Flixborough disaster. Thus the introduction of such standards helped the Institution to consolidate its position with manufacturers first, then with other engineers, and finally with the government and general public.

The earliest mention of an ethical stance on the part of the Institution appears in the minutes of the Provisional Council in 1922, reporting an address by K. B. Quinan. Quinan called for an Institution comprising 'competent men of highest integrity and character', stressing that 'high moral character was

---

[131] Reynolds, *op. cit.*, 18.

[132] For a survey of theories, see A. Abbott, *The System of the Professions: An Essay on the Division of Expert Labor* (Chicago: University of Chicago Press, 1988), pp. 3–31.

[133] J. L. Berlant, *Profession and Monopoly* (Berkeley: Univ. of California Press, 1975).

[134] A distinct set of outside pressures – such as the design of weaponry for the Viet Nam war and particular cases of whistle-blowing – were responsible for the rise of a new ethical professionalism among American engineers during the same period. See P. Meiksins, 'Engineers in the United States: a house divided', in: Peter Meiksins and Chris Smith (eds.), *Engineering Labour: Technical Workers on Comparative Perspective* (London: Verso, 1996), pp. 61–97.

equal to, if not more essential than, mere efficiency'.[135] While publicly commending and vaunting such platitudes, the Provisional Council members took no overt action to incorporate ethical standards into their constitution. The practical scope and definition of a 'high moral character' for chemical engineers was left unaddressed in both private discourse and public rhetoric.

In principle, a code of ethics was particularly relevant for chemical engineering, and could apply to issues on several levels. At the most public level, social concerns were raised by the operation of chemical plants which polluted the air and rivers. Such preoccupations preceded the recognition of chemical engineering itself. Leblanc soda manufacturing plants were copious producers of hydrogen chloride gas, and bleach and dye works, paper mills and tanneries dumped waste liquids in the rivers beside which they were purposefully sited. Lawsuits regarding local pollution date from the medieval period, and in 1864 the British government instituted official inspections of chemical plants. G. E. Davis, for example, had been employed as one such government alkali inspector to identify and prescribe action to curb excessive pollution, later specialised in water analysis and waste treatment as a consultant, and published a book documenting the destruction of the river system surrounding Manchester.[136]

A second potential locus for improved ethical standards was in the environment for operators of chemical plants. Working conditions were often among the worst in any industry, with operators exposed to a melange of acidic fumes, caustic liquids and high temperatures.[137] Conditions, again, were recognised and accepted widely by industry and government. The state, indeed, promoted this atmosphere in His Majesty's Munitions Plants of the first world war, in which female workers were transformed into yellow-skinned and greenish-haired 'canary girls' through their absorption of TNT through the skin.[138]

The organisers of the IChemE no doubt had little desire to raise issues of culpability for a profession that was struggling to assert its legitimacy and value to the public and to other engineers. Secondly, it could be argued, such moral lapses were more the fault of owners and managers of chemical plant than of the designers and salaried operators who made up the greater fraction of the IChemE membership.[139] The competing factors of company profit, public injury and governmental inaction could not easily be balanced by the professional body. A more restricted definition of ethical conduct was, however,

[135] CM 8 Mar 1922.
[136] George E. Davis, *The River Irwell and Its Tributaries* (Manchester, 1890). See also F. Stainthorp, 'The polluted Irwell a century on', *TCE*, 23 Aug 1990, 16–20.
[137] See, for example, Norman Swindin, *Engineering Without Wheels: A Personal History* (London: Weidenfeld & Nicolson, 1962), for accounts of conditions in chemical plants before the second world war.
[138] M. Rayner-Canham and G. Rayner-Canham, 'The Gretna garrison', *Chem. in Britain*, Mar 1996, 37–41; Angela Woollacott, *On Her Their Lives Depend: Munition Workers in the Great War* (Berkeley: University of California Press, 1994).
[139] Nevertheless supervisors, managers and directors comprised a sizeable 38% of the membership between 1923 and 1926.

amenable to supervision by the Institution: the professional practice of consulting engineers. Nevertheless, with self-described consultants making up only 6% of the early membership, standards of conduct did not unduly exercise the Institution in the interwar period.

The issue of ethical standards for the new profession came up occasionally, but without any resulting action. In 1929, for example, there was considerable discussion on the desirability of establishing a code of professional conduct, but Council agreed 'that the Institution had not yet been in existence for a sufficiently long time to make the drawing up of a specific code a desirable thing'.[140] The minute books are also noticeably silent on cases of professional misconduct and Institutional discipline. Indeed, there appears to have been only a single case in the interwar period, and that was more a matter of boundary-policing than internal censure.[141] Through to the 1970s, only a handful of cases of misconduct was raised. As a result, no systematic procedure was put in place to deal with such occurrences, only one going so far as the establishment of an investigatory committee.[142]

The complacency of the Institution began to change only after the second world war. The rapid expansion of chemical engineering and plant design opened new opportunities for both consulting work and misconduct by consultants.[143] In 1945 Hugh Griffiths, then President, recommended revitalising the pre-war 'register of consultants', primarily to avoid embarrassment to the Institution:

> We all know that in the past there has been a more or less artificial assumption that no one should describe himself as a consultant unless he were completely independent in the strictest possible ethical sense, and that he should also follow the same type of rules of professional conduct as a lawyer or a medical man. Those of us who have had many years contact with the consulting fraternity realise that very few engineering consultants could really make a living if these principles were strictly applied. There are a few consultants of high eminence who have undoubtedly succeeded in maintaining this very high standard of conduct, but I believe that it would be fair to say that there are many others not so independent as may be pretended. It is unnecessary to go into details to describe the various mechanisms by which the independent facade can be maintained.[144]

---

[140] CM 9 Jan 1929.

[141] In the early 1930s Council discussed the case of a non-member consultant using the qualification 'AMIChemE' after his name. The Honorary Secretary contacted the man and warned him to desist – apparently an effective action despite its hollow threat. A similar situation occurred as late as 1971, when Council discovered they were 'powerless to prevent the misuse of the term chemical engineer by firms'. CM 12 Jul 1971.

[142] Evans archive, R4: Rules for professional conduct. The cases recorded include bankruptcy, a threat to a former employer and accusation of improper expert testimony in court.

[143] His widow complained bitterly that when consulting after the first world war, 'rascals' and 'fraudulent representations by the unscrupulous' had led John Hinchley into several disastrous joint ventures. E. M. Hinchley, *John William Hinchley: Chemical Engineer* (London: Lamley & Co., 1935), pp. 41, 43–4.

[144] Letter, H. Griffiths to Joint Hon Secretaries, 2 Nov 1945, Gayfere archive box VII/2, file 'Register of Consultants (1945–1958)'.

Among such misuses were the promotion of processes, equipment or business ventures in which the consultant had a financial interest, or consultation in areas outside the consultant's domain of expertise. The old pre-first world war criticism of chemical engineers acting as representatives of equipment manufacturers could still stick. But Griffiths, himself a consultant, admitted that a register would pose problems. The Institution could not possibly categorise and accredit consultants beyond their status as Fellows, Associates or Graduates; it would be forced to trust chemical engineers to accurately cite their subjects of competence. And the question of trust could be dangerous to the Institution unless the would-be consultants had clear guidelines to follow.

There the matter rested for another thirteen years. Provoked by repeated requests from industry to recommend suitable qualified consultants, however, Council finally set up a committee to examine the whole question in 1958.[145] Noting that 'the Institution had no code of conduct in this matter', the committee used the Professional Practice rules from the Association of Consulting Engineers, the Code of Ethics of the AIChE and Institution of Civil Engineers, and the Bye-Laws of the Institution of Electrical Engineers to devise their own Code of Professional Conduct.

Ethical issues, however, were decidedly secondary in this. The committee saw its Register of Consultants as akin to an extension of the war-time Central Technical Register. It had been compiled so that information about corporate members 'may be made available readily to Government departments, industrial organisations and individuals ... to serve not only professional but also vital national interests.[146] But the basis of the Register was a specification of suitable arrangements between consultants and their clients. Six items instructed the consultant to avoid indirect fiduciary benefits and solicitation of consulting work, and to make known any links he might have with contractors, manufacturers and patent holders involved with the work. Apart from a general admonition to be guided 'by the highest principles to honour and uphold the dignity of the chemical engineering profession and the reputation of the Institution', the points of conduct referred narrowly to consulting practice rather than to the ethical responsibilities of either managers or operators of chemical plant. Thus admission to the Register of Consultants demanded an acceptance of professional responsibility beyond that of non-consultant members of the Institution, who remained unbound by a code of practice.

The question of a general code of practice for all members was raised in a separate ad hoc committee on professional conduct established in 1958 with Sir Hugh Beaver as Chairman.[147] In 1961 and 1962 the committee sent draft rules to the ICE and IEE for comment. The ICE wrote to say that it hoped to get agreement of professional conduct rules between the British engineering

---

[145] CM 19 Mar 1958.
[146] 'Register of consulting chemical engineers', typescript, 1958, Gayfere archive box VII/2, file 'Register of Consultants (1945–1958)'.
[147] Gayfere archives box VII/2, file 'Rules of professional conduct (1958–64)'.

institutions with a view to international adoption through the Conference of Engineering Societies of Western Europe and the USA (EUSEC).

Indeed, the growth of such inter-professional bodies was the deciding factor in establishing ethical policy. Participation in the CEI brought calls for a more explicit definition of a code of practice for chemical engineers, because member institutions agreed to uphold standards of qualification, competence and conduct.[148] While the first two were well regulated by the IChemE, the third was not. In 1972 Council decided to add professional code of conduct rules such as the CEI had established. These demanded, for example, that 'a chartered engineer shall uphold the dignity standing and reputation of his profession ... have regard to the public interest including public health and safety ... accept and discharge with integrity his responsibility to his employer or client and to his employees ... and exercise professional skill and judgement'.[149] Any such codes of practice remained impotent, however, without means of enforcement. This finally became possible, at least in theory, with the introduction of a revision to the by-laws of the Institution stating that any member who flouted rules on conduct would be expelled.[150]

Flixborough and its consequences for the profession

A link between ethical conduct and plant safety was not always clear-cut. There was an inevitable tension between manufacturers who wanted inexpensive product (and who hired chemical engineers to design and operate their plants) on the one hand and public acceptance of such plants on the other. Yet chemical engineers had wider social responsibilities than curbing pollution alone, argued IChemE president C. E. Spearing in 1963. He cited the example of a petroleum plant design costed at £33M which increased to £50M after safety features were added. The revised plan was rejected because of its cost, and the original design worked economically. Spearing called for chemical engineers to 'grasp their opportunities with firmness and courage' to 'become leaders of the present age' by sensibly trading-off excessive safety provisions to ensure the viability of contracting. Thus could employment and cost-effective products be maintained.[151]

Such public sentiments became unfashionable after 1 June, 1974, when the Nypro chemical plant at Flixborough, Lincolnshire exploded.[152] The sudden

---

[148] *TCE*, Dec 1965, CE308.

[149] CM 31 May 1972.

[150] Modified by-law (43).

[151] C. E. Spearing, 'Cost, courage and calculated risk in chemical engineering', *Chem. & Indus.*, 4 May 1963, 745. This compromise between safety and profitability appeared in several subsequent publications, often as a graph. There developed a clear difference between British and American approaches to pollution, however. The US set legal limits for contaminants, based on measurement; the British adopted the policy of employing 'the best possible means' of control, which could give a more positive impetus for seeking technical solutions.

[152] *The Flixborough Disaster: Report of the Court of Inquiry* (London: HMSO, 1975); J. Cullen, 'Flixborough ten years on', *TCE* Jun 1984, 25–33; T. A. Kletz, *What Went Wrong? Case Histories of Process Plant Disasters* (Houston: Gulf Publishing Co., 1985). Nypro was formed in 1964 and, at the time of the explosion, was owned 55% by Dutch State Mines (DSM) and 45% by the National Coal Board.

loss of 28 lives, devastation of the plant and extensive damage to three nearby villages were followed by a more gradual loss of reputation and confidence for the chemical engineering profession. The public addressing of ethical issues was an important consequence.

Some three weeks after the accident the IChemE convened a special meeting of Council to consider the consequences.[153] While Council members feared that 'members of the Institution had been involved at the design stage', the body had a good record of promoting research and education into hazards.[154] One suggested that 'if the Institution did not itself take appropriate action it was likely that action would be forced upon it'. Feeling that standards much like those in the nuclear industry would be introduced, 'it was essential the Institution took positive and objective action, if only to prevent unnecessarily burdensome legislation'. Two issues were thus seen as being at stake: the need to counter constriction of professional freedom (despite its possible strengthening effects by clearly circumscribing professional identity) and a desire to protect the public image of chemical engineers as competent professionals.

Government spokespersons suggested that one outcome of Flixborough enquiries might not only concern stricter regulation of the chemical plant industry but of chemical engineers themselves, by the extension of registration or licensing to professional engineers, just as was applied to medical practitioners. The IChemE General Secretary, David Sharp, suggested that this would have a dual effect. First, it would promote the application of the code of conduct still being developed by the Institution and, second, would provide a means of enforcing such a code. Misconduct or malpractice could, for the first time, be punished by the withdrawal of membership and, with it, the right to practise chemical engineering.[155] A later study indicated, however, that professional training and membership in professional bodies seemed to have small influence on attitudes towards social responsibility.[156]

The government's Health and Safety Executive (HSE) judged that chemical engineers were inadequately trained in safety issues. The Institution responded that the problem was not that of either the Institution or of British chemical engineering training. Some of the personnel at Flixborough, claimed then General Secretary A. M. McKay defensively, had held Dutch technical qualifications 'unlikely to have satisfied this Institution's requirements for Corporate membership', and the Model Degree Scheme demonstrated the long-standing commitment of the Institution to broad-based training.[157] But public relations

---

[153] CM 25 Jun 1974.

[154] The first conference had been held in 1960 and repeated at frequent intervals. In 1971 there had been an international conference on Loss Prevention, and in co-operation with the European Federation of Chemical Engineering a second international conference had, ironically, been held in the month before the Flixborough accident.

[155] D. H. Sharp, 'From the General Secretary', *TCE* Mar 1975, 208.

[156] B. Harvey, 'Flixborough five years later', *TCE* Oct 1979, 697–8.

[157] A. M. McKay, *TCE* Feb 1976. The General Works Manager and Managing Director were chemical engineers with Dutch qualifications, but they had 'no mechanical engineering training or qualifications'. A chartered chemical engineer and chartered fuel engineer were plant managers for neighbouring regions of the facility. The plant's chartered mechanical Works Engineer had recently left its employ. *Court of Inquiry, op. cit.*, 3–4.

were sensitive: an IChemE Council member and member of the Flixborough Court of Enquiry indicated that 'any pronouncement by the Institution could be most embarrassing'. The Institution consequently drafted a three-part brief outlining its policies of training, admission and qualification in case members were called upon to represent it.[158] Plant safety and loss prevention (in the widest and most ambiguous sense of the term) became a subject of notably greater attention afterwards, particularly in the three years immediately after Flixborough.[159] The Institution began to take an active role in such issues. It organised conferences; published proceedings and monographs by working parties; organised short courses; and, modified its Model Degree Scheme.[160] The Model Degree, in 1982, required that safety components be mandatory in any chemical engineering course accredited by the IChemE (unlike its American counterpart). The subject remains more prominent in British university curricula than in America. The lack of Institutional activity regarding occupational health and safety, however, came under criticism.[161]

Indeed, the issue of environmental pollution, increasingly in the public mind since the publication of Rachel Carson's *Silent Spring* in 1962, and opposition to the perceived dangers of the British nuclear programme, galvanised by the formation of the Campaign for Nuclear Disarmament (CND) in 1958 forced chemical engineers for the first time to seek to shape and amplify their ambiguous and retiring social identity. Rather pointedly, the IChemE adopted a new 'clean, green' logo in 1990 to suggest modernity and sensitivity to environmental issues. Such corporate imagery was emblematic of an outward-looking stance not seen from the organisation since the 1920s. The intended audiences for such portrayals were not practising engineers, but rather industry, government and the general public. Similarly, since 1980 the Institution had been describing itself as 'an independent voice for the profession, as it is a voluntary association of individuals bound by a strong code of ethics'.[162] Indeed, the newly drafted regulations required members to have due regard for the working and living environment. The Institution, in turn, offered to support a member 'whistle-blowing' by going to an external body if unable to convince his or her company to deal with an environmental problem it had created.[163] Into the 1980s, however, concerns were raised about the Institution's 'vulnerability' to potential

---

[158] CM 29 Oct 1974; 10 Apr 1975.

[159] T. A. Kletz, 'Safety today', *TCE* March 1983, 47. For coverage of the same issues, see also coverage of the 19 Nov 1984 liquid petroleum gas inferno at the Mexico City Pemex plant, the 3 Dec 1984 Bhopal methyl isocyanate leakage disaster, and the 26 Apr 1986 Chernobyl nuclear reactor explosion: D. Lihou, 'Why did Bhopal ever happen?', *TCE* Apr 1985, 18–23; 'Pemex: the forgotten disaster', *TCE* Oct 1985, 16–22; *TCE* Nov 1986 (several articles). Loss prevention was one of the most prominent subjects for articles in *TCE* through the late 1980s.

[160] V. C. Marshall, 'Accrediting the safety professional', *TCE* May 1983, 392–3; B. M. Hancock, 'Safety and loss prevention: the IChemE's involvement', *TCE* May 1983, 99. See also *Loss Prevention Bulletin*, first published in Jan 1974 and then as a bi-monthly from 1980.

[161] V. C. Marshall, *TCE* Dec 1983, 3.

[162] *TCE*, Mar 1980, 1.

[163] 'At the council table', *Diary & News* Mar 1992. As of the late 1990s, however, this provision appears never to have been exercised.

attacks on its social responsibility, particularly regarding an official stance on nuclear energy, nuclear fuel reprocessing and relations with South Africa.[164]

OTHER DIMENSIONS OF THE PROFESSIONAL IMAGE

One of the respects in which ethics came ostensibly to the fore in the Institution was concerning union participation. This was also arguably the most personal expression of professional identity for practising chemical engineers. Through the 1970s, the influence of trades unions became more profound for them and for the British engineering professions as a whole. For a time, the disparities between a 'labour-based' and 'professional' identity became explicit.[165] Pressures to unionise arose from two directions. Internally, professional engineers were being adversely affected by falling real salaries and a deep slump in the chemical industry during the first half of the decade. Externally, the Trades Union Congress (TUC) was promoting the universalisation of the closed shop, and with it came the development of more formalised procedures of industrial democracy. The IChemE Council had recommended to members in 1971 that they join a union to protect their personal interests. The CEI similarly recommended that its members' institutions encourage professional engineers to join appropriate professional unions,[166] although at least one later president saw a conflict between union affiliation, professional ethics and employer allegiances.[167]

Ostensibly, the demographic profile of chemical engineers appeared well-matched to the traditional membership of unions. At the end of the 1980s, although 70% of all university entrants came from families of social classes 1 and 2 (professional and semi-professional workers, according to government ranking), about 40% of chemical engineering students were drawn from social classes 3 and 4.[168] Yet the IChemE even in the late Seventies had fewer trade union members than any other CEI body, partly because it had the fewest members employed in the public sector which traditionally had been the first to organise.[169] The relatively recent science-based origins of the profession also tended to divorce it from crafts-based engineering professions and the unions that developed with them.[170] The IChemE Council resisted suggestions that

---

[164] Letter, D. C. Freshwater to council, 25 Jun 1986. A position statement on nuclear energy was consequently prepared. Evans archive, R8: Freshwater Working Party Relationships With Other Institutions.

[165] For the 'labour-based' perspective, see for example James B. Jeffreys, *The Story of the Engineers, 1800–1945* (New York: Lawrence & Wishart, 1945) and Keith Middlemas, *Power, Competition and the State* (Basingstoke: Macmillan, 1990).

[166] R. J. Horlick, 'Trade unions and the professional engineer', *TCE* Apr 1976, 253–4.

[167] J. Solbett interview with Johnston, 16 Dec 1997.

[168] 'Chemical engineering's timebomb', *Chem. in Britain 25* (1989), 978.

[169] T. J. Evans, 'Professional chemical engineers and trade unions', *TCE* Jan 1979, 9–10. According to a 1977 survey, 25% of chemical engineers versus 44% of general chartered engineers were members of a trade union. Union influence was also fragmented: no individual union claimed more than 6% of IChemE members in its membership.

[170] P. Whalley and C. Smith, 'Engineers in Britain: a study in persistence', in: Meiksins and Smith, *op. cit.*, 27–60.

the Institution alter its Charter and reorganise as a labour union for pragmatic reasons. It argued that existing unions would not welcome a destabilisation of the status quo, and that the cross-industry distributions of its members might make a single union inappropriate. Pressures for an explicit union stance were lessened, too, by the Institution's intermediate status. The IChemE was unique in being the only body affiliated with both the CEI and Council of Science and Technology Institutes (CSTI), thus straddling the engineering/science divide.[171] In 1980 alone, for example, the Institution had consulted over a dozen times with the Royal Institute of Chemistry (RIC) and the CSTI, particularly over health and safety matters. Nevertheless, contact with the RIC was not particularly close, being tinged by implicit accusations of 'snubbing' of the chemical engineering profession.[172]

A millennial identity

The evolution of the debate concerning institutional unification left chemical engineers largely untouched and indifferent except as far as it related to their accreditation and employment opportunities. In parallel with the moves towards merger with other bodies on financial grounds or for engineering unification, the IChemE continued to scout related occupations for new members and new territory. The Institution set up an Employment Opportunities Task Force in early 1979 to promote recognition of chemical engineers in 'fringe' (quickly re-dubbed 'non-traditional') industries.[173] In 1983, UK graduates were widely dispersed in industry. About one-quarter found employment in the chemical and allied industries; a fifth in the engineering industries; a tenth in atomic energy and public utility; about 7% each in the food and oil industries, finance and government-related posts. Another fifth found jobs in miscellaneous industries, services and consulting.[174] In contrast to the situation two decades earlier, the BP Group and ICI divisions were the principal employers, but a dozen other British companies employed over 200 chemical engineers each, and nearly two dozen others employed at least 50.[175]

In other respects the end of the century showed a lack of clear trends. University enrolment in chemical engineering declined from the Eighties, university grants were reduced and departments were rationalised. In 1992, over 8% of new chemical engineers were unemployed six months after graduation. Their recruitment by companies was, in fact, halved between 1991 and 1993.[176] While there was a large surplus of chemical engineers in the severe economic

---

[171] T. J. Evans, 'The CEI/CSTI Group of the meeting of heads of professional bodies', *TCE* Aug/Sep 1980, 527.

[172] Evans archive, R1: Royal Society of Chemistry Jan 1984–Dec 1990.

[173] At that time, employment of the profession was again depressed owing to the second oil crisis of late 1978, triggered by Iranian fundamentalism and the Iran–Iraq war.

[174] G. S. G. Beveridge, 'The politics of change, or the moulding of today's Institution', *TCE*, May 1985.

[175] T. Evans, 'Common interest', *Diary & News* Oct 1989, 24.

[176] *TCE* 24 Sep 1992, 7 and 19 Aug 1993, 5. Sources: national statistics records and the CIA.

climate of 1991, for example, the Chemical Industries Association decided to promote science and chemical engineering for schoolchildren to counter the drop in chemical engineering graduates, an action mirroring that of the Institution during the downturn of employment in the early 1960s.[177]

The Council repeatedly questioned the Institutional identity. Beveridge, in 1985, saw it as a 'focal point and communications agency' to advance the science of chemical engineering. This 'learned society' function had grown significantly since the 1960s, when income from publications and conferences began to become a significant source of Institutional income.

Two years later, Rolf Prince again analysed the way ahead for what had become 'a vast, quite complex organisation'.[178]

The role of members, who now contributed merely 20% to the IChemE income, was questioned.[179] The 'customers' of the Institution were the industries employing members, argued some Council members, not the members themselves.[180] For their part, members stated that qualifications and access to professional titles were the chief reasons for belonging to the Institution.[181] Thus the institutional and professional identities had cleaved.

The Institution itself became more responsive to the needs of individual members as the cyclical nature of the chemical and related industries dampened employment prospects. In 1987 the Institution launched an employment agency in conjunction with the Institution of Civil Engineers, the first foray into job

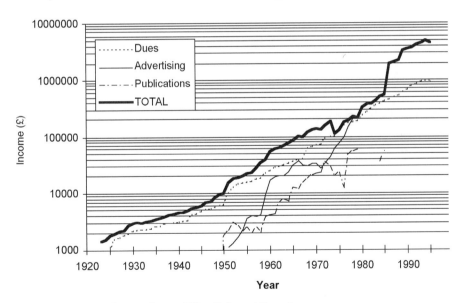

Figure 8-3 IChemE income. Source: IChemE Annual Reports.

[177] CIA, 'Graduate supply and demand', Dec 1991.
[178] R. Prince, 'Successes and the way ahead', Annual Report for 1987.
[179] The majority of Institution income by this time was from advertising and conference revenue.
[180] 'At the council table', *Diary & News* Jul 1990, 8.
[181] T. Evans, 'The attitude survey – more grist to the mill', *Diary & News* Aug 1993, 8.

finding by the Institution since its prewar Appointments Bureau.[182] Efforts were made, too, to encourage member participation: committee and council members were sought from 'active' subject group members and respondents to surveys.[183] By the early 1990s, strategic planning had come to occupy an important place in council activities – a distinct departure from the traditional 'response to crisis' mode of the Institution. The goals of the IChemE, its guiding principles and a clear definition of both chemical engineering and its customers were painfully thrashed out.[184] One outcome was the re-organisation of the Institution in 1996 to streamline the Council and to add three specialist Boards (Qualifications, Communications and Services, and Technical) reporting to it. The harder economic climate of the late 1990s also forced a reduction in the IChemE staff for the first time.[185]

Amidst the institutional liaisons and membership evolution that character-ised the last two decades of the century, professional visibility remained an unattained goal. On its seventy-fifth anniversary, the IChemE Annual Report rehearsed old themes by observing that there was 'never a better opportunity for chemical engineering to break through into public consciousness'. The Institution was adapting its self-image and that of a membership no longer entirely at the mercy of external events, but the two were still embedded within a complex ecology of professions.

[182] *Diary & News* Jun 1987, 16 In 1991 this was joined by a Graduate Employment Task Force which developed schemes to help newly qualified chemical engineers to seek and find employment.
[183] 'Welcome to member power', *Diary & News* Jun 1996, 1, 8,
[184] 'Strategic statements provoke a thrill', *Diary & News* Oct 1994, 8.
[185] From a single assistant secretary in 1923, staff reached 10 in the mid 1950s, 35 by the mid 1970s, 50 in 1985 and 75 by 1990. Except for the 1990 figure, this represented an almost constant 310 to 330 IChemE members per member of staff. IChemE Annual Reports; 'Rugby scrum', *TCE* 19 Nov 1998, 24.

CHAPTER 9

CONCLUSIONS: DISTILLING AN IDENTITY

Why did chemical engineers evolve as they did in Britain? What were the enduring qualities that persistently directed their trajectory? Can the emergence of what became the 'chemical engineer' over a troubled century reveal broader characteristics of new technical professions? Our study shows that the successful creation of a professional identity relied upon a whole series of events, both planned and contingent.

Making sense of chemical engineers is complicated by the tenacious segmentation of the specialism. Statistics of all varieties – the number of engineers employed by specific industries; the fraction of female workers; the proportion of members supporting Institution initiatives; their geographical distribution, educational and career backgrounds – such indicators seldom exceed 20% of the working population of the IChemE. This, admittedly, is true to varying degrees for many professions. For chemical engineering, however, there has been a particularly long-lived invisibility to 'outsiders' and a weak sense of cohesion from within. The identity of the profession was consistently diffuse, tenuous, unclassifiable – a good example of what Abbott has described as the 'fluctuating and geographic' nature of such social structures. The point of alternating identities – even for the practitioners themselves – is made strongly by the disparity between census figures and university graduation numbers for chemical engineers.[1] The lack of clarity of the professional borders made the jurisdictions of chemical engineering more vulnerable.[2] The multiple voices, fluid constituencies and lack of meaningful majorities also make historical generalisation misleading and dangerous.

The heterogeneous character of the specialism, and its mutating representations, is nevertheless itself a recognisable attribute. And other qualities, particularly the early identification of higher education as a distinguishing trait, demarcate this profession from those that came before it and which grew alongside it. The strategies and exigencies of the chemical engineering profession

---

[1]   See Appendix.
[2]   Abbott, *The System of the Professions: An Essay on the Division of Expert Labor* (Chicago: University of Chicago Press, 1988), pp. xv, 56.

are arguably an archetype for twentieth century technical groups, dissimilar from both the nineteenth century evolutionary pattern of the 'big three' engineering institutions and from the more recent inchoate professions such as software engineering. This 'in between' quality of chemical engineering is intrinsic to it and an important factor in its success.[3] And the persistent skirting of extremes makes the subject an excellent probe of conventional historical and sociological explanations.

A NEW KIND OF TECHNICAL PROFESSION

Nevertheless, it would be wrong to suggest that chemical engineers did not form a distinct social group, in the sense of sharing elements of a common identity and distinguishing themselves from seemingly similar workers such as industrial chemists. Their collective claims regarding distinct analytical methods helped make their profession viable.

But chemical engineers did not experience division of labour along the same lines nor as pervasively as did earlier technical professions. They emerged from a tradition of usually educated but occupationally undifferentiated workers. They differed from the chemical technologists and technicians, organised by John Hinchley and others in the 1910s, in subtle ways – for example by more often straddling social categories (very often within a single career – from charge hand, to plant designer, to works manager). While an unusually large fraction came from middle-class backgrounds, their Institution collected members indiscriminately from all these occupational strata. And if, in the interwar period, the IChemE was hesitant to promote a technician grade for fear of degrading the status of its members, after the war the Institution identified as 'chemical engineers' those receiving training as diverse as an HND or research degree. Their employers and clients were similarly heterogeneous. Chemical engineers came to be seeded promiscuously among firms both small and large. In an environment in which chemical engineers scrambled for occupational niches, career mobility and migration were high; not one niche became their exclusive domain. Thus hierarchies are difficult to discern.

Formal training was consistently important. Education was established early on as the key attribute and principal tool of professional differentiation. As detailed in Chapters 4 and 6, chemical engineering programmes at colleges and universities were successful in creating an identifiable discipline. Building a potential system of abstract knowledge that could be applied to particular cases strengthened the claims of the profession, and was something from which the Institutions of Gas, Mining and Nuclear Engineers never benefited. Although there was moderate resistance from neighbouring chemistry departments, the few outright battles for jurisdiction served as set-pieces to hone the definitions of the nascent discipline. Close ties with chemical science were

---

[3]   A similar point has been made, in the context of the sociology of scientific knowledge, about those responsible for the synthesis of ammonia in Germany, in Timothy Lenoir, *Instituting Science: The Cultural Production of Scientific Disciplines* (Stanford: Stanford University Press, 1997), Chap. 8.

responsible for this academic basis, and set it apart from near contemporaries such as Production Engineering and Fuel Engineering. Indeed, the disciplinary perspective was crucial in distinguishing what were not dissimilar occupations. The organisation of production engineers, for instance, had a strong spark fanned by Taylorist and other trans-Atlantic rationalisation schemes; fuel engineers had a strong crafts-based history stretching back to mid-Victorian gasworks. Early chemical engineers, by contrast, embodied the themes of industrial economy (with a strong national flavour), the deep-rooted traditions of the British chemical industry, and a historically close association with academic chemists. The adoption of university education as a desirable prerequisite for membership in the young IChemE was therefore a means to mark out chemical engineers as a coherent group in a way that mere jobs could not. Teachers of the subject pressed this disciplinary exclusiveness at every opportunity. The post second world war integration of science moved chemical engineering beyond the technical colleges into university curricula, and forged new specialist departments from kindred materials-based disciplines such as oil and fuel engineering.

THE ROLE OF STATE PATRONAGE

This 'backwards profession' – in the sense of being more successfully seeded in academe than industry – could not grow without broader support. Lacking a class-based cohort or even an uncontested occupational niche, chemical engineers sought sponsorship and alliances promiscuously. Their original champion – the first world war Ministry of Munitions – set a precedent for state involvement in technical occupations, and introduced a long-standing dalliance within the Institution of Chemical Engineers for corporatist alliances. The state proved to be an fickle sponsor, however. The closest links were forged during times of war (both 'hot' and 'cold'), when chemical engineers were imbued with a particular kind of competence: the ability to design large-scale plants on short time scales, and to tune and operate them to high efficiency. These attributes were less sought in times of peace, because other economic factors often overrode their importance. This was particularly true of government involvement with the profession, which was concerned primarily with national competitiveness. The government played a negative role before the second world war, too, by adopting a policy of siting large petroleum refineries outside the UK. While chemical engineers did work at refineries such as the one in Abadan, they had less opportunity to use this industry as a stepping-stone as did their American counterparts.

Activists within the IChemE struggled to take advantage of the sudden detours in societal and political trends that characterised the century by adapting their profession's identity to them. The post first world war enthusiasm for 'reconstructionism', according to which British industry would be reorganised along the lines of wartime corporatist structures, endured just long enough to promote the creation of the CEG and IChemE. By the mid Twenties, this impetus had been mutated into 'rationalisation', a less congenial arrangement in which chemical engineers strove to consolidate an overt role as engineers of efficiency.[4] A decade later, and through the next war, the Institution pragmatically played the card of supplying technical workers attuned to the rapid accommodation of production to suit the state's needs. After the war, this role altered only slightly to support the not dissimilar economic warfare of international trade. And from the 1960s and beyond the IChemE attempted, with considerable success, to proactively support initiatives by successive Labour and Conservative governments to reorganise the technical professions in Britain.

THE BATTLE FOR INDUSTRY

According to Abbott, the goals of all professions reduce to gaining authority over a circumscribed class of work, and the status and perquisites surrounding it. Jobs were thus the bottom line for practising chemical engineers. Paradoxically, employment was the most implicit of the challenges and actions of the IChemE, yet the most germane to the careers of its members.

The IChemE employed a variety of tactics to seduce industrialists into employing its members. The most immediate was by direct association. Even at its formation, the Institution incorporated a high proportion of managers and directors among its Councils, committees and members. Secondly, the training standards and professional accreditation offered by the Institution held out the promise of a higher quality of technical worker, and one better suited to employers' needs. Thirdly, chemical engineers could be portrayed – at times, at least – as ideal intermediaries straddling the worlds of government, academe, design office and plant. For some industries, they could argue that they embodied the most suitable combination of skills: their chemical and physical knowledge was essential for understanding the processing of, and designing and operating plants for, explosives, nitrogen fixation, coal products, polymers and petrochemicals. Nevertheless, these attractions gained few adherents among major industrialists between the wars.

The winning over of industry was, in the end, dictated mainly by economics. The postwar take-up of chemical engineers was pressed forward by a shortage of technical labour and the threat of receding international markets. The increased population of chemical engineers by the late 1950s coincided with increased commercial competition not dissimilar to the military competition of the world wars. ICI, in particular, was forced to restructure its operations, adapting its historical hodgepodge of grafted corporate cultures to accept this more financially-aware and intellectually trained species of engineer. Such foundational changes of major employers provided, at last, a measure of recognition and stability in employment.

---

[4]   On the political place of interwar engineering, see Craig R. Littler, *The Development of the Labour Process in Capitalist Societies* (London: Heinemann, 1982), Chapter 8.

THE CENTRAL THEME: BOUNDARY DISPUTES AND A CONTENTIOUS IDENTITY

The question of professional recognition was pervasive and never fully resolved. The inherent mixture of career backgrounds and elusive authority was responsible for the broad characteristics of the evolution of the British profession. Establishing a 'space' in the ecology of professions was a long-standing concern of the Institution. The tactics varied with time and circumstance. Two among them – the targeting of education as a distinguishing attribute, and the courting of an influential sponsor – are reflected in the recurring pattern of negotiation and accommodation with other bodies. Among a number of tactics described in general terms by Abbott, let us mention several below that are particularly relevant to British chemical engineers.

Cleaving from existing organisations

The act of formation of the Institution was substantially one of reproduction and mutation from the Chemical Engineering Group of the SCI. The CEG, formed only four years before the IChemE, was seen by several of its ambitious organisers as a convenient vehicle for the creation of a more autonomous accrediting body. The role of the CEG, too, reflected their faith in promoting the discipline of chemical engineering before too strenuously attempting to found a profession. The CEG served to focus the nucleus of interest within the Society of Chemical Industry from which the stronger professional claims could later be made.

Development of alliances

The role of the CEG as a 'stepping stone' to professionalism partly explains the close and enduring relationship between it and the IChemE. The system of alliances developed by the Institution, however, went beyond chemical engineering proper. The Provisional Council, aware of the new Institution's professional fragility, sought affiliations widely. It would, it decided, 'avoid clashing in any way' with other Institutions and 'afford all the assistance in its power to other learned Societies ... and qualifying Institutions'.[5] Within three years of its formation, for example, members of the Institution had been granted reduced fees by the Institution of Fuel Economy Engineers, and came to similar arrangements with the Institutes of Fuel and Gas.[6] So topical was the subject that Sir Alexander Gibb made the co-ordination of professional interests the subject of his Presidential address in 1929.[7]

The sharing of territory concerned not only professional domains but extended to the very literal matter of physical accommodation. The Chemical Engineering Group had shared office space with the IChemE since 1924. In

---

[5] Discussion paper, IChemE Prov. Council, 14 Mar 1923.

[6] CM 10 Feb 1926.

[7] A. Gibbs, 'The co-ordination of engineering institutions and societies', Presidential Address, IChemE, 20 Mar 1929.

1927, the IChemE Council sounded out the Institution of Mechanical Engineers regarding the possibility of accommodation for the Institution as well as some 'loose system of affiliation'.[8] Rebuffed by the Mechanicals, the IChemE Council agreed the following year to offer tenancy to the Diesel Engine Users' Association; the DEUA was still sharing office and library space nine years later.[9] In 1952 the British Association of Chemists offered to share its premises with the IChemE, but the Institution decided to co-habit with the Society of Chemical Industry, an arrangement that endured until the mid 1970s.

Intellectual territory was shared too, and common interests demonstrated, through the relatively safe mechanism of joint meetings. The earliest of these was the 1925 joint meeting with the American Institute of Chemical Engineers, Society of Chemical Industry and Chemical Engineering Group.[10] Through the 1940s the Institution held joint technical meetings with the Institute of Petroleum, Gas and the Institutions of Mechanical, Electrical and Civil Engineers. From the 1950s, there were also joint conferences with Society of Instrument Technology on automatic control in the process industries.

The most effective alliances, however, were with the larger institutions of the Mechanicals, Electricals and Civils. The gradual elevation of the IChemE to what was sometimes described as the 'Big 3½' in the late 1950s, following its Royal Charter, and then to the 'Big 4' during the Finniston and Fairclough inquiries, has been detailed in Chapters 9 and 11. Its consolidation as a 'major' profession responsible for a 'primary technology' was a phenomenon of the post second world war period.

Appropriating territory

With Royal Charter in hand, the Institution found itself on the other side of the professional fence. Its Councils were quick to defend and expand its borders. Indeed, the first suggestion put forward when an Institutional motto was sought in the early 1960s was 'Qui non proficit deficit' (translated by contemporaries in territorial terms as 'He who does not advance loses ground').[11] Yet the profession was careful to avoid direct challenges to the jurisdiction of others.

---

[8]   CM 9 Feb 1927.

[9]   CM 14 May 1930, 15 Apr 1936. The organiser was F. A. Greene, who from 1929 was the Honorary Treasurer of both the IChemE and DEUA. The DEUA had, in 1927, about 300 members compared to the Institution's 400 to 460.

[10] Direct membership in both the IChemE and its American counterpart was uncommon, seldom exceeding 1% of the membership, but cordial links at the level of the organisations themselves were developed through joint meetings. The 1925 meeting was followed by an American meeting and tour of the north-east US and Canada in 1928, a visit by the Americans to the 1936 Chemical Engineering Congress in London, and a 1947 meeting in Detroit. Gayfere archives, box XIV/1 'IChemE Joint Activities', folders 'AIChE Visit 1925', 'AIChE Visit 1928', 'AIChE/IChemE Nov 1947'.

[11] CM 18 Jul 1962 and TCE, June 1964, CE122. The final choice, suggested by J. M. Pirie, was 'Findendo Fingere Disco' ('I learn to make by separating') – a motto with professional implications.

Its subordination to the 'big three' until the postwar years was difficult to discern; conflicts were largely avoided.

As a chartered institution, the IChemE was able to protest against the granting of charters to other, weaker, professional bodies, just as the ICE and IEE had in 1949 protested against it. In 1959, for example, the IChemE successfully objected to the petition of the Institution of Production Engineers (which had been formed in 1921, before the IChemE itself) for a Royal Charter.[12]

The colonisation of 'vacant' intellectual territory is well illustrated by the case of nuclear engineering, where the Institution found itself siding with its new allies even more directly, as discussed in Chapter 7. Other subjects were similarly claimed. Harold Hartley argued in 1955 that mining engineering should be separated from mineral engineering, which included all processes of ore treatment and extractive metallurgy, and which involved intimately chemical engineering techniques. And discipline mirrored practice: he recommended that some universities establish departments of mineral processing and extractive metallurgy working closely with departments of chemical engineering.[13] Significantly, each of these cases of attempted dominance by the IChemE was over Institutions that had a weaker claim to a disciplinary foundation; none was university-based.

Yet other subjects were portrayed as 'naturally' ceding to chemical engineers owing to their emphasis on the process and the importance of both chemical and physical knowledge. The influx of chemical engineers into environmental engineering from the late 1960s seems to have occurred without significant contestation of territory.[14]

Division of interests

Not all subjects could be so easily embraced. Hartley had cited atomic energy, the treatment of ores and food processing as up-and-coming chemical engineering specialisms.[15] But while the relatively new subject of biochemical engineering – in the form of antibiotics manufacturing – was a reasonably straightfor-

[12] In 1962, when the IProdE prepared a fresh petition, its President wrote to the current IChemE President, Frank Morton, to seek his reaction. Upon Morton's advice, Council agreed not to oppose further petition by the IProdE. Their Royal Charter was granted in 1964. For more on the IProdE, see H. B. Watson, *Organizational Bases of Professional Status: A Comparative Study of the Engineering Profession* (PhD thesis, U. London, 1976), pp. 215–7.

[13] *The Times*, 7 Feb 1955. This professional division was recognised also by mining and metallurgical engineers: the Institution of Mining and Metallurgy (IMM), which attempted a merger with the 'coal mining oriented' Institution of Mining Engineers in 1988, abandoned it in 1990. On the other hand the IMM, while 'a multidisciplinary organisation since its inception', did not see its members as being in any sense akin to chemical engineers. Even petroleum engineers, historically connected closely with the IChemE, comprised few members within the IMM. See A. J. Wilson, *The Professionals: The Institution of Mining and Metallurgy 1892–1992* (London: Institution of Mining and Metallurgy, 1992), Chap 17.

[14] N. W. Schmidtke, 'Chemical and civil engineers – a symbiotic relationship in environmental engineering', *TCE* Dec 1982, 458–9.

[15] H. Hartley, 'Chemical engineering', *Financial Times* May 19, 1955.

ward extension to the work of chemical engineers, food processing was not. Through the early 1950s, in particular, the Society of Chemical Industry promoted closer integration of chemical engineering and the food industry.[16] Companies such as APV, which manufactured plant for both industries, maintained separate chemical engineering and food processing groups. These showed little collaboration despite sharing the company's expertise in distilling plant.[17] Yet the Process Plant Association, formed in 1971 by a merger between the Tank and Industrial Plant Association, the BCPMA and the Food Machinery Association, made much of its food processing connections. The first chairman noted that the building of food plants demanded 'the employment of the large chemical engineering contractors who have built up their reputations in the petroleum and allied fields'.[18] But by 1974 an IChemE discussion paper admitted that food processing was one of the 'non-traditional' industries that the profession had failed to penetrate.[19]

While such intellectual and occupational space was actively contested or divided, other territory was just as quickly ceded. Labour issues were long rejected as being beyond the scope of the Institution. When the Association of Scientific Workers invited the IChemE in 1945 to a meeting on 'Science and World Progress' at which atomic energy would be discussed, the IChemE representative said that the meeting 'would be rather political in character', and Council decided to take no further action. Similarly, when the World Federation of Scientific Workers sought co-operation from IChemE in 1953 in circulating a questionnaire on the working conditions of members, Council declined to help 'owing to the Memorandum of Association'.[20]

By the end of the 1950s such territorial renegotiation had had a mixed reception. An editorial in *Chemistry & Industry* criticised chemical engineering as a 'diffuse' discipline still 'in the hands of the chemist and thermodynamicist' and for not having more actively colonised engineering as a cognate specialism.[21] But Herbert Cremer noted the difficulty of finding an uncontested intellectual space: defining chemical engineering for three professional engineers, he described how a chemical engineer performed the scaling up from laboratory 'including all the mechanical and electrical engineering and much of the civil engineering work as well. "Ah!" said my inquisitors, "there is something wrong there – they were usurping the functions of *our* classes of engineer, who themselves should deal with such matters".' Thus Cremer felt that he had failed

---

[16]   See, for example, papers in the 1954 volume of Chem. & Indus.
[17]   G. A. Dummett to Johnston, 2 Apr 1997; G. A. Dummett, *From Little Acorns* (London: Hutchison Benham, 1981).
[18]   'Process Plant Association', *Chem. & Indus.*, 25 Sep 1971, 1075; 'BCPMA annual dinner', *Chem. & Indus.* 6 Nov. 1971, 1299.
[19]   R. J. Kingsley, 'The future of the Institution', paper F.115, 2 Sep 1974 (submitted to all members), p. 9. T. Evans, General Secretary from 1976, saw the failure of the IChemE to attract membership from personnel in the off-shore oil industry as another serious lost opportunity. Interview, Evans with Johnston, 23 Feb 1998.
[20]   CM 9 Dec 1953.
[21]   'Is chemical engineering too diffuse?', *Chem. & Indus.* 23 Jul 1960, 945.

to vindicate the profession and brought down the accusation of 'blacklegging'. 'In one breath these good people criticized our ability to exercize the function of 'real' engineers and at the same time protested when we had proved our ability to do so'.[22]

The wish to extend their disciplinary identity has been repeatedly voiced by chemical engineering educators, perhaps because they most actively police the academic (and, by extension, professional) borders. Prof. A. R. Ubbelohde (1907–88) of Imperial College, for example, praised his large research school which promoted new extensions to the discipline by engendering 'vigorous hybridisation and interaction' with other fields.[23]

The contestation of territory is continual, however. Perhaps the most recent example is the substitution of chemical engineers for civil engineers in the British water industry in the 1990s. Technical and financial factors drove this occupational change. Water quality legislation led to commitments by water companies to monitor contaminants with better instruments. There was a move away from the conventional civil engineering solution of improving and cascading filters, and towards the chemical engineering solution of process plants. The financial driver was the conversion of water companies from public ownership to public limited companies, which emphasised profits and short amortisation times for technical improvements. In combination, these pressures aided an expansion of chemical engineering into the water quality domain.

Seeking amalgamation

The protracted institutional discussions from the late 1960s with the SCI, and later with engineering bodies as part of the Finniston and Fairclough inquiries, indicate the increasing urgency for interaction in a domain denuded of resources. The consideration of inter-institutional co-operation, mergers and amalgamations occupied an increasing fraction of the energy of IChemE administrators.

ARCHETYPAL IMAGERY AND ITS CHAMPIONS

The British chemical engineering profession advanced by such manoeuvres, amongst which a few seminal actions can be distilled. With them are associated individuals who played defining roles as enthusiasts of incisive ideas and as instigators of policies. Their individual aspirations were translated into collective action. Kenneth Quinan used his position in what could have been anonymous wartime administration to set new standards and openness in scientific design and the operation of chemical plant. William Macnab, from Ministry of Munitions official to backroom organiser, was the most ardent propagandist for the profession. Linking chemical engineers to the interests of state and industry, he helped to popularise an image that swept in chemical engineering

22 H. W. Cremer, 'Chemical engineering: fact or fiction?', *Chem. & Indus.* 14 Jan 1960, 31–3.
23 A. R. Ubbelohde, 'Chemical engineering and science-based education', *Chem. & Indus.* 31 Jul 1971, 878–80.

organisations on the coat-tails of post first world war government policy. John Hinchley, educator and tenacious organiser, should be remembered as the advocate for the successful 'occupational eugenics' policy of the IChemE: to make the profession viable by producing not merely a *distinct*, but an *improved*, and academically-based, class of chemical engineer linked closely to a growing discipline. A. J. V. Underwood almost single-handedly identified threats, created confrontations, and wrested professional status from mundane bureaucratic classifications during his own tenure as Honorary Secretary. In particular, he distinguished and contrasted the interests and expertise of chemical engineers and chemists and promoted stronger affinities with an identity based on engineering. And Sir Harold Hartley was far and away the most active imperialist of the profession, seeking to expand and usurp occupational territory during a period of buoyant growth. Yet beyond these decisive constructions by a few architects, the broad sweep of Institutional activities can be described as limited responses precipitated by external events. By the 1960s, the major 'ideological battles' – the pressing of jurisdictional claims of the profession – had been fought. Subsequent evolution focused on knitting the rends that had formed in creating a space, but now from a position of relative strength as one of the major British engineering professions. There were fewer obvious heroes, and a greater number of effective participants, in such bridge-building activities. An Institutional momentum, and strong ties with other organisations, increasingly directed the course of the IChemE and constrained the identity of the chemical engineer.

While admitting such generalisations we must puncture some ubiquitous myths. G. E. Davis was not the 'father' of chemical engineering in any historically meaningful sense. A peripatetic chemical worker, consultant, successful author and sometime lecturer, he certainly pressed publicly for the distinctive title 'chemical engineer' and tried to associate with it a new way of conceptualising work. He failed, though, to produce a direct line of descent for his ideas that was recognised by his near-successors. Similarly, the common story of ICI as a monolithic giant which excluded chemical engineers does not bear close examination. ICI was one of the principal employers of IChemE members from its first decade; some of its managers promoted the specialism by adopting explicit employment practices and through their activity in IChemE functions; and, for other Divisions of this company of varied technical cultures, the categorisation of its technical workers into mutually exclusive 'chemists' and 'engineers' mirrored the practice of government departments. The most subtle and seductive myth is perhaps the notion that 'it had to be this way' – that chemical engineering is inherently more adaptable than other professions, and so more competitive. The chemical engineer was socially constructed. The successes of the specialism in the workplace, in teaching institutions and among other Institutions were qualified and hard-won.

WHO SPEAKS FOR CHEMICAL ENGINEERS? THE ROLE OF THE IChemE IN IDENTITY-SHAPING

There were thus multiple paths for development of the profession. To its credit, the Institution of Chemical Engineers was responsible largely for the existence and present status of the profession in Britain and several other countries. The IChemE – unlike organisations for the would-be nuclear engineering profession, for instance – never found its role as representative of the profession challenged. Its early association with educational accreditation is largely responsible for this. The IChemE implicitly negotiated the sharing out of certain forms of control with other organisations. Thus the Chemical Engineering Group retained its character as a 'learned society' and developed its social role as an arena for 'networking' between senior members of the chemical industry. The IChemE shared research sponsorship with the ABCM, and technical publishing of periodicals with commercial publishers. Nevertheless, the 'public' role of the Institution masked private efforts to garner and sustain influence. The interwar period saw an impotent Institution seldom able to convince employers of their need for its special class of workers. The buoyant postwar expansion of industry and technical education secured employment for chemical engineers, but equally threatened the authority of the IChemE by the 'decentration' caused by employer- and government-led research programmes and regional member organisation. Subsequent decades witnessed a vacillation in the strength of the authoritative voice of the Institution – weakened by member complacency, but later revived by its central role in discussions on engineering unification. The incongruity of simultaneous dilution and dominance that characterised the CEI and Finniston activities promoted a new definition of the chemical engineer. General Secretary Trevor Evans defined the 'professional chemical engineer' as being synonymous with 'the attainment of Corporate Membership of the Institution of Chemical Engineers', or 'those progressing towards their attainment of Corporate Membership'. Such a one-to-one relationship between membership, technical competence and professional identity had been desired but never claimed by the early architects of the profession. The definition of 'professionalism' had shifted, as well:

> ... I believe that the point made in the Institution's evidence to the Finniston Inquiry is right; a profession must be circumscribed and there must be formal and public signals of inclusion within the body. Membership of a professional body also imposes on its members a code of ethical conduct and disciplinary regulations (and control of the profession consequently requires the power to exclude members as well as to accept them).[24]

As discussed in Chapter 8, the explicit link to ethics had been rejected, down-played or ignored by the founders of the Institution and early chemical engineers. Such concerns were largely imported from the CEI.

The claim that the IChemE broadly represented chemical engineers in Britain and beyond had a limited currency, and could be argued only with difficulty over the Institution's history. Until the late 1950s, there was a poor overlap between the professional identity and occupational realities – in effect the 'internal' and 'external' identities were incongruous. However, it is equally true

---

[24]   T. Evans, IChemE Symposium Series No. 53, 'Supply of Professional Chemical Engineers' (1980).

that after that time the Institution moulded British chemical engineering publicly and achieved a position of some status next to more established professions.

It is necessary to emphasise, though, that all such conclusions must be qualified. The 'homogeneity' of the Institution and its profession is largely a mirage. Just like the outside influences to which it responded, the IChemE was a mutating collection of interests; many identities could be, and were, constructed. Beginning with the early dichotomy of consultants versus former government workers, and progressing to the later Presidents drawn alternately from business or academic spheres, the Institution has represented a disparate collection of emphases. Nevertheless, the 'official voice' of the IChemE was usually clear and undivided, if mutable. The evolving power of the internal interests also were a positive force in the success of the profession: by emphasising those aspects of their identity most in demand, chemical engineers could fluidly adjust their collective portrayal for employers.

POSITION ON THE SCIENCE/ENGINEERING FRONTIER

The profession trod, with considerable success, the narrow line between 'science' and 'engineering'. It attracted members, in the early days, almost equally apportioned between the two camps. The 'engineers' within the Institution soon were assimilated as Chemical Engineers; the equivalent absorption of chemist members was nearly complete by the end of the second world war. Because of the more explicit jurisdictional contentions with chemists, the Institution's greatest successes – the National Registry categorisation, the Royal Charter, and the grasping of a prominent voice in responding to government moves towards engineering profession amalgamation – all resulted from alliances with engineering organisations rather than alignment with scientific bodies. However, the long-winded and infertile moves towards merger of the engineering institutions caused a turning back towards learned body and scientific activities in more recent years.

CONTINGENCY VS 'NATURAL' EVOLUTION

In this analysis, it is important to recognise the interplay between occupational aspirations, technical realities and historical trajectories. While our focus on chemical engineers as a species of technical specialist suggests a continuity and evolution of identity, it is important to recognise that other groups saw not the 'rise of a profession' but merely the episodic appearance and disappearance of a contentious category of worker.

The emergence of chemical engineers was neither 'natural' nor 'inevitable'. They were not a profession destined by technological trends. The different national experiences of chemical engineering professions are evidence enough of this. Rather than being predetermined by technological progress, chemical engineers were equally shaped by historical accidents and active promotion. Consider the question of the geographical placement of the sites where chemical engineers worked. Why, for example, did G. E. Davis in the 1870s, K. B. Quinan during the first world war, Christopher Hinton during and after the second

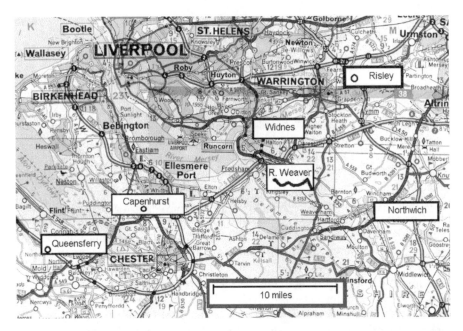

Figure 9-1 Location of some of the major chemical engineering sites in North-West England. Except for Quinan's 1915 Queen's Ferry munitions factory in Clwyd, North Wales, the others – Widnes, location of Davis's attempt to form a Society of Chemical Engineers in 1880 and later home of ICI's General Chemicals Division; the Davis Chlorine Processing Co of the 1890s on the River Weaver; Brunner, Mond's Northwich Alkali Division and a first world war State factory; Hinton's Risley Industrial Group headquarters from 1946, and the second world war and UKAEA sites at Capenhurst – are located in Cheshire, England.

world war and the UKAEA thereafter site major plants within the same ten-mile radius in the North-West of England? There were certainly 'natural' advantages: the existence of salt marshes had encouraged salt production in Cheshire during the early nineteenth century; later, access to coal for fuel, and water supplies for cooling processes and for dispersing their waste products, made chemical plants economical there.

Existing support facilities for rail and shipping access also played a part, as did the low concentration of major towns to be inconvenienced by pollution or accidents, yet giving access to an adequate population of workers. The very momentum of such siting decisions encouraged further development. Water for waste dilution and cooling had been important for choosing Billingham at the mouth of the River Tees on the north-east coast at the end of the first world war but, despite its vulnerable military position during the next war, the extensiveness of facilities there over-ruled re-siting. Hinton's industrial nuclear headquarters and first reactors were built on the former wartime factory sites with which he was familiar; they, in turn, had been based on locations claimed by Kenneth Quinan for his own war-work.[25] This geographical nexus collected

---

[25] The 300 acre Queen's Ferry guncotton facility, for example, was a derelict industrial site in 1914.

generations of chemical workers and chemical engineers, and the working culture that developed there shaped their profession.

THE NATIONAL CONTEXT

The weight of documentary evidence shows that the thesis of chemical engineering as an American phenomenon, exported only imperfectly and tardily to Britain and other European countries, is untenable. The British context of the specialism reveals its dependency on indigenous political policy, local industrial culture and academic opportunities. For these very reasons, the British profession was able to export its version of chemical engineering in no more than a piecemeal fashion to its former colonies and dependants, where other forces were in play. Canadian chemical engineers associated largely with chemists (in the Chemical Institute of Canada); in the antipodes, the most fertile alliances were with engineers (in the Institutes of Engineers of Australia and of New Zealand, respectively); and in India and South Africa, indigenous organisations became popular owing to the more explicitly political factors of independence and apartheid, respectively. Thus socially recognised technical occupations, and the bodies of knowledge on which they are based, can have an overwhelmingly national flavour. The similarity of the chemical engineering discipline and practice today is as much the result of the globalisation of knowledge and the economy as to any innate coupling between 'natural' disciplines and technological 'progress'.

WIDER FINDINGS

What then does this study tell us about the institutional politics of those technical occupations that have claimed professional status in Britain during the twentieth century? Obviously the analysis of just a single would-be profession is, in general terms, merely suggestive rather than conclusive. Nevertheless, our findings support the view that British engineers – and perhaps other technical occupations as well – should be regarded as being among the 'professions of capital'. During and after the first world war, for instance, the leaders of the inchoate profession clearly thought that the path to social and economic authority lay in claiming a place for chemical engineers high in the hierarchy of some kind of a corporate industrial order. The basis for this claim lay in the particular technical expertise in the economic manufacture of chemical products and their associated plants that chemical engineers said was not to be found among chemists and other kinds of engineers. The same refrain was picked up during and after the second world war, and subsequently during periods of industrial trauma: recessions from the late 1950s and pollution crises from the 1970s.

Our study also confirms the relevance for the professionalisation of technical occupations of the phenomenon of 'corporate bias' that Keith Middlemas has skilfully diagnosed as operating more generally in the industrial politics of Britain throughout this century. Technical occupations that see themselves as developing in a symbiotic relationship with industrial employers have to deal with the continuing problem that these employers do not always share common views with the occupation. They are, moreover, unlikely as a group to welcome the disruption to the existing industrial order implied by the jurisdictional claims of a would-be profession. Hence the rudimentary profession has to construct a consensus, although by definition it does not itself yet possess the necessary social and cultural resources with which to do so. The state can act as an important surrogate agent in this process. The British state, via the powerful Ministry of Munitions and Ministry of Reconstruction, carried through a reorganisation of the relationships between chemical firms and their technical staffs that was sufficiently radical to permit 'chemical engineers' a realistic chance of occupying a key place in the new industrial landscape after the first world war. Administrators of the IChemE were mindful of this opportunity for sponsorship, and made good use of it as peripheral consequences of subsequent government actions: by latching onto the Central Register as a opportunity for recognition during the second world war; through aggressive lobbying for teaching provision during the expansionist period after it; and, by becoming an active voice in the CEI, Finniston enquiry and Fairclough proposal.

Despite courting state patronage, chemical engineering has never attained a stable image in the public arena. From 'engineers of efficiency' after the first world war – their historical peak of visibility – to an unappealing hybrid or species of chemist between the wars, to a kind of applied scientist in postwar nuclear facilities, to scapegoats for the pollution crises two decades later, the identity of this chameleon-like profession has been ambitiously claimed by its protagonists, and often crudely categorised or ignored by others. Through this series of successes and failures a particular mode of evolution becomes apparent. The coalescence of a viable profession required active manipulation on social, technical and political levels. Chemical engineering is a heterogeneous and nationally distinct profession to which chance and opportunism have contributed as much to success as have foresight and design.

APPENDICES

Table A-1 Early organisers of chemical engineering in Britain

Name	CEG Provisional Committee (July 1918)	CEG first Council (Mar. 1919)	IChemE Provisional Committee (Nov. 1921)	IChemE Provisional Council (Jan. 1923)	IChemE first Council (Jan. 1923)	CEG Council (1923)	Other (at time of participation)
1 E. A. Alliott	Member	Council				Council	Officer, ABCM (1920) and BCPMA (1920), 'AMIME, Engineer and chemical engineer'
2 Dr E. F. Armstrong		Vice Chairman					Organiser (1916), ABCM. 'DSc, FRS, FCGI, FIC'
3 A. J. Broughall						Council	
4 D. Brownlie			Member				Fellow, SCI.
5 Dr H. J. Bush	Member	Council					Officer (1920), BCPMA. 'MSc, PhD, Chemical engineer'
6 Dr C. Carpenter					Vice President		Sat with Lord Moulton on first Committee for Explosives Supply, 1914; President (1916), SCI; organiser (1916), Chairman (1918 Vice President (1919), ABCM. Resigned VP of IChemE Feb 1924.
7 Prof. F. G. Donnan						Council	
8 Sir A. Duckham			Chairman	Chairman	President		'KCB, MICE'
9 A. C. Flint			Ass't. Secretary	Ass't Secretary	Ass't Secretary		[died 1924]
10 C. S. Garland	Member	Council	Co-opted Jan 1922	Council	Council	Vice Chairman	FIC; Council member, SCI; Unionist MP (1922–3); Founder (1917), British Association of Chemists, Officer, National Union of Manufacturers. 'ARCSc, BSc, AIC, Chemical engineer'
11 C. J. Goodwin	Member	Council	Member			Council	Hinchley had worked for father. '(Capt.) ACGI, BSc (Eng), AMInstCE, Consulting chemical engineer'
12 W. T. Gee							
13 F. A. Greene			Member				
14 Dr H. C. Greenwood	Member	Council					Advisor, Min. of Munitions. '(Lieut), DSc, FIC, Chemical engineer' [Died Nov 1919]
15 H. Griffiths		Council					Chemist, Min. of Munitions. 'ARCSc, BSc'
16 Dr E. Hill		Council					'ARCSc, FIC, DSc'
17 Prof. J. W. Hinchley	Chairman	Chairman	Convenor	Hon. Secretary	Hon. Secretary	Council	Organiser (1911), Association of Chemical Technologists; President, Institution of Chemical Technologists; organiser (1917), British Assoc. of Chemists. 'WhSc, ARSM, FIC, Professor of Chemical Engineering, Imperial College'

	Name						Notes
18	C. S. Imison					Council	'ARCSc, BSc, DIC'
19	? Howroyd					Council	
20	H. F. V. Little	Council		Member			'AMInstCE, Consulting engineer'
21	A. J. Liversedge	Council					
22	W. Macnab		Member		Council	Council	FIC; Council, SCI; Sat on Committee for Explosives Supply with Lord Moulton; senior advisor to Ministry of Munitions
23	J. MacGregor		Member				
24	A. E. Malpas					Council	
25	Lord Moulton						Head of Dept. of Explosives Supply, WWI; President, ABCM (1918) [died Mar 1921]. 'GBE, KCB, LLD, MA, FRS'
26	D. M. Newitt		Member				Former student of Hinchley.
27	Dr W. R. Ormandy	Council	Member		Council	Council	President, Institution of Automobile Engineers; Council, Institute of Petroleum Technologists. 'DSc, FIC, MIPetTech Consulting chemical engineer'
28	P. Parrish					Council	
29	H. J. Pooley					Council	
30	F. M. Potter					Council	Council (1911), Association of Chemical Technologists.
31	K. B. Quinan			(attended)	Vice President		Gen. Manager, Cape Explosives Co; Chief Advisor, Explosives Supply.Chief Advisor, Min. of Munitions; Companion of Honour.
32	J. A. Reavell	Council	Member	Member	Council	Chairman	Officer (1923), BCPMA, Council, SCI. 'MIME' MD of furnace engineering firm.
33	S. M. Ridge		Member	Member			
34	F. H. Rogers	Hon. Treasurer	Member	Hon. Treasurer	Hon. Treasurer	Hon. Treasurer	'MIME, Consulting engineer and patent agent'
35	Dr E. W. Smith		Member	Member		Council	
36	H. Talbot	Hon. Secretary		Member	Council	Hon. Secretary	Dep. Director of Gauges, Min. of Munitions 'ARCSc, BSc, Chemical engineer'
37	S. J. Tungay		Member			Council	Min. of Munitions; lecturer at Imperial College.
38	S. G. M. Ure					Council	
39	J. H. West					Council	
40	W. J. U. Woolcock		Vice Chairman			Council	Secretary of Pharmaceutical Society of G.B.; Coalition Liberal MP (1918–21); Parliamentary Private Sec'y to Minister of Munitions (1919–21); General Manager (1918–28), ABCM.

INSTITUTIONAL STATISTICS

COMPARATIVE MEMBERSHIP GROWTH

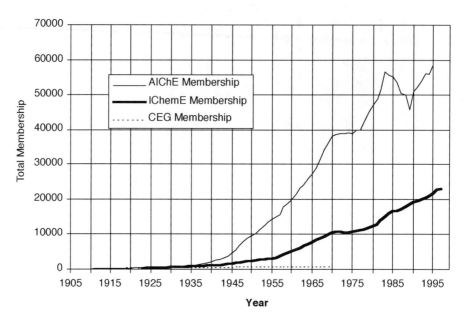

Figure A-1   Membership of chemical engineering organisations (all grades).   Sources: Annual Reports, IChemE; T.S. Reynolds, *Seventy-Five Years of Progress: A History of the American Institute of Chemical Engineers 1908-1983* (New York: AIChE, 1983); AIChE reports; *Proc. CEG.*

IChemE MEMBERSHIP BY GRADE

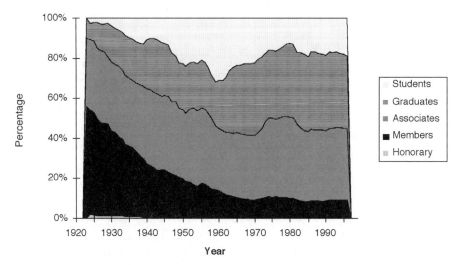

Figure A-2   Distribution by membership category.   Source of raw data: IChemE reports.

'Corporate' membership (comprising 'Associate Members' and 'Members') grew steadily from the birth of the IChemE, although 'Members' made up a decreasing fraction of the total.

In 1972 the membership grades were extended to more closely reflect practice in other engineering institutions.[1] 'Associate Members' were renamed 'Members' and 'Members' were renamed 'Fellows'. The new title 'Associate' referred to a technician-class member. 'Companion' members were also introduced as a class having an interest in chemical engineering but no recognised qualifications.

---

[1]   The Civils, for instance, had instituted such name changes in 1968, after having created a 'Graduate' class in 1952.

FEMALE MEMBERSHIP

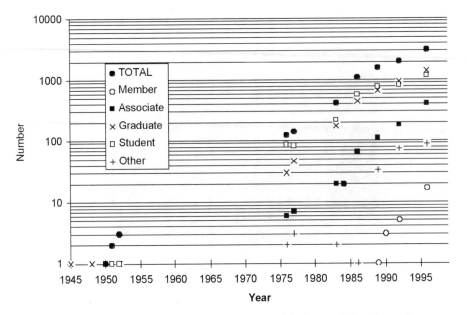

Figure A-3 IChemE female membership (pre-1972 categories). Source: IChemE annual reports; application forms.

PROFESSIONAL AFFILIATIONS

Members of the early Institution frequently were also members of several other bodies; nearly three-quarters had at least one such affiliation. These were distributed almost equally between chemistry- and science-related bodies and engineering or trade organisations (e.g. the Institute of Petroleum, IMechE, Coke Oven Manufacturers' Association and BCPMA). The most common chemistry connections were with the SCI (over 30% of corporate members), the Chemical Society (about 17%) and the British Association of Chemists (5%).

Through the 1930s, new members had fewer outside affiliations, but these remained dominated by the SCI, CS and BAC.[2] The Institution of Petroleum Technologists temporarily claimed a noticeable fraction – 10% of new members. After the second world war this downward trend in outside affiliations continued. Perhaps because of narrower training and a heightened sense of community, only 20% of new IChemE members belonged to other organisations by the mid 1950s.[3] Their sense of identity appeared, however, to be biased in terms of a chemical *profession* and an engineering *occupation*. In a 1958 survey,

---

[2]   SCI: 26%, CS: 21%, and BAC: 9% of new corporate members for the period 1930–9. Note that, because a single member could have multiple memberships, these figures cannot be related directly to the fraction of members holding no other memberships.

[3]   SCI: 11%, CS: 6%, BAC: 1% for the period 1950–6.

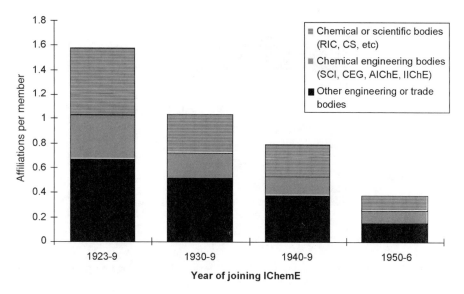

Figure A-4 Average number of professional affiliations for new corporate members, defined as (number of listed memberships of a given type)/(number of members). Source of raw data: Mackie/Roberts sampling of IChemE application forms.

the Institution found that those of its members having outside connections were largely allied with a chemical professional body, not an engineering body, and that a 'negligible' number belonged to more than two professional institutes. Links with engineering trade organisations, though, remained common.[4] By the late 1990s, a mere 7% of corporate members were also corporate members of another engineering body, and the fraction was still falling.[5]

MEMBERSHIP BY EMPLOYMENT SECTOR

There are broadly discernible trends in the industries in which chemical engineers worked. Five industrial sectors showed large changes over the life of the IChemE: the broad category of 'chemical process' industries; mining/metals/foundries; coal gas/coal tar/coke oven; oil/petroleum; and process plant typically included about one-third to one-half of new members. The largest specific industry between the wars was the coal gas/coal tar/coke oven industry. While this declined from the 1920s, it employed a burst of new members immediately after the second world war when coal research was briefly supported more generously by companies and government following nationalisation of the coal-mining industry. Employment in the nuclear industry peaked in the 1960s. Jobs in the plastics, food and consulting sectors varied more irregularly from decade to decade. Among the other sectors employing a handful of new members were

---

[4]   Sampled IChemE application forms for corporate members, and IChemE 1958 statistics, Gayfere archive, box VII/2.

[5]   T. Evans data.

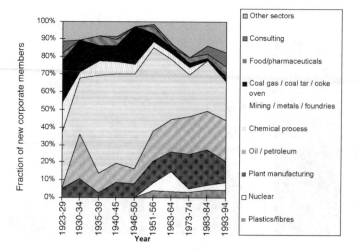

Figure A-5 Membership by employment sector. Source of data: Mackie/Roberts and Johnston sampling of membership application forms.

cement works; food manufacturing; paper mills; synthetic fibres such as viscose and rayon; paints; pharmaceuticals; leather.

QUALIFICATIONS VERSUS TIME

Three observations can be made of Figure A-6. First, there is a general trend towards degree-educated corporate members. Second, advanced degrees appear to be more common among new corporate members during times of high unemployment. And third, the slight increase in the fraction of non-degree holders during the 1960s may reflect the increase in holders of HNCs during this period. The almost total dominance of degree holders after the 1970s reflects the success of the CEI definition of 'engineer' as degree-based.

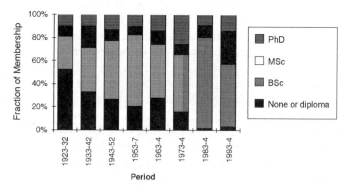

Figure A-6 Qualifications of new corporate members. Source of data: Mackie/Roberts and Johnston sampling of IChemE membership application forms.

INSTITUTIONAL ORGANISATION

By 1998 there were a dozen active regional branches in the UK (Scotland, Northern, North Western, Irish, Yorkshire, South Humberside, Midlands,

Table A-2 Regional branches of the ICheme

Year	Group or Branch
1945	North-Western Branch (Manchester); preceded by G&S section meetings.
1949	Midlands 'area', S. Africa and India interest groups (unorganised)
1950	Midland Group (Birmingham; F. H. Garland, chair). Australia Advisory Panel
1951	Scunthorpe, S. African and Indian meetings and committees.
1952	S. African Branch (Johannesburg, March 1952); Scunthorpe Group (Nov 1952).
1954	Scottish Group
1956	South Wales and Monmouthshire Group (July 1956)
1958	North-Eastern Group (Oct 1958; King's Coll, Newcastle-Upon-Tyne)
1961	South-Eastern Branch (Jan 1961, London); Scottish Branch (Feb 1961); Victorian Group (Melbourne, Australia, Oct 1961).
1962	Yorkshire Branch (Leeds, April 1962); Irish Committee; Indian Panel.
1963	New South Wales Group (Sydney, May 1963).
1964	South Wales & Monmouthshire Branch (Feb 1965); Northern Branch (Billingham; March 1965).
1966	Irish Group (April 1966)
1967	Now 'LONDON and South-Eastern Branch'
1968	Now 'Scunthorpe Group' (South Humberside Group); Queensland Group (Nov, 1968).
1969	New Zealand National Committee formed.
1970	South Australian Group (March, 1970)
1972	New Zealand Chemical Engineering Group (Christchurch, 1972).
1973	Western Australian Group (Dec, 1973).
1974	South-West England Branch (Bristol, Oct. 1974). Monmouthshire renamed 'South Wales Branch'.
1975	European Branch
1977	Irish Branch
1994	Singapore Branch

South Wales, South Western, South Central, East Anglian and London & South East) and three elsewhere (Europe, Singapore and Australia).

Subject Groups were formally approved by Council in 1977. All had open membership, although non-members of the Institution were encouraged to become 'Companion Members'.[6] Twenty years later, some 5 300 members – about three-quarters of them IChemE members, representing 16% of the total IChemE membership – were associated with one of the twenty-one groups. Non-IChemE member participation ranged from nearly 70% (for the Energy Conversion group) to a mere 4% (Computing Club).

In 1999 the groups were: Applied Catalysis; Applied Rheology; Biochemical engineering (BESG); Coal Utilisation; Computer Aided Process Engineering (CAPE); Education; Energy Conversion Technology; Environmental Protection (EPSG); Fluid Mixing Processes; Fluid Separation Processes; Food & Drink; Materials: Use and Development (MUD); Oil and Natural Gas Production (SONG); Particle Technology; Pharmaceuticals, Toiletries and Cosmetics (PTC); Process Control; Project Management; Process Economics and Cost Engineering (PENCE); Safety and Loss Prevention; Solids Drying; and Water. Some groups, such as Coal Utilisation, were on the decline (between 1994 and 1997, membership in Coal Utilisation had dropped by more than half).

---

[6]   For details and model constitution, see *TCE*, Aug 1977, 551.

TEACHING STATISTICS

CHEMICAL ENGINEERING PROGRAMMES AND CHAIRS IN BRITISH TEACHING INSTITUTIONS

In the interwar period, most chemical engineering programmes were non-degree courses, even when supported by a chair in chemical engineering. Such courses could be presented as full-time, part-time day or part-time evening classes, and had various products: post-graduate diplomas; preparation for the IChemE Associate Membership Examination; HNCs (after 1950), or HNDs (after 1965).

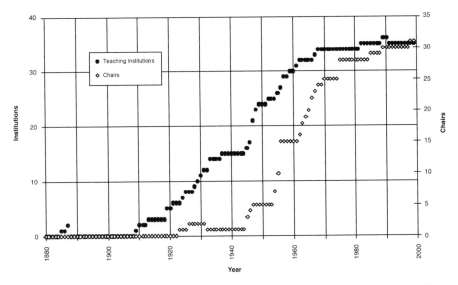

Figure A-7 British university departments offering chemical engineering courses and chairs. Note: courses are included based on their contemporary, not modern, descriptions. Sources (not exhaustive): Anon., 'Chemical engineering courses in Britain', *Quart. Bull. IChemE* Jul 1953, xxxii–xxxv; Anon., 'A survey of output of chemical engineers in Great Britain', *TCE* Apr 1958, A37–A41; IChemE Annual Report 1960; IChemE, Chemical Engineers for the Future (Rugby, 1989); S. R. Tailby, 'Early chemical engineering education in London and Scotland', in: W. F. Furter (ed.) *A Century of Chemical Engineering* (New York: Plenum, 1982), pp. 65–126; F. R. Whitt, 'Early teachers and teaching of chemical engineering', *TCE* Oct 1969, CE356–CE360 and Oct 1971, 370–374; W. K. Hutchison, 'Industry, science and the chemical engineer', *Chem. & Indus.* 6 May 1961, 567–72; Minutes of IChemE council, Education Committee and Accreditation Committee; Gayfere archive, box VI/1; inaugural lectures.

ANNUAL PRODUCTION OF CHEMICAL ENGINEERS IN BRITAIN

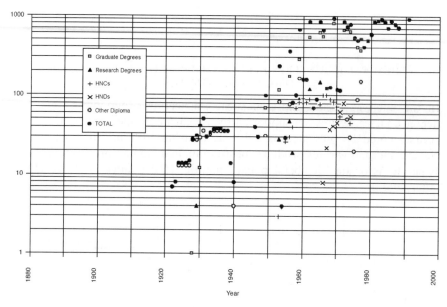

Year

Figure A-8   Annual production of chemical engineers in Britain. Sources (not exhaustive): R. T. W. Hall, 'Supply and demand for chemical engineers', *TCE* May 1978, 388–90; D. B. Purchas, 'Who joins the Institution?', *TCE* May 1978, 329; A. J. Biddlestone & J. Bridgwater, 'From mining to chemical engineering at the University of Birmingham', and D. C. Freshwater, B. W. Brooks and A. Foord, 'Loughborough – The development of a department', both in: N. A. Peppas (ed.), *One Hundred Years of Chemical Engineering* (Dordrecht: Kluwer, 1989), pp. 237–44 and 245–61; IChemE, *Chemical Engineers for the Future* (Rugby, 1989); J. Lamb, Surrey; B. Azzopardi, Nottingham; C. Jones, Heriot-Watt; H. Briers, Loughborough; U. Newcastle; P. J. Bailes, Bradford; Anon., 'A survey of output of chemical engineers in Great Britain', *TCE* Apr 1958, A37–A41; Gayfere archive box VI/1; Minutes of IChemE Education Committee.

OCCUPATIONAL STATISTICS

Prior to 1971 chemical engineers were not singled out for analysis in the Qualified Manpower tables. Subsequent figures are the numbers of practitioners between 18 and 69 years of age employed in Britain under category 3.9.1 (Chemical engineering not elsewhere classified). This excludes unemployed persons, and 'hidden' chemical engineers classified as managers, unspecified engineers or other types of specialist. The relation between these figures and those of Figure A-8 (Annual production of chemical engineers in Britain) and Figure A-1 (IChemE membership) is unclear. The small census figures, compared to both annual graduates and IChemE membership, suggest that they are under-reported by specifying another employment description – indirect evidence for a weak occupational identity.

Table A-3 Census statistics for chemical engineers

Male	Higher degree	Male First degree	GCE 'A' level or HNC/HND	Higher degree	Female First degree	GCE 'A' level or HNC/HND	Total
1971[7]	157	1318	138	3	7	10	1633
1981[8]	205	1205	159	6	23	3	1601
1991[9]	272	1200	213	23	76	16	1800

---

[7] *Census 1971 Great Britain: Qualified Manpower Tables* (London: HMSO, 1976).

[8] *Census 1981: Qualified Manpower Great Britain* (London: HMSO, 1984).

[9] *Census 1991: Qualified Manpower Great Britain* (London: HMSO, 1994).

SALARIES

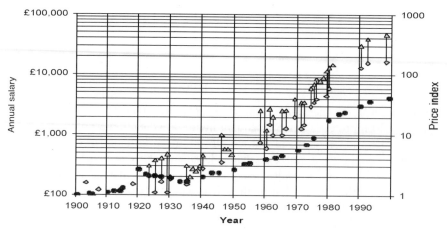

Figure A-9   Median salaries for persons engaged in 'chemical engineering' work (triangles), compared to the UK Price Index (circles). Sources: Castner-Kellner Co. records, CRO DIC/BM 20/101;
Albright & Wilsons records, MS 1724 Box 50 Agreements; D. J. Oliver, 'Chemical industry pay
and conditions 1930 to 1939' (unpublished typescript, Jan 1997; PRO AB 17/231, *Careers in Nuclear
Engineering* (Harwell, 1958); A. Cluer, 'Careers in chemical engineering' (typescript, 1960); IChemE
salary surveys; job advertisements, *Quart. Bull. IChemE* and *TCE*; Sir Frederick Warner interview
with C. Cohen, 10 May 1996. Chemical engineering salaries until at least the 1970s remained
broadly similar to those of a skilled fitter, according to figures collated by Oksana Newman and
Allan Foster, The Value of a Pound: Prices and Incomes in Britain 1900–1993 (New York: Gale
Research International, 1995). The Price Index figures (derived from Newman and Foster p. 305)
are normalised to 1900 = 1.

THE LITERATURE OF CHEMICAL ENGINEERING

The array of journals, popular magazines and books on chemical engineering
played an important role in defining chemical engineers themselves. As
Anderson has written of nationalism, such literature helped to define an imagined community in which 'most members will never know most of their fellow
members, meet them, or even hear of them, yet in the minds of each lives the
image of their communion'; they came 'to visualize in a general way the
existence of [others] like themselves through print language'.[10] The advent of
illustrated advertisements, magazines and books made chemical engineers
themselves tangible.

The books and periodicals of chemical engineering are also a major source
of information for historians of the subject. They provide a 'public' view of the
discipline and the profession, revealing glimpses of how its intellectual currents
were stirred by contemporary concerns. More abstractly, the number and
variety of titles reveal the changing recognition and scope of the subject.

Such sources do not, though, provide an unclouded window to the past.
Texts do not always mirror activities in contemporary chemical engineering,

---

[10]   Benedict Anderson, *Imagined Communities* (London: Verso, 1991), pp. 6, 77.

when most today are written by academics having relatively little industrial experience.[11] The disparity between the chemical engineering literature and practice existed, too, in the late nineteenth century when, despite an expanding bulk chemicals industry, commercial secrecy had a dampening effect on publication. In an atmosphere of technical concealment when informative books were sparse, periodicals performed the complementary task of identifying, rallying and serving a working community of chemical plant designers and operators.

Structure, content and role

From the turn of the century, the texts of chemical engineering performed several functions. Initially, books such as G. E. Davis's *A Handbook of Chemical Engineering* (1901, 1904) and Jacob Grossman's *The Elements of Chemical Engineering* (1906) had an implicit proselytising and propagandising role in attempting to define the scope and content of the discipline. These books structured their content in the same way as their authors categorised their subject. For Davis, chemical engineering was best communicated as a series of processes based on units of plant. The chapters of his book consequently discussed operations and equipment such as separating, absorbing and transporting. For Grossman, by contrast, the subject was to be understood as a large-scale analogue of laboratory chemistry. His chapters correspondingly promoted this mapping (e.g. 'The Beaker and Its Technical Equivalents', 'The Air Bath and Its Technical Equivalents').

In the years during and after the first world war, during which the chemical engineering profession was struggling, there were few further attempts to focus or redirect the emphases of the nascent British discipline through books. Instead, authors concentrated on details of practice; Hinchley's *Chemical Engineering: Notes on Grinding, Sifting, Separating and Transporting Solids* (1914) was typical. British texts concentrated on works describing plant design and operation for practising engineers, and apparently found a ready market. Hinchley's book was one of several adapted from articles in *The Chemical World* between 1913 and 1919.[12]

But the connection between the authors of such works and chemical engineering organisations is not close, illustrating the lack of close cohesion between the occupation, discipline and profession at that time. Among 22 authors of the Benn Bros Chemical Engineering Library of the early Twenties, only eight developed connections with the IChemE or the Chemical Engineering Group.[13] Authors of other books of the period appear to have been operating within other spheres. S. S. Dyson, for example, having worked some fifteen years for

---

[11]   D. C. Freshwater, 'The development of chemical engineering as shown by its texts', in Nicolaos A. Peppas ed., *One Hundred Years of Chemical Engineering* (Dordrecht: Kluwer Academic Publishers, 1989), pp. 15–25.

[12]   Others included texts by William Porter Dreaper (1913), Edgar Jobling (1916) and James Arthur Crowther (1919).

[13]   Authors who became IChemE members included E. A. Alliott, C. Elliott, W. E. Gibbs, H. Griffiths, P. Parrish, A. I. Robinson, N. Swindin and S. J. Tungay.

G. E. Davis, much of it as editor of his *Chemical Trade Journal*, founded another journal, *Chemical Engineering & the Works Chemist*, in 1911 and published two rather unsuccessful texts.[14] While Davis's son Keville and Norman Swindin, another employee, became early members of the IChemE, Dyson never applied.[15]

General British handbooks of chemical engineering were sparse in the interwar period. Among the exceptions (none of which was particularly popular) were Hugh Griffiths, *The General Principles of Chemical Engineering Design* (London, 1922), J. C. Olsen, *Unit Processes and Principles of Chemical Engineering* (London, 1932) and Harold Tongue, *A Practical Manual of Chemical Engineering* (London, 1939). American books during the interwar period were targeted at a broader readership and, unlike the British products, included a significant number of teaching texts and works on engineering economics (nevertheless, American texts appeared only from 1923, indicating the common intellectual bases for the two national disciplines).[16] Besides the national differences in content, titles published on chemical engineering continued to be more numerous in America than in Britain, typically in a ratio of eight to one.[17] Among the most influential texts from American publishers were W. H. Walker, W. K. Lewis and W. H. McAdams, *The Principles of Chemical Engineering* (1923), W. L. Badger and W. L. McCabe, *Elements of*

---

[14]   *Chemical Works: Their Design, Erection and Equipment* (London, 1912), written with S. S. Clarkson, was advertised in his periodical; *A Manual of Chemical Plant* (London, 1916), published in 13 parts, ended abruptly after Chapter 1 of Volume 2.

[15]   On George Keville Davis (1881–1935), see N. Swindin, *Engineering Without Wheels* (London: Weidenfeld & Nicolson, 1962), 40–2; IChemE AF 202. George Davis's brother Alfred gave up joint management of the Journal and partnership in their chemical engineering practice in 1905. Keville continued his father's consultancy and Manchester Technical Laboratory until the 1920s. On his death, editorship of *The Chemical Trade Journal* was passed to others of the Davis clan, his cousin Eric N. Davis and J. N. Davis.

[16]   On the most active branch of American publishing, see S. D. Kirkpatrick, 'Building the literature of chemical engineering', *Chemical Engineering* 59 (July 1952), 166–73.

[17]   Some 200 British and 1300 American-published texts were identified at the US Library of Congress and 1400 at the British Library, the most complete repositories of English-language books. The US Library of Congress and British Library subject category of 'chemical engineering' was searched (using three catalogues at the British library: Retrospective, Current General and Current Scientific, Technology and Business). Titles were also itemised in the catalogues of major academic libraries, and in specialist bibliographies, notably W. P. Dreaper (ed.), *Textbooks of Chemical Research and Engineering* (London, 1913), *A Reference List of Bibliographies: Chemistry, Chemical Technology and Chemical Engineering* (New York, 1924), and K. Bourton, *Chemical and Process Engineering Unit Operations: A Bibliographic Guide* (London: MacDonald, 1967). No single library collection is entirely representative of the subject either nationally or internationally. The rate and thoroughness of library acquisitions depends on the collection policy and budget allocated to a subject, which may change as the subject becomes more established or more important to the institution concerned, or on the purpose for which the collection is intended. Similarly, librarians' definitions (or subsequent re-definitions) of 'chemical engineering' determine which books in the collection become so categorised. Such graphs are thus a 'cultural composite' of the decisions of authors, publishers, book purchasers, librarians and information professionals.

Chemical Engineering (1936) and J. H. Perry, *Chemical Engineers' Handbook* (1934).

Even in that country, however, chemical engineering titles were sometimes altered to suit a broader market: thus a book planned as *Mechanical Control of Chemical Engineering Operations* appeared as *Industrial Instruments for Measurement and Control* (1941); the manuscript *Chemical Engineering in Industry* became *The Chemical Process Industries* (1945); and *Materials of Construction for Chemical Engineering Equipment* was published as *Materials of Construction for Chemical Process Industries* (1950).[18]

British books on the subject were, in fact, almost entirely absent between 1940 and 1950 owing to wartime and postwar constraints. Hugh Griffiths, on a visit to India in 1949, reported that 'everywhere the same remark is passed: "Why are there no British textbooks on chemical engineering?"'[19] From the Fifties, however, British book publishing again began to expand. The faculty of the new department of chemical engineering at Cambridge published a half dozen texts during the decade.[20] The most influential of these, the two-volume *Chemical Engineering* by J. M. Coulson and J. F. Richardson, sold over a thousand copies per year through the 1950s, 1600/yr in the Sixties and 2400/yr into the 1990s. Published by Pergamon Press, at the time launching a new

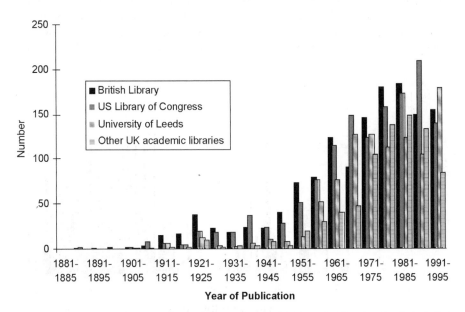

Figure A-10 Chemical engineering titles (all countries of origin) in major libraries. Books included in the classification 'chemical engineering' are those assigned by the respective libraries.

---

[18] Kirkpatrick, *op. cit.*, 171.
[19] Letter, H. Griffiths to IChemE Hon. Registrar, 3 Nov. 1949, Gayfere archive, box III/1, 'Branches – India'.
[20] John Bridgwater, *Fifty Years Young: Products and Processes – The Future of Chemical Engineering* (Inaugural lecture, Cambridge: Cambridge University Press, 1996).

chemical engineering journal, the Coulson & Richardson text replaced the American standards of Walker, Lewis & McAdams, Perry and Badger & McCabe in many British syllabi.[21] Nevertheless, some British authors (e.g. R. Edgeworth Johnstone and M. W. Thring, *Pilot Plants, Models, and Scale-Up Methods in Chemical Engineering*, 1957) preferred to publish for the much larger American audience that Coulson and Richardson failed to attract. The most ambitious work completed in this period was the twelve-volume H. W. Cremer, T. R. Davies, S. B. Watkins (eds.), *Chemical Engineering Practice* (London: Butterworths, 1956–65), an uneven collection of articles by a variety of contributors useful chiefly as a reference work. British publications from this time began to show a distinct concentration on biochemical engineering and plant safety.

Similarly, texts could occasionally serve as signposts marking out new intellectual territory, or certifying a new-found status for the specialism. The design of chemical reactors, for example, achieved a higher rank as catalysis and transport processes became better understood.

Periodicals

Where chemical engineering texts focused on the issues of intellectual foundations, standard practice and teaching, periodicals communicated occupational, technical and professional news to practitioners and students. Reflecting the changing conditions of chemical engineers, the number of periodicals increased

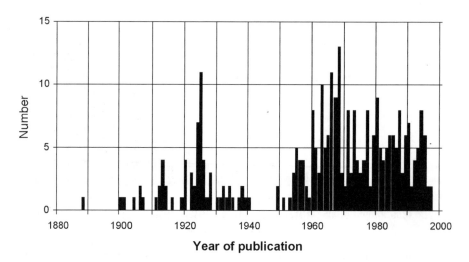

Figure A-11 Chemical engineering books published in Britain. Source: catalogues of British Library, National Library of Congress and University of Leeds for 'chemical engineering'.

---

[21]   From the late 1960s the IChemE itself became the principal publisher of chemical engineering books, initially through Pergamon Press and mainly in the form of conference proceedings. In the last two decades of the century such publications began to dominate chemical engineering publishing in Britain.

rapidly in the periods following each of the two world wars. Each had a distinct readership and subject content. The range of British periodicals included *The Chemical Trade Journal* (Manchester, 1887–1966) and *The Journal of the Society of Chemical Industry* (1882–1950). The first periodicals in Britain to use the term *chemical engineer* in the title, and thereby claim a new readership, were the renamed *The Chemical Trade Journal & Chemical Engineer* (from 1908) and *Chemical Engineering and the Works Chemist* (1911–1925?). The latter was the official organ of the Association of Chemical Technologists, organised in part by John Hinchley, until it was replaced in 1911 by the *Journal of Chemical Technology*. All three published patent reviews and news items concentrating on occupational issues and new technical developments. The *Proceedings of the Chemical Engineering Group* (1919–74) carried papers by members of the Group and advertisements. It included a brief one or two page annual report on the activities of the CEG, but otherwise had no editorial, news or non-technical content.

The *Transactions of the Institution of Chemical Engineers* (1923–1982) served as the principal repository for technical papers delivered through the IChemE.[22] Members of the IChemE and AIChE also shared their journals during the interwar years. Because the American journal averaged twice as many papers (30 vs 15 for the *Trans. IChemE* over its first decade) it was a major source of current knowledge for IChemE members. The IChemE *Transactions* were continued by *Chemical Engineering Research and Design* (1983–) and later augmented by *Process Safety and Environmental Protection* and *Food and Bioproducts Processing*. For information other than research papers, the Institution published *The Quarterly Bulletin of the Institution of Chemical Engineers* (1924–56), continued as *The Chemical Engineer* (1956–). This incor-porated news and technical features of broad interest. Other contemporary periodicals sympathetic to chemical engineering were *The Chemical Age: A Weekly Journal devoted to Industrial and Engineering Chemistry* (1919–) and *Chemistry and Industry* (1923–), the official news organ of the IChemE and other organisations and published by the SCI.

University periodicals had a limited circulation and were devoted to local news and activities. They included *The Journal of the Ramsay Society of Chemical Engineers* (UCL, 1953?–), *The Journal of the Imperial College Chemical Engineering Society* (1945–61?), *Birmingham University Chemical Engineer* (1949–76?), the Victoria University of Manchester *Mutech Chemical Engineering Journal* (1958–?), and the *Loughborough University of Technology Chemical Engineering Society Journal* (1965–72).

Commercial publications (as distinct from the publications of institutions) appeared after the second world war. *Chemical Engineering Science* (1951–)

---

[22] On the first technical paper published – an 82 page tour de force taking up most of the first volume – by the Head of Research of the first IChemE President's firm, see T. C. Finlayson, 'Industrial Oxygen', *Trans. IChemE* 1 (1923), 1–88; W. Aitkin, 'Industrial Oxygen', *TCE* 11 Jun 1998, 16–18. The other three papers of that first volume were by M. B. Donald, based on his MSc work at MIT.

was the first venture of the new Pergamon Press in October 1951. The initial contributions were from Netherlands and the U.K., with Editorial and Advisory Boards drawn from chemical engineers in Western Europe, but following the appointment of an American Editorial Board, the journal operations moved to the US in 1956.[23] Other commercial competitors were *British Chemical Engineering* (1956–72),[24] continued by *Process Technology International* (1972–), *Chemical Engineering Communications* (1973–), *Computers and Chemical Engineering: an International Journal* (Oxford: Pergamon, 1979–) and *Process and Chemical Engineering* (Cambridge: RSC, 1982–).

Such British periodicals competed for attention in a growing international market. While the sole English-language alternatives initially were the *Transactions of the American Institute of Chemical Engineers* (1908–46), continued by *Chemical Engineering Progress* (1947–) and the *Journal of Industrial and Engineering Chemistry* (1909–22), other national publications appeared after the second world war.[25] Since then, commercial journals such as *Chemical Engineering* (New York: McGraw-Hill, 1946–), *The Journal of Chemical and Engineering Data* (1956–), *International Chemical Engineering* (1961–94) and *Chemical Engineering Communications* (1973–84) have carried technical communications. Other commercial periodicals gaining readership from the early 1970s included *Chemical and Process Engineering* (1964–71), followed by *Process Engineering and Plant Control* (1971–73), *The Chemical Engineering Journal* (Lausanne: Elsevier, 1970–) and *Process Engineering* (1973–).[26] At the end of the century, the UK is second only to the USA in the number of chemical engineering periodicals.

---

[23] J. Bridgwater, 'History of a research journal, Chemical Engineering Science', in N. Peppas ed., *One Hundred Years of Chemical Engineering* (Dordrecht: Kluwer, 1989), pp. 39–46.

[24] J. O. S. MacDonald, 'Chemical engineering publications', *TCE* (Jul/Aug 1964), CE183-CE186.

[25] E.g. *Vegyipari Szakirodalmi Tajekoztato* (Chemical Engineering Abstracts, Budapest, 1949–); *Canadian Journal of Chemical Engineering* (1957–); *Chemical Engineering World* (Jodhpur, 1966–); *Journal of Chemical Engineering of Japan* (1968–); *Teoreticheskie osnovy khimicheskoi tekhnologii* (Theoretical Foundations of Chemical Engineering, 1968–); *Latin American Journal of Chemical Engineering and Applied Chemistry* (1971); *Huagong Jixie* (Chemical Engineering and Machinery, Langzhou, 1974–); *Chemical Engineering in Australia* (1976–); *Chemical Engineering and Technology* (Weinheim, 1978–); *Korean Journal of Chemical Engineering* (1983?–).

[26] M. Spear, 'All our yesterdays and a few tomorrows', *Process Engineering* 76 (1995), 41–72.

GEORGE E. DAVIS – *A HANDBOOK OF CHEMICAL ENGINEERING* (FIRST EDITION, 1901)

The work of George E. Davis is often referred to but not many people will have had a chance to read the pages of his pioneering text, *A Handbook of Chemical Engineering*.

George E. Davis attempted to establish a professional body for chemical engineers as early as the 1880s. It is a matter of coincidence that this book on The Institution of Chemical Engineers should be written almost exactly 100 years after the appearance of that first George E. Davis Handbook. The facsimile pages that follow are a selection taken from a copy of Volume I of the first edition of the Handbook held in the IChemE Library. Marginal notes on the pages are in Davis' own hand and were made in preparation for revisions to be made in the Second Edition.

A Handbook

OF

Chemical Engineering.

CHAPTER I.

Introduction.

The functions of the Chemical Engineer are very generally misunderstood. It has often been thought that the chemical contractor, who erects plant, either from well-worn designs in his possession, or from the designs of others, must of necessity be styled a Chemical Engineer; others have deemed the term correctly applied to the ironfounder, the mason, the coppersmith, or the builder who has supplied work for use in connection with the chemical industries; but a little thought will soon enable us to see that both these propositions are untenable.

The ironfounder may have been supplying castings all his life, and the coppersmith building stills, digesters, or vacuum pans, but he may not have been made acquainted with the various purposes to which it is intended to put them, so that unless he possesses a very extended knowledge of the many operations over which manufacturing chemistry extends, he can hardly lay claim to be called a Chemical Engineer. From the manufacturer's point of view it is certainly to his advantage to recognise an intermediary between himself and the contractor, in the same way as the architect and surveyor is recognised by both builder and house-owner, as, even if the professional man does make a mistake sometimes (and he would scarcely be human if he did not), the evil will be more than counterbalanced by the good offices of securing a careful perusal of plans and execution of the job in a workmanlike manner. Of course where the proprietor of a chemical establishment is in himself a good engineer and chemist, or where his general manager has a good knowledge of these subjects, the professional man from the outside will not be so much in requisition ; but it is astonishing how "familiarity breeds contempt," and how often an outside opinion will enable matters to be seen in a totally different aspect.

The Chemical Engineer has been a creation of necessity, brought about by the wonderful advances made in recent years by manufacturing chemists, and the activity of capital in exploiting invention after invention so soon as it is made known. If we study the specifications of patents granted in this country during the past fifty years, we cannot fail to observe the change that has come over chemical inventions. Crude ideas have given way to carefully detailed manipulations, and descriptions of apparatus are now usually so complete as to leave nothing to be desired in this respect. The mechanical aspect of manufacturing chemistry is still advancing with very rapid strides, so that the skill of the engineer is being more and more sought after and employed. This has been the outcome of the exploitation of some of the finer processes of the chemical manufacture, notably in the aniline colour industry, though we must be careful not to give undue credit to this branch alone, as the heavy chemical trade has within the past twenty years undergone almost a complete revolution : and it is not unreasonable to suppose that many processes, tried and abandoned in times gone by, would have had a greater measure of success if they had been installed with some of the beautiful appliances of the present time

The gradual development of certain processes seems a matter worthy of comment, as it is most probable that the want of success in their early days was due to a lack of knowledge of Chemical Engineering. Chemists were not engineers, and engineers were not chemists, and the experience of the one was of but little assistance to the other, and the fact must not be ignored that the engineer with a little knowledge of chemistry generally was more successful than the chemist with a smattering of engineering, which goes a long way to prove that the engineering knowledge was the more valuable of the two.

It may be that the retardation of practical success in the direction of several chemical processes has in the past been due in a great measure to the arbitrary division of chemistry into "practical" and "theoretical," this leading many otherwise well informed persons to infer that there is no harmony between theory and practice, whereas, had they thought deeply enough, they would have appreciated the fact that theoretical chemistry *per se* gives only a portion of the information necessary for the successful carrying out of chemical operations on the large scale. Practical chemistry is, to a large degree, dependent upon theoretical chemistry, but not in any greater degree than upon physics and mechanics, and to attempt the solution of problems in manu-facturing chemistry without the aid of these sister branches of science would now-a-days only be commenced by the ignorant. The idea that theory and practice will not work together is as old, or older, than the Greek philosophy, and though the idea is dying fast it cannot be said that it is actually defunct. Theory is based, or should be based, on known facts marshalled and generalised, and it helps us in many manufacturing troubles by showing us the probable reason why good results or bad results are obtained; while practice demonstrates pretty clearly what we should do to obtain the best results, or to obviate existing errors—each has its *rôle*, and it is well they should not be confounded.

All practical questions involve the use of some fixed datum, selected after careful experiment, and it is the province of Chemical Engineering to engage in this work so far as it relates to manufacturing chemistry, and to supply figures which, when woven in with pure theory, show that some desired result is attainable, or on the other hand demonstrate its impracticability.

But in addition to the theorems of chemistry, physics, and mechanics, there is another species of knowledge essential to success in manufacturing chemistry. This knowledge springs from observation of good and bad materials, and of various descriptions of workmanship It is acquired during the actual transaction of business where materials have to be provided, and workmen employed, not only in constructing the apparatus, but in its manipulation afterwards, so that the necessary operations can be satisfactorily carried out. If we call this knowledge "practical experience," we shall find the first aid comes in from Theoretical Chemistry, which tells us of the nature of the raw materials to be operated upon and the various mutations they are likely to undergo; Mechanics comes forward and teaches how to combine and arrange our apparatus; while Physics will most probably supply the remainder of the required information.

Chemical Engineering differs from general engineering chiefly on account of the destructive character of the materials operated upon, and the very nature of the processes themselves require such careful study from a purely chemical point of view, that theoretical and applied chemistry of a high order should form a necessary part of the curriculum of the Chemical Engineer. It is well known that this view is not shared in some quarters, and a slight smattering of chemistry tacked on to an engineering student is sometimes held

244 *STEAM PRODUCTION AND DISTRIBUTION.*

283 lb. of coal were burned per hour, and 2,603 lb. of water evaporated during the same time. To put 50,000 units of heat to the steam escaping from this boiler would raise its temperature :—

$$\frac{50,000}{2,603 \times \cdot 475} = 40^\circ$$

0·475 being the specific heat of steam under constant pressure, so that if the usual boiler pressure carried be 50 lb., the temperature of which, as shown by Table 77, is 298°F., the temperature after superheating must be 338°F., which is higher than that of the waste gases escaping from the boiler setting, showing the operation of superheating in this instance to have been impossible, as the waste gases were only 324°F. This shows us the difference between heat units and temperature. If we admit 184°F. as the lowest practicable temperature in the waste gases necessary to produce a good draught in the chimney, we have 140°F. in excess of this in the Philadelphia trial, and as 283 lb. of fuel were burned per hour we have :—

$$283 \times 24 \times 140 \times \cdot 238 = 226,309$$

heat units escaping per hour in excess of that actually required to produce a draught, and yet insufficient to produce superheating, as anyone will be able to see that, should the temperature of the exit fuel gases fall below that of the steam, cooling and condensation of the latter must inevitably take place.

When a superheater is placed in the boiler flue the products of combustion will be cooled very considerably during the process of heating the steam. In the foregoing instance the superheat has been assumed to require 50,000 units per hour, and as the specific heat of air is 0·238, the following calculation will show the cooling of the gases by the superheating of the steam :—

$$283 \times 24 = 6,792 \text{ lb. per hour.}$$

$$\frac{50,000}{6,792 \times \cdot 238} = 30^\circ\text{F.}$$

thus reducing the temperature of the exit flue gases from say 500°F. to 470°F. No account is taken here of the water coming over with the steam, which would require a further and considerable number of heat units.

The fifth condition for economical steam raising, viz., to utilise all hot condensed water from steam traps, pipes, etc., should not require mention; yet how often do we find this pure hot distilled water finding its way to the nearest drain.

To keep the steam boiler free from scale is the next important condition that must be observed. As a rule this matter does not receive sufficient consideration, but a little thought upon the case nearly touch the shaft, and the rings of blades on the shaft lie between those on the case, as shown in the figure (109), and Fig. 110 further shows the form of blade which is used, the smaller and larger circles being cross-sections through the shaft and cylinder respectively.

The steam enters at J (Fig. 109) through a ring of fixed guide blades, and is projected in a rotational direction upon the next ring of moving blades, imparting to them a rotatory motion; it is next thrown back upon the succeeding ring of guide blades, and so on in succession until the whole of the fixed blades and all the moving blades have been passed, the reaction at each series of blades increasing the rotational force. The energy to give the steam its high rotary velocity at each successive ring is supplied by the fall in pressure, the steam expanding gradually by small increments. At the left side of the spindle are groved pistons or "dummies," D, E, and F, the object of which is to prevent end-thrust, and they also act as a practically steam-tight joint. The governing of the motor is accomplished as follows :—Steam is admitted to the turbine

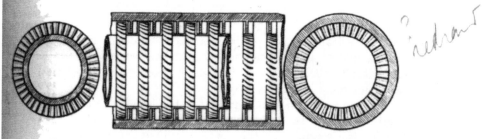

FIG. 110.—FORM OF BLADE IN THE PARSONS STEAM TURBINE.

in a series of gusts by the periodic opening and closing of a double beat valve operated by means of a steam relay in mechanical connection with the turbine shaft. Where a dynamo is being driven, the duration of each gust is controlled by an electric solenoid which is connected as a shunt to the field magnets, the core of the solenoid being hung from a long lever. The fulcrum of this lever is periodically moved up and down by means of a link connecting it with the eccentric, which receives its motion from the worm on the sleeve coupling, the short end of the lever controlling the valve of the steam relay. Where no dynamo is being driven the speed can be controlled with an ordinary centrifugal governor.

The bearings of this motor are all under a head of oil, the oil being continually circulated by means of a pump, and as no lubricant is used in the cylinder the exhaust steam is free from oil and other impurities. This is no small advantage, as the condensed water can be used again in the steam boiler without filtration or other purification. With the exception of the bearings, the Parsons

FRUE VANNERS. 31

.The details of the power required for a ten stamp wet-crushing gold mill, capable of working from 15 tons to 18 tons of gold ore per day of 24 hours, may safely be put down as follows :—

One Blake Rock Crusher	6 horse-power	
Two ore feeders	—	
Ten stamps, 750 lbs., 90 drops	12	,,
Four Frue Vanner Concentrators	2	,,
One grinding pan, 3 ft. diameter ../ ...	3	,,
One settler	3	,,
Allow for friction	4	,,

                                              30 horse power.

A twenty stamp mill, with eight Frue vanners, would only require 46 horse-power.

Ore Concentrating.—The Frue vanners deserve mention in this place. On page 26 the operation of levigation has been briefly described, which, in effect, is the separation of finely-divided particles from the coarser particles by means of a current of water. "Vanning" is hardly this, as the main object of the operation is the

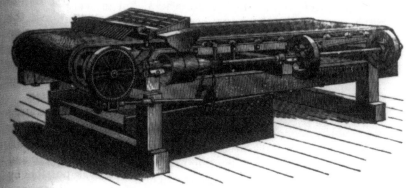

FIG. 26.—THE FRUE VANNER.
(Messrs. Fraser and Chalmers, Ltd.)

separation—more or less complete—of minerals which differ not only in gradation of size, but also in specific gravity. The vanner was the invention of the late Mr. W. B. Frue, the manager of the Silver Islet mines, on Lake Superior, and was applied by him to the concentration of low grade silver ores as early as 1873. Many forms of the vanner, or concentrator, are now made, but the reader must be contented with a description of the Frue vanner, as made by Messrs. Fraser and Chalmers, an illustration of which is shown by Fig. 26.

This concentrator, or Frue vanner, consists in the main of a continuous rubber belt, with a rubber flange on each edge. The

48   THE APPLICATION OF HEAT AND COLD.

	No. 1. Per cent.	No. 2. Per cent.
Moisture	0·33	0·49
Volatile matter	2·25	1·31
Fixed carbon	80·54	97·46
Sulphur	0·60	0·72
Ash	6·28	10·02
Structure, cells	52·94	50·04
Structure, coke	47·06	49·96
Specific gravity	1·697	1·890
Heat units (per lb.) ...	13,540	12,937

No. 1 is a light coke with medium porosity, and will give a quick, intense heat.

No. 2 is a strong coke and will hold up a heavier burden of iron ; the porosity is lower—*i.e.*, the coke is more dense—consequently a stronger blast is required.

A strong, heavy coke will give a steady, continued heat."

A few words may now be said upon liquid fuels. In times of fuel scarcity some manufacturers have endeavoured to rely upon liquid fuel of various kinds as a substitute for coal. Some of these attempts have lasted longer than others, but in every case that has come under the author's notice, the users have only been too glad to discontinue the use of liquid fuel at the earliest possible moment.

It must not be supposed that the discontinuance of the use of liquid fuel was in any way due to its failure from a physical or mechanical standpoint. It is a fiscal question entirely, and when heat can be raised more cheaply from liquid fuel than with coal, manufacturers will revert to the former. It is, therefore, necessary to know the economics of these substitutions.

In Great Britain at the present time the only liquid fuels available for large scale heating are, gas-tar, coke-oven and blast-furnace oils, and coal-tar creosote, though in other countries crude petroleum and the semi-liquid astatki, or petroleum residue, may form a subject for serious consideration. Crude petroleum, consisting of 86 parts (by weight) of carbon and 14 parts of hydrogen, is a very good substance for comparison. If we follow the methods described on page 234 of Vol. I., we shall find that a substance of this composition gives a theoretical heating power in British thermal units as follows :—

	%	Per lb. of oil			B.T.U.
Carbon	86·0	...	0·86 × 14544	=	12507
Hydrogen ...	14·0	...	0·14 × 62032	=	8684
					21191
Less water ...	126·0	...	1·26 × 1147		1445
					19746

162 *SEPARATING SOLUBLES FROM INSOLUBLES.*

the sulphate of soda has been dissolved, or when manganese liquor has been neutralised, more often than not these insoluble particles have to be separated. There are several methods of doing this, but perhaps the most general way is by sedimentation or subsidence. The liquid is allowed to flow into large tanks where the suspended matters fall to the bottom and form a layer of greater or less thickness, entangling some solution amongst its particles.

It will be naturally asked whether the dimensions of the settling tanks have any influence upon the rate of settling. They have; but whether each insoluble substance has its own specific rate of settling, and this too in varying menstrua, the author cannot say, but it seems likely. In this direction the following table will be of use to the designer as an aid if not an absolute guide. Four hundred gallons of manganese mud at 170° F., suspended in chloride of calcium solution of .24° Tw, were placed simultaneously in five vessels, two feet, three feet, four feet, five feet, and six feet diameter respectively, and the amount of settling noticed. The figures in the next table show the amount of clear liquor in gallons yielded by each in the given times recorded in the first column. The same liquid placed in a tall litre jar in the laboratory, yielded in 24 hours 500 c.c. of clear solution.

TABLE **12.**

SHOWING THE SETTLING IN VESSELS OF VARYING DIMENSIONS.

	2 ft.	3 ft.	4 ft.	5 ft.	6 ft.
	feet.	feet.	feet.	feet.	feet.
Depth of liquor	20	9	5	3·3	2·2
In 15 minutes	59	73	61	64	47
In 30 minutes	76	96	77	84	71
In 45 minutes	95	115	102	104	96
One hour	105	132	128	136	129
One hour, 15 minutes	112	144	144	152	140
One hour, 30 minutes	120	160	153	168	140
One hour, 45 minutes	128	162	164	172	
Two hours	132	168	169	176	
Two hours, 15 minutes	136	174	180	184	
Two hours, 45 minutes	148	185	184	184	
Three hours, 15 minutes	156	192	189		
Three hours, 45 minutes	160	194	189		
Four hours, 15 minutes	165	196			
Six hours	181	200			
Seven hours	186	200			
Eight hours	189				
Nine hours	194				
Ten hours	200				
Twelve hours	200				

It may be of interest here to state that a series of four subsidence tanks of three-eighths steel plate, each tank being 16ft. square by 7ft. deep, being one long tank with three partitions, so as

BIBLIOGRAPHY

NOTE ON ARCHIVAL SOURCES

This study employed archival research, examination of published material, and, where appropriate and possible, interviews with historical actors. The interviewees are listed in the Preface.

The written historical materials of relevance fall into several classes. Archives at engineering and scientific institutions, regional record offices and a number of British universities were consulted. The most crucial were the records of the Institutions themselves, some of which contain large amounts of previously unexamined material. Not surprisingly, the largest collection was found at the headquarters of the Institution of Chemical Engineers in Rugby, Warwickshire.

The most accessible of the IChemE materials is a virtually complete set of application forms from the origins of the Institution to the present day. The applications, which include curriculum vitae, letters of reference and supporting documentation, number some 75 000 as of 1999 and occupy over 250 linear feet of shelf space. An analysis of the changing profile of membership was important for understanding an Institution which consistently lacked identifiable majorities; significant cohorts – whether consultants, workers in the munitions industry, women members, chemical engineers in the postwar atomic factories or Commonwealth members – were historically limited to one or two 'tenths'.

Minute books also provided a continuous record of Institutional activities. Minutes ranging from the 1921 Provisional Committee to the 1999 council were available, as were those of major sub-groups such as the Education and Nominating committees. And, the Institution houses valuable but, as yet, unarchived collections of documents. The most useful of these, the 'Gayfere archive', consists of 19 boxes of files pruned from the records accumulated by the Institution during its half century in London. These are selective – omitting, in most cases, correspondence, for example – and terminate in the late 1950s to early 1960s. Nevertheless, these records flesh out the terse minute books and annual reports and, in some cases, reveal otherwise unrecorded Institutional affairs. The four folders of papers of Sir Frederic Nathan, second President of the IChemE, provided an extra-Institutional view of the emergence of chemical engineering. Two boxes of material selected by J. B. Brennan, the first General Secretary, from now presumably discarded records, and a much more voluminous store of 120 boxes of records accumulated by Trevor Evans, General Secretary from 1976, fill in blanks. In particular, discussion papers for Council members document official business.

The IChemE library proved to be a repository for valuable, if largely forgotten, materials. These included diaries, papers and a book collection of George E. Davis; first world war reports collected by Sir Harold Hartley; books from the libraries of S. G. M. Ure and A. J. V. Underwood; and, early chemical engineering volumes donated by members and publishers. Here, too, were found symposia, reports and particularly useful periodicals: the *Quarterly Bulletin of the IChemE*, *The Chemical Engineer*, *Diary*, *Transactions of the IChemE* and Annual Reports.

Thus, the Council minute books, letters, Home Papers and unpublished reports were an adequate, if not extensive, source of information; by contrast, membership applications, lists and related documentation provided a wealth of prosopographical (i.e. collective biographical) data; annual reports, salary, employment and membership surveys yielded considerable information.

Beyond the IChemE, historical materials were studied at some two dozen locations. Nevertheless, historical material concerning chemical engineering as an occupation proved particularly meagre. Organisations employing chemical engineers (or failing to do so) hold information of relevance, but have been only sampled owing to the size of the task. Moreover, such sources seldom explicitly distinguish chemical engineering tasks or employees; contemporary evaluation of their importance can be difficult to gauge. Exceptions were the records of Albright & Wilson Ltd at the Birmingham City Library, and the records of companies that merged to become ICI (notably Castner Kellner and Brunner Mond) located at the Cheshire Record Office. Other revealing sources were the company histories written by W. E. Bryden (George Scott & Sons), G. A. Dummett (APV), J. A. Reavell (Kestner), W. J. Reader (Unilever and ICI) and A. M. Pettigrew (ICI).

In all too few cases have key chemical engineers and participants in the Institution deposited archives. The earliest located is that of G. E. Davis at the Science Museum, London. More voluminous collections are the papers and memoirs of Sir Christopher Hinton at the Institution of Mechanical Engineers in London; the Sir Harold Hartley collection at the Churchill College Archive Centre of the University of Cambridge; and, some 75 boxes of papers deposited by Norman Swindon at the archives of Loughborough University. The Public Record Office (PRO) at Kew holds records of K. B. Quinan, C. Hinton and N. L. Franklin.

Institutional records were studied at the headquarters of the Society of Chemical Industry, London (minute books); the PRO (government-related records concerning factories, research organisations, education, labour and munitions); and the Institutions of Mechanical and Electrical Engineers. The histories of chemical engineering departments were studied at the archives of the Universities of Birmingham, Cambridge, Glasgow, Strathclyde and Manchester, and Imperial, King's and University Colleges of the University of London.

PERIODICALS EMPLOYED AS PRIMARY SOURCES

Chemical Age
Chemistry and Industry
Chemical Engineering & the Works Chemist
Chemical Trade Journal & Chemical Engineer
Diary of the Institution of Chemical Engineers
Journal of the British Nuclear Energy Conference
Journal of the British Nuclear Energy Society
Journal of the Society of Chemical Industry
Proceedings of the Chemical Engineering Group
Quarterly Bulletin of the Institution of Chemical Engineers
The Chemical Engineer
Transactions of the Institution of Chemical Engineers

OTHER PUBLISHED WORKS

Abbott, Andrew Delano, *The System of the Professions: An Essay on the Division of Expert Labor*, Chicago: University of Chicago Press, 1988.

Abrahart, E. N., *The Clayton Aniline Company Limited 1876–1976*, Manchester: Clayton Aniline Co, 1976.

Adams, R. J. Q., *Arms and the Wizard: Lloyd George and the Ministry of Munitions, 1915–1916*, London: Cassell, 1978.

Aftalion, Fred, *A History of the International Chemical Industry*, Philadelphia: University of Pennsylvania Press, 1991.

Aldcroft, Derek H., *Education, Training and Economic Performance 1944 to 1990*, Manchester: Manchester University Press, 1992.

Aldcroft, Derek H., *The Inter-War Economy: Britain, 1919–1939*, London: B. T. Batsford, 1970.

Alder, Ken, *Engineering the Revolution: Arms and Enlightenment in France, 1763–1815*, Princeton: Princeton University Press, 1997.

Alliott, Eustace A., *Centrifugal Dryers and Separators*, London: Ernest Benn, 1926.

Anderson, Benedict, *Imagined Communities:Reflections on the Origin and Spread of Nationalism*, London: Verso, 1991.

Antwerpen, Franklin J., and Sylvia Fourdrinier, *High Lights: The First Fifty Years of the American Institute of Chemical Engineers*, New York: AIChE, 1958.

Armytage, Walter Harry Green, *A Social History of Engineering*, London: Faber & Faber, 1970.

Arnold, Lorna, *Windscale 1957*, London: Macmillan, 1992.

Arrowsmith, H. *Pioneering in Education for the Technologies: The Story of Battersea College of Technology. 1891–1962*, Guildford: Univ. of Surrey, 1966.

Aronowitz, Stanley, *The Politics of Identity: Class, Culture, Social Movements*, New York: Routledge, 1992.

Aspray, William, *Engineers as Executives: An International Perspective*, New York: IEEE Press, 1995.

Association of British Chemical Manufacturers, *Report of the British Chemical Mission on Chemical Factories in the Occupied Areas of Germany*, London: ABCM, 1919.

Atkinson, B., *Research and Innovation for the 1990s: The Chemical Engineering Challenge*, Oxford: Pergamon, 1986.

Badger, Walter Lucius and Warren Lee McCabe, *Elements of Chemical Engineering*, New York: McGraw-Hill, 1936.

Baines, A., and F. R. Bradbury and C. W. Suckling, *Research in the Chemical Industry: The Environment, Objectives and Strategy*, Amsterdam: Elsevier Publishing Co, 1969.

Ball, D. F., *Process Plant Contracting Worldwide*, London: Financial Times Business Information, 1985.

Baumler, Ernst, *A Century of Chemistry*, Dusseldorf: Econ Verlag, 1968.

Becher, Tony (ed.), *Governments and Professional Education*, Milton Keynes: Open University Press, 1994.

Bergengren, Erik; transl by Alan Blair, *Alfred Nobel – The Man and His Work*, London: Thomas Nelson and Sons, 1962.

Berlant, V. L., *Profession and Monopoly*, Berkeley: Univ. of California Press, 1975.

Bernhard, C. C., E. E. Crawford & P. Sorbom (eds.), *Science, Technology and Society in the Time of Alfred Nobel*, Oxford: Pergamon Press, 1982.

Bickel, Lennard, *Rise Up to Life: A Biography of Howard Walter Florey Who Gave Penicillin to the World*, London: Angus Robertson, 1972.

Blackman, Jules, *Chemicals in the National Economy*, Washington, DC: Manufacturing Chemists' Assoc, 1964.

Bott, T. Reg, *Beating Pollution: Chemical Engineering*, Reading: Educational Explorers, 1973.

Bourton, Kay, *Chemical and Process Engineering Unit Operations: A Bibliographical Guide*, London: MacDonald, 1967.

Bowden, Mary Ellen & John Kenly Smith, *American Chemical Enterprise: A Perspective on 100 Years of Innovation to Commemorate the Centennial of the Society of Chemical Industry (American Section)*, Philadelphia: Chemical Heritage Foundation, 1994.

Bradbury, F. R. and B. G. Dutton, *Chemical Industry: Social and Economic Aspects*, London: Butterworths, 1972.

Braunholtz, Walter T. K., *The Institution of Gas Engineers: the First Hundred Years, 1863–1963*, London: Institution of Gas Engineers, 1963.

Brennan, John Basil, *The First Fifty Years: A History of the Institution of Chemical Engineers 1922–1972*, Rugby: unpubl. manuscript, 1972.

British Petroleum Company Ltd, *Our Industry: An Introduction to the Petroleum Industry for the Use of Members of the Staff*, London: BP, 1958.

Broadberry, S. N., *The British Economy Between the Wars: A Macroeconomic Survey*, Oxford: Basil Blackwell, 1986.

Brodie, John Alfred, *Australia's Engineering Milestones: The Chemical Industry*, 1987.

Brophy, Leo P., Wyndham D. Miles and Rexmond C. Cochrane, *The Chemical Warfare Service: From Laboratory to Field*, Washington, DC: Office of the Chief of Military History, 1959.

Brown, John K., *The Baldwin Locomotive Works 1831–1915: A Study in American Industrial Practice*, Baltimore: Johns Hopkins University Press, 1995.

Bucciarelli, Louis L., *Designing Engineers*, Cambridge, Mass: MIT Press, 1994.

Bud, Robert, *The Uses of Life: A History of Biotechnology*, Cambridge: Cambridge University Press, 1993.

Bud, Robert F., and Gerrylynn K. Roberts, *Science Versus Practice: Chemistry in Victorian Britain*, Manchester: Manchester University Press, 1984.

Budapest Technical University, *Hundred Years of the Faculty of Chemical Engineering, Technical University Budapest, 1871–1971*, Budapest: Budapest Technical University, 1972.

Burk, Kathleen, *War and the State: The Transformation of British Government, 1914–1919*, London: George Allen & Unwin, 1982.

Burn, Duncan, *The Structure of British Industry*, Cambridge: Cambridge University Press, 1958.

Calder, Angus, *The Myth of the Blitz*, London: Pimlico, 1991.

Camilleri, Joseph A., *The State and Nuclear Power: Conflict and Control in the Western World*, Seattle: University of Washington Press, 1984.

Carlisle, Rodney P., and Joan M. Zenzen, *Supplying the Nuclear Arsenal: American Production Reactors, 1942–1992*, Baltimore: Johns Hopkins University Press, 1996.

Cartwright, A. P., *The Dynamite Company: The Story of African Explosives and Chemical Industries Ltd*, London: MacDonald, 1964.

Castner–Kellner Ltd, *Fifty Years of Progress 1895–1945: The Story of the Castner–Kellner Alkali*

Company Told to Celebrate the Fiftieth Anniversary of Its Formation, Manchester: Castner–Kellner, 1945.

Cathcart, Brian, *Test of Greatness: Britain's Struggle for the Atomic Bomb*, London: John Murray Ltd, 1994.

Cawson, Alan (ed.), *Organized Interests and the State: Studies in Meso-Corporatism*, London: Sage Publications, 1985.

Coetzee, Frans and Marilyn Shevin-Coetzee (eds.), *Authority, Identity and the Social History of the Great War*, Providence: Berghahn Books, 1995.

Clark, Hans T. et al. (eds.), *The Chemistry of Penicillin: report on a collaborative investigation by American and British chemists under the joint sponsorship of the Office of Scientific Research and Development, Washington, D.C., and the Medical Research Council, London*, Princeton: Princeton University Press, 1949.

Clark (ed.), John, *Technological Trends and Employment 2: Basic Process Industries*, Aldershot: Gower, 1985.

Clark, Ronald W., *The Birth of the Bomb: The Untold Story of Britain's Part in the Weapon That Changed the World*, London: Phoenix House, 1961.

Cole, Margaret, *The Story of Fabian Socialism*, London: Heinemann, 1961.

Cole, George Douglas Howard, *Great Britain in the Post-War World*, London: Victor Gollancz, 1942.

Coleman, Donald Cuthbert, *Courtaulds: An Economic and Social History* (2 vols), Oxford: Clarendon Press, 1969.

Connor, Ralph A; D. Churchill Jr; R. H. Ewall; C. Heimsch; W. R. Kerner, *Chemistry: A History of the Chemistry Components of the National Defense Research Committee 1940–1946*, Boston: Little, Brown & Co., 1948..

Corbett, A. H., *The Institution of Engineers Australia: A History of Their First Fifty Years 1919–1969*, Institution of Engineers, Australia, 1973.

Corfield, Penelope J., *Power and the Professions in Britain 1700–1850*, London: Routledge, 1995.

Corley, Thomas Anthony Buchanan, *A History of the Burmah Oil Company 1886–1924*, London: Heinemann, 1983.

Coulson, John Metcalfe and John Francis Richardson, *Chemical Engineering*, Oxford: Pergamon, 1954.

Court of Inquiry, *The Flixborough Disaster*, London: HMSO, 1975.

Crawford, Stephen, *Technical Workers in an Advanced Society: The Work, Careers and Politics of French Engineers*, Cambridge: Cambridge University Press, 1989.

Cremer, Herbert William and Trefor Davies (eds.), *Chemical Engineering Practice*, London: Butterworths, (12 vols), 1956–65.   ·

Croucher, Richard, *Engineers At War 1939–1945*, London: Merlin Press, 1982.

Davis, George Edward, *A Handbook of Chemical Engineering* (2 vols), Manchester: Davis Bros., 1901–4.

Delanty, Gerard, *Inventing Europe: Idea, Identity, Reality*, Basingstoke: Macmillan, 1995.

Dewar, George Albemarle Bertie, *The Great Munitions Feat*, London, 1921.

Donnelly, James Francis, *Chemical Education and the Chemical Industry in England from the Mid-Nineteenth to the Early Twentieth Century*, unpubl. PhD thesis, University of Leeds, 1987.

Dummett, George Anthony, *From Little Acorns: A History of the A.P.V. Company Limited*, London: Hutchison Benham, 1981.

Dutton, William S., *Du Pont: One Hundred and Forty Years*, New York: Charles Scribner's Sons, 1951.

Dyson, S. S., *A Manual of Chemical Plant*, London: Dover, 1916.

Edgerton, David, *Science, Technology and the British Industrial 'Decline' 1870–1970*, Cambridge: Cambridge University Press, 1996.

Edyvean, R. G. J. (ed.), *Centenary of the Sheffield Technical School*, Sheffield: University of Sheffield, 1986.

Elder, Albert L. (ed.), *The History of Penicillin Production*, New York: AIChE, 1970.

Engineering Council, *A Guide to the Engineering Institutions*, London: The Engineering Council, 1990.

Eyre, John Vargas, *Henry Edward Armstrong, 1848–1937: The Doyen of British Chemists and Pioneer of Technical Education*, London: Butterworths, 1958.

Fairchild, Byron, and Jonathan Grossman, *United States Army in World War II: The War Department: The Army and Industrial Manpower*, Washington, DC, Dept of the Army, 1959.

Ferrier, R. W., *The History of the British Petroleum Company Vol I: The Developing Years 1901–1932*, Cambridge: Cambridge University Press, 1982.

Field, Robert, *Chemical Engineering: Introductory Aspects*, Basingstoke: Macmillan Education, 1988.

Finniston, Sir Montague, *Engineering our Future: Report of the Committee of Inquiry into the Engineering Profession*, London: HMSO, 1980.

Forrest, John Samuel (ed.), *The Breeder Reactor: Proceedings of a Meeting at the University of Strathclyde 25 March 1977*, Edinburgh: Scottish Academic Press, 1977.

Fox, Robert, and Anna Guagnini (eds.), *Education, Technology and Industrial Performance in Europe, 1850–1939*, Cambridge: Cambridge University Press, 1993.

Furter, William F. (ed), *A Century of Chemical Engineering*, New York: Plenum Press, 1982.

Furter, William F. (ed), *History of Chemical Engineering*, Washington: American Chemical Society, 1980.

Galloway, R., *Education, Scientific and Technical*, London: Trubner, 1881.

Gerstl, Joel Emery and Stanley Peerman Hutton, *Engineers: The Anatomy of A Profession. A Study of Mechanical Engineers in Britain*, London: Tavistock Publications, 1966.

Giddens, Anthony, *Modernity and Self-Identity: Self and Society in the Late Modern Age*, Cambridge: Polity, 1991.

Gispen, Kees, *New Profession, Old Order: Engineers and German Society, 1815–1914*, Cambridge: Cambridge University Press, 1989.

Glover, Ian A., and Michael P. Kelly, *Engineers in Britain: A Sociological Study of the Engineering Dimension*, London: Allen & Unwin, 1987.

Gowing, Margaret Mary, with assistance by Lorna Arnold, *Independence and Deterrence: Britain and Atomic Energy 1945–1952* (2 vols), London: Macmillan, 1974.

Gowing, Margaret Mary, *Britain and Atomic Energy 1939–1945* (2 vols), London: Macmillan, 1964.

Grant, Allan, *Steel and Ships: The History of John Brown's*, London: Michael Joseph, 1950.

Grieves, Keith, *Sir Eric Geddes: Business and Government in War and Peace*, Manchester: Manchester University Press, 1989.

Griffen, Francis James, *The History of the Society of Chemical Industry*, London: unpublished typescript, no date.

Griffiths, Hugh, *The General Principles of Chemical Engineering Design*, London: Benn Brothers, 1922.

Grossman, Jacob, *The Elements of Chemical Engineering*, London: Charles Griffin, 1906.

Grove, Jack William, *Government and Industry in Britain*, London: Longmans, 1962.

Haber, Ludwig Fritz, *The Chemical Industry During the Nineteenth Century: A Study of the Economic Aspect of Applied Chemistry in Europe and North America*, Oxford: Clarendon Press, 1958.

Haber, Ludwig Fritz, *The Chemical Industry 1900–1930: International Growth and Technological Change*, Oxford: Clarendon Press, 1971.

Haber, Ludwig Fritz, *The Poisonous Cloud: Chemical Warfare in the First World War*, Oxford: Clarendon Press, 1986.

Hague, Douglas C., *The Economics of Man-Made Fibres*, London: Gerald Duckworth & Co., 1957.

Hall, H. Duncan and C. C. Wrigley, *Studies of Overseas Supply*, London: HMSO, 1956.

Hall, H. Duncan, *North American Supply*, London: HMSO, 1955.

Hall, Tony, *Nuclear Politics: The History of Nuclear Power in Britain'*, Harmondsworth: Penguin, 1986.

Hannah, Leslie, *The Rise of the Corporate Economy*, London: Methuen, 1976.

Hannah, Leslie, *Engineers, Managers and Politicians: The First Fifteen Years of Nationalised Electricity Supply in Britain*, London: Macmillan, 1982.

Hardie, David William Ferguson and J. Davidson Pratt, *A History of the Modern British Chemical Industry*, Oxford: Pergamon Press, 1966.

Hargreaves, E. L., and Margaret Mary Gowing, *Civil Industry and Trade*, London: HMSO, 1952.

Harris, Ralph and Arthur Seldon, *Advertising in Action*, London: Andre Deutsch, 1962.

Harrison, Godfre, *Alexander Gibb: The Story of an Engineer*, London: Geoffrey Bles, 1950.

Harrison (ed.), Tom, *War Factory: A Report by Mass-Observation*, London: Victor Gollancz, 1943.

Hartley, Harold, *Report of the British Mission Appointed to Visit Enemy Chemical Factories in the Occupied Zone Engaged in the Production of Munitions of War*, London: Ministry of Munitions, 1919.

Hayes, Peter, *Industry and Ideology: IG Farben in the Nazi Era*, Cambridge: Cambridge University Press, 1987.

Hayman, Roy, *The Institute of Fuel: The First 50 Years*, London: Institute of Fuel, 1977.

Hewlett, Richard G., and Jack M.Holl, *Atoms for Peace and War 1953–1961: Eisenhower and the Atomic Energy Commission*, Berkeley: University of California Press, 1989.

Hinchley, John W., *Chemical Engineering: Notes on Grinding, Sifting, Separating and Transporting Solids*, London: J. & A. Churchill, 1914.

Hinchley, Edith M., *John William Hinchley: Chemical Engineer*, London: Lamley & Co, 1935.

Hinton, Christopher, *Engineers and Engineering*, Oxford: Oxford University Press, 1970.

Holloway, David, *Stalin and the Bomb: The Soviet Union and Atomic Energy 1939–1956*, New Haven, CT: Yale University Press, 1994.

Homburg, Ernst, Anthony S. Travis and Harm G. Schröter (eds.), *The Chemical Industry in Europe, 1850–1914. Industrial Growth, Pollution, and Professionalization*, Dordrecht: Kluwer, 1998.

Hopkins, Harry, *The New Look: A Social History of the Forties and Fifties in Britain*, London: Secker & Warburg, 1963.

Hornby, William, *Factories and Plant*, London: HMSO, 1958.

Hounshell, David A., *From the American System to Mass Production, 1800–1932*, Baltimore: Johns Hopkins University Press, 1984.

Hutchison, Sir Kenneth, *High Speed Gas: An Autobiography*, London: Duckworth, 1987.

Hutt, William Harold, *Plan for Reconstruction: A Project for Victory in War and Peace*, London: Kegan Paul, Trench, Trubner & Co, 1943.

Hutton, Stanley, and Peter Lawrence, *German Engineers: The Anatomy of A Profession*, Oxford: Clarendon, 1981.

IChemE, *Model Form of CONDITIONS OF CONTRACT FOR PROCESS PLANTS Suitable for Lump-Sum Contracts in the United Kingdom*, London: IChemE, 1968.

IChemE, *Presidents – The Institution of Chemical Engineers*, undated, unpublished illustrated volume.

Imperial Chemical Industries, *Ancestors of an Industry: ICI*, [n.p.]: Kynoch Press, 1950.

Institution of Petroleum Technologists, *Petroleum: Twenty-Five Years in Retrospect, 1910–1935*, London: Institution of Petroleum Technologists, 1935.

Ives, Kenneth James, *Sixty Years of Sanitary Science: The Department of Municipal Engineering, University College, London, 1897–1957*, London: The Chadwick Trust, 1958.

Jefferys, James B., *The Story of the Engineers 1800–1945*, New York: Lawrence & Wishart, 1945.

Jenkins, Richard, *Social Identity*, London: Routledge, 1996.

Johnson, Paul, *Twentieth Century Britain: Economic, Social and Cultural Change*, London: Longman, 1994.

Jordan, Grant, *Engineers and Professional Self-Regulation: From the Finniston Committee to the Engineering Council*, Oxford: Clarendon Press, 1992.

Kimball, Bruce A., *The "True Professional Ideal" in America: A History*, Cambridge MA: Blackwell, 1992.

Kirkpatrick, Sidney D. (ed.), *Twenty-Five Years of Chemical Engineering Progress: 1908–1933*, New York: Van Nostrand, 1933.

Kletz, Trevor A., *What Went Wrong? Case Histories of Process Plant Disasters*, Houston: Gulf Publishing Co., 1985.

Kletz, Trevor A., *Myths of the Chemical Industry, Or 44 Things a Chemical Engineer Ought NOT to Know*, Rugby: IChemE, 1984.

Kuhn, James W., *Scientific and Managerial Manpower in Nuclear Industry*, New York: Columbia University Press, 1966.

Larson, Magali Sarfatti, *The Rise of Professionalism: A Sociological Analysis*, Berkeley: University of California Press, 1977.

Layton, Jr., Edwin T., *The Revolt of the Engineers: Social Responsibility and the American Engineering Profession*, Cleveland: Press of Case Western Reserve University, 1971.

Lenoir, Timothy, *Instituting Science: The Cultural Production of Scientific Disciplines*, Stanford: Stanford University Press, 1977.

Littler, Craig R., *The Development of the Labour Process in Capitalist Societies: A Comparative Study of the Transformation of Work Organization in Britain, Japan and the USA*, London: Heinemann, 1982.

Lloyd George, David, *War Memoirs of David Lloyd George*, London: Ivor Nicholson & Watson, 1933.

Locke, Robert R., *The End of the Practical Man: Entrepreneurship and Higher Education in Germany, France and Great Britain, 1880–1940*, London, Jai Press, 1984.

Lomas, J., *A Manual of the Alkali Trade. including the Manufacture of Sulphuric Acid. Sulphate of Soda and Bleaching Powder*, London: Crosby, Lockwood, 1886.

Longhurst, Henry, *Adventure in Oil: The Story of British Petroleum*, London: Sidgwick and Jackson, 1959.

MacDonald, Keith M., *The Sociology of the Professions*, London: Sage, 1995.

Macnab, William, *Preliminary Studies for H.M. Factory, Gretna and Study for an Installation of Phosgene Manufacture*, London: HMSO, 1920.

Malatesta, Maria, transl. by Adrian Belton, *Society and the Professions in Italy, 1860–1914*, Cambridge: Cambridge Univ Press, 1995.

Manchester Association of Engineers, *One Hundred Years of Engineering in Manchester: The Centenary of the Manchester Association of Engineers 1856–1956*, Manchester: MAE, 1956.

Marwick, Arthur, *The Home Front: The British and the Second World War*, London: Thames & Hudson, 1976.

Mayne, Richard, *Postwar: The Dawn of Today's Europe*, London: Thames & Hudson, 1983.

Mayo, Frank, *The Beginning of Fawley Oil Refinery*, unpublished manuscript, 1996.

McClelland, Charles E., *The German Experience of Professionalization: Modern Learned Professions and Their Organizations From the Early Nineteenth Century to the Hitler Era*, Cambridge: Cambridge Univ Press, 1991.

McMahon, A. Michal, *The Making of a Profession: A Century of Electrical Engineering in America*, IEEE Press: New York, 1984.

Meadows, Donella H., and D. L. Meadows, J. Randers, W. W. Behrens III, *Limits to Growth: A Report for the Club of Rome's Project on the Predicament of Mankind*, London: Pan Books, 1972.

Meiksins, Peter, and Chris Smith (eds.), *Engineering Labour: Technical Workers in Comparative Perspective*, London: Verso, 1996.

Mendelsohn, Everett, Merritt Roe Smith and Peter Weingart (eds.), *Science, Technology and the Military*, Dordrecht: Sociology of the Sciences Yearbook 1988 (2 vols), 1988,

Mensforth, Eric, *Family Engineers*, London: Ward Lock, 1981.

Meynaud, Jean (transl by Paul Barnes), *Technocracy*, London: Faber & Faber, 1968.

Michael, T. H. Glynn, and B. T. Newbold, L. W. Shemilt and A. W. Tickner., *Chemical Canada 1970–1995*, Ottawa: The Chemical Institute of Canada, 1995.

Middlemas, Robert Keith, *Politics in Industrial Society: The Experience of the British System since 1911*, London: Andre Deutsch, 1979.

Middlemas, Robert Keith, *Power, Competition and the State*, Basingstoke: Macmillan (3 vols), 1986–90.

Miles, Frank Douglas, *A History of Research in the Nobel Division of I.C.I.*, Glasgow: ICI Nobel Division, 1955.

Ministry of Labour, *The Whitley Report, Together with the Letter of the Minister of Labour Explaining the Government's View of its Proposals*, London: Ministry of Labour, 1917.

Ministry of Munitions, *H. M. Factory, Gretna: Description of Plant and Process*, London: HMSO, 1918.

Ministry of Munitions, *List of Staff and Distribution of Duties*, September, 1918.

Ministry of Munitions, *Mr. Lloyd George's Farewell Address to the Staff of the Ministry*, London: Harrison & Sons, 1916.

Ministry of Munitions, *Report on the Statistical Work of the Factories Branch: Statistical Information*, London: HMSO, 1919.

Ministry of Munitions, *Service Charges on H. M. Factories*, London: Department of Explosives Supply, 1918.

Ministry of Reconstruction, *Reconstruction Problems 1: The Aims of Reconstruction'*, London: HMSO, 1918.

Ministry of Reconstruction, *Reconstruction Problems 5: New Fields for British Engineering*, London: HMSO, 1918.

Ministry of Reconstruction, *Reconstruction Problems: 28: Scientific Business Management*, London: HMSO, 1919.

Ministry of Reconstruction, *Reconstruction Problems 31: Trusts, Combines and Trade Associations'*, London: HMSO, 1919.

Ministry of Reconstruction, *Industrial Councils: The Whitley Scheme*, London: Ministry of Reconstruction, 1919.

Molle, Willem, and Egbert Wever, *Oil Refineries and Petrochemical Industries in Western Europe: Buoyant Past, Uncertain Future*, Aldershot: Gower, 1983.

Morgan Crucible Company Ltd., *Battersea Works 1856–1956*, London: Morgan Crucible Co. Ltd., 1956.

Morris, Peter John Turnbull, Colin A. Russell and John Graham Smith (ed.), *Archives of the British Chemical Industry, 1750–1914: A Handlist*, London: BJHS, 1988.

Morris, Peter John Turnbull, and William Alec Campbell and Hugh L. Roberts, *Milestones in 150 Years of the Chemical Industry*, London: Royal Society of Chemistry, 1970.

Mowat, Charles Loch, *Britain Between the Wars, 1918–1940*, London: Methuen & Co, 1955.

Mulder, Karel F., *Choosing the Corporate Future: Technology Networks and Choice Concerning the Creation of High Performance Fiber Technology*, PhD thesis: University of Groningen, 1992.

Müller, Detlef K., Fritz Ringer and Brian Simon (eds.), *The Rise of the Modern Educational System: Structural Change and Social Reproduction 1870–1920*, Cambridge: Cambridge University Press, 1989.

Multhauf, Robert P., *The History of Chemical Technology: An Annotated Bibliography*, New York: Garland, 1984.

National Economic Development Office, *Manpower in the Chemical Industry: A Comparison of British and American Practices*, London: HMSO, 1967.

Newitt, Dudley M. (ed.), *Chemical Engineering and Chemical Catalogue*, London: Leonard Hill, 1930.

Newman, Oksana and Allan Foster (compilers), *The Value of a Pound: Prices and Incomes in Britain 1900–1993*, New York: Gale Research International, 1995.

Newman, Otto, *The Challenge of Corporatism*, London: Macmillan, 1981.

Noble, David F., *America By Design: Science, Technology and the Rise of Corporate Capitalism*, New York: Knopf, 1979.

Noble, David F., *Forces of Production: A Social History of Industrial Automation*, Oxford: Oxford University Press, 1986.

Norris, James D., *Advertising and the Transformation of American Society, 1865–1920*, New York: Greenwood Press, 1990.

Nuclear Energy in Britain, London: HMSO, 1969.

Ossipoff, George, *The Institution of Chemical Engineers in Australia: A Historical Chronology 1960–1994*, unpublished typescript, IChemE Australia, 1997.

Pannell, John Percival Masterman, *An Illustrated History of Civil Engineering*, London: Thames and Hudson, 1964.

Parke, Victor E., *Billingham – the First Ten Years*, Durham: Imperial Chemical Industries Ltd, 1957.

Parker, Henry Michael Denne, *Manpower: A Study of War-Time Policy and Administration*, London: HMSO, 1957.

Political and Economic Planning, *Trade Associations: Activities and Organisation*, London: PEP, 1957.

Pelz, Donald C., and Frank M. Andrews, *Scientists in Organizations: Productive Climatse for Research and Development*, New York: John Wiley, 1966.

Peppas, Nikolaos A. (ed.), *One Hundred Years of Chemical Engineering: From Lewis M. Norton (M.I.T. 1888) to Present*, Dordrecht: Kluwer, 1989.

Perkin, Harold, *The Rise of Professional Society: England Since 1880*, London: Routledge, 1989.

Pettigrew, Andrew M., *The Awakening Giant: Continuity and Change in Imperial Chemical Industries*, Oxford: Basil Blackwell, 1985.

Pittock, Murray G. H., *The Invention of Scotland: The Stuart Myth and the Scottish Identity, 1638 to the Present*, London: Routledge, 1991.

Pollard, Sidney, *The Development of the British Economy, 1914–1990*, London: Edward Arnold, 1992.

Postan, Michael Moissey, *British War Production*, London: HMSO, 1952.

Postan, M. M., D. Hay and John Dick Scott, *Design and Development of Weapons: Studies in Government and Industrial Organisation*, London: HMSO, 1964.

Potter, John R., *Chemical Engineering: An Introduction*, London: Butterworths, 1971.

Potts, H. E., *Patents and Chemical Research*, Liverpool: Liverpool University Press, 1921.

Priestley, John Boynton, *English Journey*, Toronto: Macmillan, 1934.

Pringle, Peter, and James Spigelman, *The Nuclear Barons*, London: Sphere Books, 1982.

Reader, William Joseph, *A History of the Institution of Electrical Engineers 1871–1971*, London: Peter Peregrinus, 1987.

Reader, William Joseph, *Fifty Years of Unilever 1930–1980*, London: Heinemann, 1980.

Reader, William Joseph, *Imperial Chemical Industries: A History* (2 vols), Oxford: Oxford University Press, 1970, 1975.

Reavell, J. Arthur, et al., *The Kestner Golden Jubilee Book 1908–1958: To Commemorate 50 Years of Chemical Engineering Endeavour*, London: Kestner Evaporator & Engineering Co, 1958.

Reimann, Guenter, *Patents for Hitler*, London: Victor Gollancz, 1945.

Reuben, B. G., and M. L. Burstall, *The Chemical Economy: A Guide to the Technology and Economics of the Chemical Industry*, London: Longman, 1973.

Reynolds, Terry S., *Seventy-Five Years of Progress: A History of the American Institute of Chemical Engineers 1908–1983*, New York: AIChE, 1983.

Rose, Hilary and Steven, *Science and Society: The Chemists' War*, Harmondsworth: Penguin, 1969.

Ross, William H., *History of the Distillers' Company Limited*, London: Distillers Ltd, 1919.

Rothbaum, Melvin, *The Government of the Oil, Chemical and Atomic Workers Union*, New York: John Wiley & Sons, 1962.

Russell, Bertrand, *Principles of Social Reconstruction*, London: George Allen & Unwin, 1916.

Russell, Colin A., *Science and Social Change, 1700–1900*, London: Macmillan, 1983.

Russell, Colin A, with Noel G. Coley and Gerrylynn K. Roberts, *Chemists by Profession: The Origins and Rise of the Royal Institute of Chemistry*, Milton Keynes: Open University Press, 1977.

Scranton, Philip, *Endless Novelty: Specialty Production and American Industrialization, 1865–1925*, Princeton: Princeton University Press, 1997.

Seton-Watson, Hugh, *Nations and States: An Enquiry into the Origins of Nations and the Politics of Nationalism*, London: Methuen, 1977.

Sharp, David, and T. F. West, *The Chemical Industry*, Chichester: Ellis Horwood, 1982.

Shemilt, Leslie W., *Chemical Engineering in Canada: An Historical Perspective*, Toronto: Chemical Institute of Canada, 1991.

Shimmin, A. N., *The University of Leeds: The First Half Century*, Cambridge: University of Leeds, 1954.

Simon, Anthony, *The Simon Engineering Group*, Stockport: privately printed, 1953.

Sinclair, Bruce, *Philadelphia's Philospher Mechanics: A History of the Franklin Institute, 1824–1865*, Baltimore: Johns Hopkins University Press, 1974.

Sohon, Julian Arell and William L. Schaaf (compilers), *A Reference List of Bibliographies: Chemistry, Chemical Technology and Chemical Engineering Publishing Since 1900*, New York: H. W. Wilson, 1924.

Spitz, Peter H., *Petrochemicals: the Rise of an Industry*, New York: John Wiley & Sons, 1988.

Stevenson, John, *British Society 1914–1945*, Harmondsworth: Penguin, 1984.

Stevenson, John, and Chris Cook, *The Slump: Society and Politics During the Depression*, London: Quartet Books, 1979.

Swindin, Norman, *Engineering Without Wheels: A Personal History*, London: Weidenfeld and Nicolson, 1962.

Sydney University Chemical Engineering Association, *Chemical Engineering and Engineering Technology at the University of Sydney 1929–1988*, Sydney: University of Sydney, 1988.

Tajfel, Henri (ed.), *Social Identity and Intergroup Relations*, Cambridge: Cambridge University Press, 1982.

Taylor, F. Sherwood, *A History of Industrial Chemistry*, London: Heinemann, 1957.

Taylor, A. J. P., *The Origins of the Second World War*, Harmondsworth: Penguin, 1961.

Thomson, Harry C., and Lida Mayo, *The Ordnance Department: Procurement and Supply*, Washington, DC: Office of the Chief of Military History, 1960.

Thornton, John, *The Engineering Profession: A National Investment. A Working Party Report on the Relationship of Education to Industry*, London: Conservative Political Centre, 1978.

Threlfall, Richard E., *The Story of 100 Years of Phosphorus Making 1851–1951*, Oldbury: Albright & Wilson, 1951.

Tilden, Sir William A., *Chemical Discovery and Invention in the Twentieth Century*, London: George Routledge & Sons, 1922.

Tolliday, Steven W. (ed.), *Government and Business*, Aldershot: Edward Elgar, 1991.

Tongue, Harold, *The Design and Construction of High Pressure Chemical Plant*, London: Chapman & Hall, 1959.

Torstendahl, Rolf, and Michael Burrage (eds.), *The Formation of Professions: Knowledge, State and Strategy*, London: Sage Publications, 1990.

Travis, Anthony S., Harm G. Schröter and Ernst Homberg (eds.), *Determinants in the Evolution of the European Chemical Industry, 1900–1939: New Technologies, Political Frameworks, Markets, and Companies*, Dordrecht: Kluwer, 1998.

Tweedale, Geoffrey, *Steel City: Entrepreneurship, Strategy, and Technology in Sheffield 1743–1993*, Oxford: Clarendon Press, 1995.

U.K. Chemical Industries Review 1974–75, London: Gower Economic Publications, 1975.

UKAEA, *Harwell: Careers in Nuclear Engineering*, Harwell: UKAEA, 1958.

Vascoe, Ian, *Organising for Science in Britain: A Case Study*, Oxford: OUP, 1974.

Veblen, Thorstein, *The Engineers and the Price System*, New York: Reprints of Economic Classics, Augustus M. Kelley, Bookseller, 1965.

Vig, Norman J., *Science and Technology in British Politics*, Oxford: Pergamon Press, 1968.

Vollmer, Howard M., and Donald L. Mills (eds.), *Professionalization*, New Jersey: Prentice-Hall, 1966.

Walker, W. H., and W. K. Lewis, W. H. McAdams and E. R. Gilliland, *Principles of Chemical Engineering*, New York: McGraw-Hill, 1937. (3rd ed.)

War Office, *The History of the Ministry of Munitions*, Harrow: HMSO, 1918–22.

Warner, Marina, *The Crack in the Teacup: Britain in the 20th Century*, London: Andre Deutsch, 1979.

Warrington, C. J. S., and R. V. V. Nicholls, *A History of Chemistry in Canada*, Toronto: Sir Isaac Pitman & Sons, 1949.

Watson, Garth, *The Civils: The Story of the Institution of Civil Engineers*, London: Thomas Telford, 1988.

Watson, Hamish B., *Organizational Bases of Professional Status: A Comparative Study of the Engineering Profession*, PhD thesis, University of London D16558, 1976.

Weir, R. B., *The History of the Distillers Company: Diversification and Growth in Whiskey and Chemicals 1887–1939*, Oxford: Oxford University Press, 1995.

Westerterp, K. R., *Forty Years – European Federation of Chemical Engineering*, ACHEMA Yearbook 91, Franfurt am Main: DECHEMA, 1990.

Weyman, Geoffrey, *The Design and Arrangement of Chemical Plant*, London: Ernest Benn, 1925.

Whalley, Peter, *The Social Production of Technical Work: The Case of British Engineers*, Basingstoke: Macmillan, 1986.

Whiffen, David H., with Donald H. Hey, *The Royal Society of Chemistry: The First 150 Years*, London: Royal Society of Chemistry, 1991.

Wilkinson, Alan, *Molasses to Acid: Saltend's First 75 Years Distilled*, Hull: BP Chemicals, 1997.

Williams, Roger, *The Nuclear Power Decisions: British Policies, 1953–78*, London: Croom Helm, 1980.

Williams, Trevor Illtyd, *The Chemical Industry: Past and Present*, Menston, Yorks: EP Publishing Ltd: Menston, Yorks., 1972.

Williamson, Peter J., *Corporatism in Perspective*, London: Sage Publications, 1989.

Williamson, Peter J., *Varieties of Corporatism: A Conceptual Discussion*, Cambridge: Cambridge University Press, 1985.

Wilson, A. J., *The Professionals: The Institution of Mining and Metallurgy 1892–1992*, London: Institution of Mining and Metallurgy, 1992.

Wilson, Charles, *The History of Unilever: A Study in Economic Growth and Social Change*, London: Cassell & Co, 1954.

Wilson, John F., *British Business History, 1720–1994*, Manchester: Manchester University Press, 1995.

Winkler, John K., *The Du Pont Dynasty*, New York: Reynal & Hitchcock, 1935.

Winter, Jay M., *The Great War and the British People*, Basingstoke: Macmillan, 1986.

Woodward, Kathryn (ed.), *Identity and Difference*, London: Sage, 1997.

Woollacott, Angela, *On Her Their Lives Depend: Munition Workers in the Great War*, Berkeley: University of California Press, 1994.

Wynne, M. D., *Chemical Processing in Industry*, London: Royal Institute of Chemistry, 1970.

Yates, JoAnne, *Control Through Communication: The Rise of System in American Management*, Baltimore: Johns Hopkins University Press, 1989.

Zeitlin, Jonathan, *Between Flexibility and Mass Production: Strategic Debate and Industrial Reorganization in British Engineering, 1830–1990*, Oxford: Oxford University Press (forthcoming).

Index

Chemists and Chemistry

19. C. Reinhardt and A.S. Travis: *Heinrich Caro and the Creation of Modern Chemical Industry.* 2000   ISBN 0-7923-6602-6
20. C. Divall and S.F. Johnston (eds.): *Scaling up.* The Institution of Chemical Engineers and the Rise of a New Profession. 2000   ISBN 0-7923-6692-1

KLUWER ACADEMIC PUBLISHERS – DORDRECHT / BOSTON / LONDON